Laboratory Tests for Unsaturated Soils

The testing of unsaturated soils requires greater care and effort than that of saturated soils. Although unsaturated soil mechanics has been embraced by geotechnical engineering, engineering practice has not yet caught up as the characterisation of unsaturated soils is difficult and time-consuming, and made harder still by a lack of standards.

Laboratory Tests for Unsaturated Soils collates test procedures to cover all laboratory tests for characterising unsaturated soils. It covers the background, theory, test procedures and interpretation of test results. Each test procedure is broken down into simple stages and described in detail. The pitfalls of each test and the interpretation of the test results are explained. Test data and calculation methods are given, along with numerical examples to illustrate the methods of interpretation and to offer the presentation of typical results.

The book is especially useful for students and researchers who are new to the field and provides a practical handbook for engineering applications.

Laboratory Tests for Unsaturated Soils

Eng-Choon Leong
and
Martin Wijaya

CRC Press
Taylor & Francis Group
Boca Raton New York London

CRC Press is an imprint of the
Taylor & Francis Group, an **informa** business

First edition published 2023
by CRC Press
6000 Broken Sound Parkway NW, Suite 300, Boca Raton, FL
33487-2742

and by CRC Press
4 Park Square, Milton Park, Abingdon, Oxon, OX14 4RN

CRC Press is an imprint of Taylor & Francis Group, LLC

© 2023 Taylor & Francis Group, LLC

Library of Congress Cataloging-in-Publication Data
Names: Leong, E. C., author. | Wijaya, Martin, author.
Title: Laboratory tests for unsaturated soils / Eng-Choon Leong and Martin Wijaya.
Description: First edition. | Boca Raton : CRC Press, [2023] | Includes bibliographical references and index.
Identifiers: LCCN 2022034960 | ISBN 9781138093829 (hbk) | ISBN 9780367860585 (pbk) | ISBN 9781315105147 (ebk)
Subjects: LCSH: Soil mechanics--Laboratory manuals. | Arid soils--Testing--Laboratory manuals.
Classification: LCC TA710 .L444 2023 | DDC 624.1/5136--dc23/eng/20221011
LC record available at https://lccn.loc.gov/2022034960

ISBN: 978-1-138-09382-9 (hbk)
ISBN: 978-0-367-86058-5 (pbk)
ISBN: 978-1-315-10514-7 (ebk)

DOI: 10.1201/b22304

Typeset in Sabon
by SPi Technologies India Pvt Ltd (Straive)

Dedicated to our families

Contents

Preface

The development of unsaturated soil mechanics lagged the development of saturated/classical soil mechanics. Our understanding of unsaturated soil mechanics has vastly improved in the last four to five decades. However, the application of unsaturated soil mechanics has still not become commonplace in practice. This is hardly surprising as noted by the late Professor Geoffrey Blight in his book *Unsaturated Soil Mechanics in Geotechnical Practice* that

> unsaturated soils do not usually give rise to geotechnical problems, as long as they remain in their unsaturated state at an approximately constant water content. ... Problems arise when unsaturated soils become subject to wetting ... that settlement, collapse or heave of soil surfaces occurs, when loss of strength results in slope failures or landslides, when piping failure is initiated, and compacted earth liners leak. Without the absorption of excess water, none of these widespread problems occur.

When these "special" problems do occur, practicing engineers who may not be familiar with unsaturated soil mechanics also faced the major problem of determining the properties of unsaturated soils. Testing of unsaturated soils entails greater care, patience and effort than saturated soils. The testing of unsaturated soils has remained mostly in the research laboratories of universities. The testing of unsaturated soils is made more difficult due to a lack of standards. Often cited problems of testing unsaturated soils are cost of the testing apparatus is expensive, the tests are complicated and the test duration is very long. While there is a workaround solution to obtaining properties of unsaturated soils by using estimation, the gold standard in obtaining unsaturated soil properties as is for saturated soils is still through tests. Hence, we have endeavoured to collate our experiences on the testing of unsaturated soils together with those of others from the literature into this book to provide a convenient source of reference for students and practitioners, as well as early researchers in unsaturated soils. Where applicable, it complements existing standards for testing of unsaturated soils and augments with current developments and knowledge.

This book assumes that the user has knowledge of classical soil mechanics and the testing of saturated soils to appreciate the significance and limitations of soil tests. Nevertheless, description of each unsaturated soil test is organised into three main parts: background and theory, test method and interpretation and analysis. The first and third parts distinguished it from the usual testing manual and are meant to help the reader to quickly grasp the basic principles of the unsaturated soil test. The test procedures are broken down into stages so that readers can associate the stages with the stages in the corresponding saturated soil test. The pitfalls in each test and the interpretation of the test results are carefully explained where applicable. The distillation of only the essential information of each test will enable the newcomer to embark on unsaturated soil testing quickly. The book contains references and further reading list to encourage readers to further develop their understanding and advanced their knowledge on unsaturated soil mechanics and unsaturated soil tests.

We hope that this book will become a useful resource and testing of unsaturated soils will become more common in practice.

Acknowledgments

This book contains the experiences in the authors' journey into unsaturated soil mechanics and unsaturated soil testing. The experiences are the collective efforts of colleagues, collaborators, students, research staff and friends who have shared the journey with the authors.

We acknowledge the influence of Professor Harianto Rahardjo who brought unsaturated soil mechanics to Nanyang Technological University. We have benefited greatly from the interactions with distinguished researchers in the field of unsaturated soils (not in any order of merit): Professors Geoffrey Blight (late), Delwyn G Fredlund, Pierre Delage, David Toll, James Graham and Sandra Houston who were at Nanyang Technological University at various points in time. We are grateful for the collaboration with various individuals: Professors Rifat Bulut, Hosaam Albuel-Naga, Ria Soemitro, Snehasis Tripathy, Arun Prasad, Alexandro Tarantino Devendra N. Singh and Lyesse Laloui.

We learnt through the experiences of many postgraduate and undergraduate students, and research staff who worked on unsaturated soils: Yeo Sir Hoon, Tran Sy Kien, Chin Kheng Boon, Than Than Nyunt, Cheng Zhuoyuan, Richard Kizza, Huang Wengui, Zou Lei, Sam Bulolo, He Liangcai, Zhang Xihu, Agus S. Samingan, Tou Jen Hau, Sri Widiastuti, Julianto Cahyadi, Lee Chin Chye, Tsen-Tieng Daryl, Trinh Min Thu, Henry Krisdani, Yang Hong, Goh Shin Guan, Alfrendo Satyanaga, Deny Tami, Harnas F.R., Indrawan I.G.B., Inge Meilani, Fifi Melinda, Lee Siew Beng, Ang Sok Ser, Senthikumar s/o Alagan, Koh Xin Yun, Jasmine Soh Xiang Qin, Rayner Wee Jian De, Ong Lay Sah @ Lisa, Peh Qingli, Tan Ee Jin, Tan Wenli Kyna, Tan Wei Jian, Liu Fang Ming, Phoong Bo Jun, Wong Kah Chou, Hardy Kek, Timothy Toh Hong Ern, Feng Yuyun, Ho Lun Fa, Muhammad Ashraf Assawaf B S, Chia Wei Ning Samuel, Ng Rui Kang Keith, Ng Teik Ting, K Loshana, Jesslyn Chua Jie Si, Jerome Tiah Sze Kiat, Eng Zheng, Leong Yoke Mun, Toh Jia Ming, Cui Xiao, Liu Chenying, Tang Weilun, Chen Xingyu, Lee Wen Yan, Jasmine Bte Sadimin, Li Lingyan, Poorooye Harivansh, Khairul Anwar Bin Ramli, Jeffrey Ho Jun Hua, V Madhavan Narayanan, Jenell A Sy Hui Zhen, Tong Wan Yi, Chong Han Siang, Chan Win Kit, Muhammad Amirul Haziq Bin Yahya, Chia You Xun, and Sarah Peter.

The second author would like to specially thank all his peers who accompanied him through his Ph.D. journey at Nanyang Technological University: Dr. Abdul Halim Hamdany, Dr. Alfrendo Satyanaga, Dr. Huang Wengui, Dr. Sam Bulolo, Dr. Richard and Felicia; friends who supported his research: Bjorn Tan and Lai Shan from Wykeham Farrance; former colleagues at Kiso-Jiban for all the great time in learning in situ and laboratory testing: Triwi Hayuningrum and Vivian Sim.

We apologise if we inadvertently omitted the names of individuals who should have been in the previous list. All contributions are gratefully acknowledged.

Last but not least, we thank the staff at Taylor & Francis: Tony Moore, who got us started on writing this book, as well Frazer Merritt and Aimee Wragg, for their patience and professionalism in handling editorial matters.

Chapter 1

Introduction

1.1 HISTORICAL DEVELOPMENT

The early study of soil mechanics was not separated into saturated and unsaturated soils as evidenced by the papers presented since the First International Society for Soil Mechanics and Foundation Engineering Conference in 1936. However, the principles and theories for saturated soils started emerging from that conference, and the concept of effective stress was first presented by Terzaghi at the same conference. The establishment of the principles of effective stress led to rapid developments in saturated soil mechanics and its application to numerous geotechnical engineering problems. In hindsight, the development of saturated soil mechanics first is natural, as saturated soils involved only two phases, soil and air (dry soil) or soil and water (saturated soil). The understanding of unsaturated soils faced many challenges in the early years. First, three phases are involved; second, soil testing was still in its infancy stage of development, and finally, most serious geotechnical engineering problems at that time involved saturated clay soils. The development of saturated soil mechanics led to the design of better soil tests, equipment and instruments. Although research on unsaturated soils did not stop completely, it was only in 1959 that a concept of effective stress for unsaturated soils appeared (Bishop 1959). In 1977, Fredlund and Morgenstern proposed two stress-state variables to describe the behaviour of unsaturated soils. With the advent of data acquisition and personal computers in the 1970s, longer duration tests and feedback control systems were made possible, and this in part helped with development of testing of unsaturated soils at constant suction where a very low shearing rate has to be applied, much lower than the shearing rate of drained tests for saturated soils.

At the same time when saturated soil mechanics was developing since the First International Society for Soil Mechanics and Foundation Engineering Conference in 1936, a separate group of people was studying a peculiar soil known as expansive soils that cause problems to many building structures around the world. The expansive soils problem occurs with soils in which the initial condition was dry/unsaturated, but they swell extensively when wetted. Such an expansive soil problem was first recognised by the U.S. Bureau of Reclamation in 1938 for a foundation of a steel siphon at their Owyhee Project in Oregon (Holtz and Gibbs 1956). Interest in expansive soils grew, and the first national conference on expansive clay was held at the Colorado School of Mines in Golden, Colorado, in 1959. The first and second International Conferences on Expansive Soils were held at Texas A & M University in 1965 and 1969, respectively. This was followed by five more conferences on expansive soils with the seventh or last International Conference on Expansive Soils being held in Dallas, Texas, in 1992. It was realised that expansive soils fall naturally into the domain of unsaturated soils and since then it was subsumed under the International Conference of Unsaturated Soils where the first conference was held in 1995 in Paris, France. Since then many international and regional conferences have been regularly held to disseminate the latest research and developments in unsaturated soils.

Despite the numerous advances made in the testing of unsaturated soils, unsaturated soil testing remains largely in the research laboratories of the universities, although a few unsaturated soil tests have been standardised. Currently, there is no book solely on the laboratory testing of unsaturated soils. This book collates our experiences and those of others from the literature on testing of unsaturated soils to bring the advances made in unsaturated soil tests in research to commercial soil testing laboratories.

1.2 UNSATURATED SOILS

Unsaturated soils usually exist near the ground surface and arise mainly due to a deep groundwater table and climatic conditions. Hence, unsaturated soils are encountered widely in the world. Commonly, unsaturated soils are associated with residual soils, expansive soils and loess. Brief descriptions of residual soils, expansive soils and loess are given in the following sections.

1.2.1 Residual soils

The wet and humid climatic conditions in the tropics led to extensive physical and chemical weathering of rocks and rock formations, leading to the formation of residual soils (Figure 1.1). The degree of weathering is most extensive near the ground surface and decreases with depth. The depth of weathering can be quite variable. The weathering profile is commonly described using Little's (1969) six grades weathering profile as shown in Figure 1.2 where weathering grade I refers to fresh intact rock and weathering grade VI refers

Figure 1.1 World distribution of residual soils.

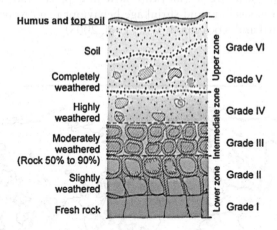

Figure 1.2 Weathering profile based on Little (1969).

to residual soil. The zones of highly weathered (grade IV) and completely weathered (grade V) which still bears the original rock structure are termed saprolite, while the zone that bears no resemblance to the original parent rock (grade VI) is termed a lateritic or residual soil. The boundary between each weathering grade is not clearly defined, and tests on residual soils can involve weathering grades IV to VI.

1.2.2 Expansive soils

Expansive soils refer to soils which swell significantly on wetting and are responsible for extensive damage to residential structures and infrastructures such as roads (Figure 1.3). Volume changes in expansive soils usually exceed 10% (Chen 1988; Nelson and Miller 1992). Damages due to expansive soils have been estimated to cost billions of dollars annually (Steinberg 1985; Dasog and Mermut 2013). The properties of expansive soils are due to the presence of swelling clay minerals, smectite or vermiculite (Rogers et al. 1993). Expansive soils are developed by geological processes which allow accumulation of predominantly silt and clay-sized particles that contained large quantities of expansive minerals (Rollings and Rollings 1996). The swelling and shrinkage of expansive soils depend on soil suction. Soil suction gives a measure of the tendency of the soil to undergo volume change when its moisture content changes with time (Cameron and Walsh 1984; Chen 1988; Nelson and Miller 1992; Bulut et al. 2001). All engineering structures on expansive soils are subjected to variations of suction at the soil surface due to climate, vegetation, drainage, site cover and watering patterns (Lytton 1977).

Hence, measuring soil suction is crucial for investigating expansive soil behaviour, and knowledge of moisture flow in expansive soils enables the estimation of the swelling and shrinkage behaviour (e.g., Arampatzis et al. 2001; Baumgartl and Kock 2004; Wray et al. 2005).

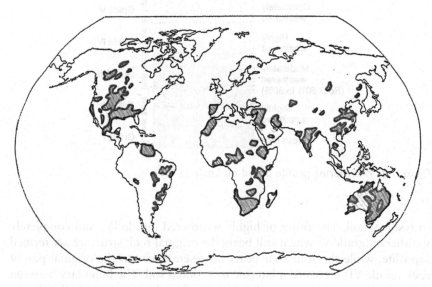

Figure 1.3 World distribution of expansive soils.

1.2.3 Loess

Loess is mainly an aeolian deposit but can also be formed by glacier grinding rocks to a fine powder and transporting them to the end of the glacier where the sediment becomes loess. The composition of loess is primarily silt-sized particles with small quantities of clay and sand particles. Loess can be found in many parts of the world (Figure 1.4), including Africa, Antarctica, Asia, central and southern Europe, northern Russia, north-western and central USA and South America (Porter 2007; Roberts et al. 2007; Rousseau et al. 2007; Zárate 2007). The most extensive occurrence of loess can be found in China. The Loess Plateau in northern and north-western China occupies a total area of about 65×10^4 km^2, which accounts for more than 6% of China's land area (Tan 1988; Derbyshire 2001; Xu et al. 2014; Li et al. 2019).

Dry loess deposits can stand vertically. However, loess is susceptible to collapse upon wetting and is associated with many geotechnical engineering problems such as wetting-induced landslides, hydroconsolidation and seismic settlement (Feda 1988; Feda et al. 1993; Dijkstra et al. 1994, 1995; Rogers et al. 1994; Derbyshire 2001; Delage et al. 2005; Yuan and Wang 2009; Xu et al. 2014). The engineering properties of loess are primarily controlled by its mineralogical composition and structure, including macroscopic texture and microstructure. Soil structure refers to both fabric and the non-frictional interparticle forces between soil particles (Lambe and Whitman 1969, Mitchell and Soga 2005). Soil fabric usually refers to the arrangement and association of particles, particle groups and pore spaces.

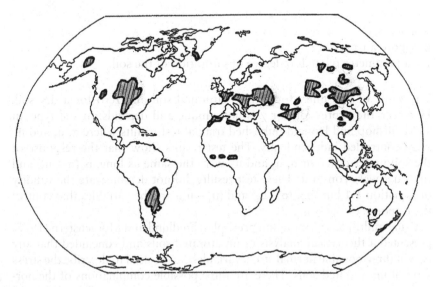

Figure 1.4 World distribution of loess.

1.3 STRESSES AND STRESS-STATE VARIABLES

In testing, the stress condition of a soil specimen in an element test such as oedometer, direct shear or triaxial test is important for interpretation. The establishment of the effective stress concept by Terzaghi (1936) has enabled the successful application of saturated soil mechanics to many geotechnical problems involving saturated soils. Hence, it is easy to understand why soil tests typically involve soil specimens in the fully saturated condition.

When a soil is fully saturated, it has two phases: soil solids and water. Both phases are incompressible, but the soil skeleton consisting of soil particles in contact (for inert particles such as sand) or in close proximity due to attractive and repulsive forces (for clay particles) reacts to the externally applied stresses. The soil particles can be re-arranged into a more compact structure with smaller voids resulting in deformation and a stronger and stiffer soil. The stresses leading to such a change are attributed to the stresses acting on the solid phase (normal stresses, σ) and the stress acting on the water phase (pore-water pressure, u_w). According to Terzaghi (1936), it is the difference between σ and u_w given by the effective stress σ' that is causing the change in state. Hence, in an element test for saturated soils, σ and u_w are applied or monitored and test results are more often interpreted based on the principles of effective stress.

Bishop (1959) proposed an effective stress for unsaturated soils:

$$\sigma' = \left(\sigma - u_a\right) + \chi\left(u_a - u_w\right) \tag{1.1}$$

where
u_a = pore-air pressure
χ = a parameter related to degree of saturation of the soil.

The value of χ is unity for a fully saturated soil and zero for a dry soil. However, the expression for χ is non-unique and depends on soil type. In 1961, Bishop and Donald published triaxial test results where σ, u_a and u_w were controlled independently. The test results show that the response of the soil under different σ, u_a and u_w were the same as long as $(\sigma - u_a)$ and $(u_a - u_w)$ were constant. These test results did not demonstrate the validity of Equation 1.1 but that $(\sigma - u_a)$ and $(u_a - u_w)$ are the variables that control the response of the soil.

Using multi-phase continuum principles, Fredlund and Morgenstern (1977) presented a theoretical analysis of unsaturated soils and concluded that any two of three possible normal stress variables can be used to describe the stress state of unsaturated soils. There are three possible combinations of the normal stress variables: (1) $(\sigma - u_a)$ and $(u_a - u_w)$, (2) $(\sigma - u_w)$ and $(u_a - u_w)$ and (3) $(\sigma - u_a)$ and $(\sigma - u_w)$. Among the three combinations, $(\sigma - u_a)$ and $(u_a - u_w)$

is the most convenient combination to apply in laboratory tests and shall be the stress-state variables used in this book.

REFERENCES

Arampatzis, G., Tzimopoulos, C., Sakellariou-Makrantonaki, M., & Yannopoulos, S. (2001). Estimation of unsaturated flow in layered soils with the finite control volume method. *Irrigation and Drainage, 50*: 349–358.

Baumgartl, Th. & Kock, B. (2004). Modeling volume change and mechanical properties with hydraulic models. *Soil Science Society of America Journal, 68*: 57–65.

Bishop, A. W. (1959). The principle of effective stress. *Teknisk Ukeblad, Norwegian Geotechnical Institute, 106*(39): 859–863.

Bishop, A. W. & Donald, I. B. (1961). The experimental study of partly saturated soil in the triaxial apparatus. In *Proceedings of the Fifth International Conference on Soil Mechanics and Foundation Engineering, Paris*, Vol. 1, pp. 13–21.

Bulut, R., Lytton, R. L., & Wray, W. K. (2001). Soil suction measurements by filter paper. In *Expansive Clay Soils and Vegetative Influence on Shallow Foundations*, ASCE Geotechnical Special Publication No. 115 (eds. C. Vipulanandan, M. B. Addison, and M. Hasen), ASCE, Reston, Virginia, pp. 243–261.

Cameron, D. A. & Walsh, P. F. (1984). The prediction of moisture induced foundation movements using the instability index. *Australian Geomechanics*, No. 8. 5–11.

Chen, F. H. (1988). *Foundations on expansive soils*, 2nd ed. New York: Elsevier.

Dasog, G. S. & Mermut, A. R. (2013). Expansive soils and clays. In Bobrowsky, P. T. (Ed.), *Encyclopedia of natural hazards*. Encyclopedia of Earth Sciences Series. Dordrecht: Springer. https://doi.org/10.1007/978-1-4020-4399-4_124

Delage, P., Cui, Y. J., & Antoine, P. (2005). Geotechnical problems related with loess deposits in Northern France. In *Proceedings of International Conference on Problematic Soils*, Eastern Mediterranean University, Famagusta, N. Cyprus, pp. 25–27.

Derbyshire, E. (2001). Geological hazards in loess terrain, with particular reference to the loess regions of China. *Earth Science Review, 54*(1): 231–260. https://doi.org/10.1016/S0012-8252(01)00050-2

Dijkstra, T. A., Rogers, C. D. F., Smalley, I. J., Derbyshire, E., Li, Y. J., & Meng, X. M. (1994). The loess of northcentral China: Geotechnical properties and their relation to slope stability. *Engineering Geology, 36*(3–4): 153–171. https://doi.org/10.1016/0013-7952(94)90001-9

Dijkstra, T. A., Smalley, I. J., & Rogers, C. D. F. (1995). Particle packing in loess deposits and the problem of structure collapse and hydroconsolidation. *Engineering Geology, 40*(1–2): 49–64. https://doi.org/10.1016/0013-7952(95)00022-4

Feda, J. (1988). Collapse of loess upon wetting. *Engineering Geology, 25*(2–4): 263–269. https://doi.org/10.1016/0013-7952(88)90031-2

Feda, J., Boháč, J., & Herle, I. (1993). Compression of collapsed loess: Studies on bonded and unbonded soils. *Engineering Geology, 34*(1–2): 95–103. https://doi.org/10.1016/0013-7952(93)90045-E

Fredlund, D. G. & Morgenstern, N. R. (1977). Stress state variables for unsaturated soils. *Journal of Geotechnical Engineering Division*, ASCE, 1039GT5: 447–466.

Holtz, G. W. & Gibbs, H. J. (1956). Engineering properties of expansive soils. *Transactions of the American Society of Civil Engineers, 121*(1): 641–663. https://doi.org/10.1061/TACEAT.0007325

Lambe, T. W. & Whitman, R. V. 1969. *Soil mechanics.* John Wiley & Sons, Inc. New York, USA.

Li, X.-A., Li, L., Song, Y., Hong, B., Wang, L., & Sun, J. (2019). Characterization of the mechanisms underlying loess collapsibility for land-creation project in Shaanxi Province, China—a study from a micro perspective. *Engineering Geology, 249*: 77–88. https://doi.org/10.1016/j.enggeo.2018.12.024

Little, A. L. (1969). The engineering classification of residual soils. In *Seventh International Conference on Soil Mechanics and Foundation Engineering,* ISSMFE, Mexico, Vol. 1, pp. 1–10.

Lytton, R. L. (1977). Foundations in expansive soils. In Desai, C. S. & Christian, J. T. (Eds.), *Numerical methods in geotechnical engineering.* New York: McGraw-Hill, pp. 427–458.

Mitchell, J. K. & Soga, K. 2005. *Fundamentals of soil behavior,* 3rd ed. Hoboken, NJ: John Wiley & Sons, Inc.

Nelson, J. D. & Miller, D. J. (1992). *Expansive soils: Problems and practice in foundation and pavement engineering.* New York: Wiley.

Porter, S. C. (2007). Loess records: China. In Elias, S. A. (Ed.), *Encyclopedia of quaternary science.* Amsterdam, The Netherlands: Elsevier, pp. 1429–1440.

Roberts, H. M., Muhs, D. R., & Bettis, E. A., III (2007). Loess records: North America. In Elias, S. A. (Ed.), *Encyclopedia of quaternary science.* Amsterdam, The Netherlands: Elsevier, pp. 1456–1466.

Rogers, C. D. F., Dijkstra, T. A., & Smalley, I. J. (1994). Hydroconsolidation and subsidence of loess: Studies from China, Russia, North America and Europe. *Engineering Geology, 37*: 83–113. https://doi.org/10.1016/0013-7952(94)90045-0

Rogers, J. D., Olshansky, R., & Rogers, R. B. (1993). Damage to foundations from expansive soils. *Claims People, 3*(4): 1–4.

Rollings, M. P. & Rollings, R. S. (1996). *Geotechnical materials in construction.* London, UK: McGraw-Hill Publishing Co.

Rousseau, D.-D., Derbyshire, E., Antoine, P., & Hatté, C. (2007). Loess records: Europe. In Elias, S. A. (Ed.), *Encyclopedia of quaternary science.* Amsterdam, The Netherlands: Elsevier, pp. 1440–1456.

Steinberg, M. L. (1985). Controlling expansive soil destructiveness by deep vertical geomembranes on four highways. *Transportation Research Records, 1032*: 48–53.

Tan, T. K. (1988). Fundamental properties of loess from northwestern China. *Engineering Geology, 25*: 103–122.

Terzaghi, K. (1936). The shear strength of saturated soils. In *Proceedings of the First International Conference on Soil Mechanics and Foundation Engineering,* Cambridge, MA, Vol. 1, pp. 54–56.

Wray, W. K., El-Garhy, B. M., & Youssef, A. A. (2005). Three-dimensional model for moisture and volume changes prediction in expansive soils. *Journal of Geotechnical and Geoenvironmental Engineering, 131*(3): 311–324.

Xu, L., Dai, F., Tu, X. et al. (2014). Landslides in a loess platform, North-West China. *Landslides, 11*: 993–1005. https://doi.org/10.1007/s10346-013-0445-x

Yuan, Z. X. & Wang, L. M. (2009). Collapsibility and seismic settlement of loess. *Engineering Geology*, *105*: 119–123. https://doi.org/10.1016/j.enggeo.2008.12.002

Zárate, M. (2007). Loess records: south America. In Elias, S. A. (Ed.), *Encyclopedia of quaternary science*. Amsterdam, The Netherlands: Elsevier, pp. 1466–1479.

FURTHER READING

Fredlund, D. G. (2006). Unsaturated soil mechanics in engineering practice. *Journal of Geotechnical and Geoenvironmental Engineering*, *132*(3): 286–321.

Houston, S. L. (2019). It is time to use unsaturated soil mechanics in routine geotechnical engineering practice. *Journal of Geotechnical and Geoenvironmental Engineering*, *145*(5): 02519001.

Chapter 2

Basic definitions, test environment and general apparatuses

2.1 INTRODUCTION

In all soil mechanics textbooks, phase relationships appear at the beginning of the book, as it is the most elementary way to describe a soil where the relationship of one phase of the soil is related to other phases of the soil in terms of mass and volume. It is also the phase relationships that recognise that soil in its most general state is unsaturated. Some phase relationships used in unsaturated soils were borrowed from other disciplines, and these have been included in this chapter for the benefit of those not familiar with unsaturated soil mechanics. In addition, a glossary of terms uncommon to saturated soil mechanics is given in Appendix A.

Unsaturated soils present challenges to the current norm of sampling, storage and sample preparation. Unlike saturated soils where it is sufficient to maintain the soil in a fully saturated condition during sampling, storage and sample preparation, the meaning of degree of saturation of unsaturated soils needs to be considered, as the degree of saturation of an unsaturated soil represents a transient condition which changes with the climatic condition. The purpose of obtaining the soil sample must be made known, which dictates the procedures needed for sampling, storage and sample preparation. If the purpose is to determine the in situ moisture content and suction, it is important to obtain the soil sample at its natural moisture condition and to ensure that there is no change in its moisture condition during sampling, storage and sample preparation. Being unsaturated, a soil exhibits higher strength and lower ductility. Hence, an unsaturated soil sample is harder to sample and more susceptible to cracks and breakage during sampling. Sampling at depth becomes more challenging.

2.2 PHASE RELATIONSHIPS

It is common to represent a soil as three phases where each phase has a mass and volume as shown in Figure 2.1. Generally, the mass of air is assumed to be negligible in relation to the other phases.

DOI: 10.1201/b22304-2

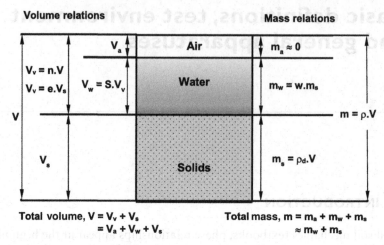

Figure 2.1 Volume-mass relations.

2.2.1 Porosity

Porosity n is defined as the ratio of volume of voids V_v to the total volume V, and is commonly expressed in percent:

$$n = \frac{V_v}{V}(100\%)$$ (2.1)

According to Equation 2.1, n can range from 0% to 100%, but in the loosest soils, the minimum value of n is about 10%.

2.2.2 Void ratio

Void ratio e is defined as the ratio of the volume of voids V_v to the volume of soil solids V_s:

$$e = \frac{V_v}{V_s}$$ (2.2)

According to Equation 2.2, e can range from 0 to greater than 1. In the densest soil, e typically ranges between 0.1 and 0.2.

2.2.3 Degree of saturation

Degree of saturation S is defined as the ratio of the volume of water in the void space to the volume of the voids and is usually expressed in percent:

$$S = \frac{V_w}{V_v}(100\%)$$ (2.3)

According to Equation 2.3, S can range from 0% for dry soil to 100% for a saturated soil and is between 0% and 100% for an unsaturated soil.

2.2.4 Gravimetric water content

Gravimetric water content w is defined as the ratio of mass of water M_w to the mass of soil soils M_s and is usually expressed in percent:

$$w = \frac{M_w}{M_s}(100\%)$$ (2.4)

Gravimetric water content w is more commonly referred to as water content in geotechnical engineering.

2.2.5 Volumetric water content

Volumetric water content θ_w is defined as the ratio of the volume of water V_w to the total volume of a soil V:

$$\theta_w = \frac{V_w}{V}$$ (2.5)

When the soil is saturated, i.e., $S = 100\%$, V_w will be equal to V_v, and the saturated volumetric water content θ_s will be numerically equal to the porosity n.

2.2.6 Soil density

There are two commonly used soil densities, i.e., total density and dry density. Total density of a soil is defined as the ratio of the total mass of the soil M and the total volume of the soil V:

$$\rho = \frac{M}{V}$$ (2.6)

Total density ρ is also referred to as bulk density, and when the soil is fully saturated ($S = 100\%$), it is usually referred to as the saturated density ρ_{sat}.

As total density ρ does not take into account the degree of saturation of the soil, it is more useful to use dry density ρ_d to express the quantity of soil solids that are packed into a unit volume of soil as a measure of its compactness. The dry density ρ_d is defined as the ratio of the mass of soil solids M_s and the total volume of the soil V:

$$\rho_d = \frac{M_s}{V}$$ (2.7)

2.2.7 Volume-mass relationships

Using Equations 2.1–2.7, other useful volume-mass relationships can be derived as summarised next:

$$S \cdot e = w \cdot G_s \tag{2.8}$$

$$\rho = \frac{1+w}{1+e} \rho_s = \frac{G_s(1+w)}{1+e} \rho_w = \frac{G_s + S \cdot e}{1+e} \rho_w \tag{2.9}$$

$$\rho_d = \frac{1}{1+e} \rho_s = \frac{G_s}{1+e} \rho_w \tag{2.10}$$

$$\theta_w = \frac{S \cdot e}{1+e} = S \cdot n = \frac{w \cdot G_s}{1+e} = \frac{S \cdot w \cdot G_s}{S + w \cdot G_s} \tag{2.11}$$

2.3 ROLE OF AIR

In the previous phase relationships, the air phase is "invisible". However, in the testing of unsaturated soils, the air phase is always present and in some tests leads to "inconveniences" and in other tests leads to poor test results. There are three different situations where the role of air has to be considered in the testing of unsaturated soils:

Situation 1:

Whenever there is water and air, water is present in the air as water vapour. In a closed environment where air is above water, water molecules leave the water surface as water vapour and re-enter the water surface eventually reaching equilibrium when the number of water molecules leaving and entering becomes equal at the saturated vapour condition. According to Dalton's law, the water vapour in the air has a partial pressure, and at the saturated vapour condition, the partial pressure of the water vapour is equal to the saturation vapour pressure. Equations to estimate the saturated vapour pressure of water are given in Appendix B. The saturated vapour pressure depends on the temperature (Figure 2.2). Evaporation takes place when there are more water molecules leaving than entering the water surface, and the partial pressure of water vapour in the air is below the saturation vapour. Condensation takes place when there are more water molecules entering than leaving the water surface, and the partial pressure of water vapour in the air exceeds the saturation vapour pressure. Hence, the partial pressure of the water vapour in the air represents the degree to which the air is saturated with water vapour, and the ratio of the partial pressure of the water vapour in the air to the saturated vapour pressure is called relative humidity.

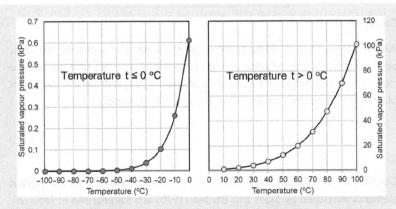

Figure 2.2 Saturated vapour pressure of water based on Huang (2018).

Condensation is observed in low suction tests (high relative humidity) when there is a drop in temperature. If the condensate falls onto the soil specimen, suction of the soil specimen may be affected. Usually, evaporation occurs during unsaturated soil tests due to two reasons: (1) air pressure above atmospheric pressure supplied to the test apparatus (closed chamber) is usually dried before entering the test apparatus, and (2) the soil specimen is exposed to ambient conditions where relative humidity is low. The suction of a soil increases as it dries. The effect of (1) on the soil specimen is dependent on the air space around the soil specimen and the air voids in the soil specimen. It is negligible when the ratio of the volume of air space to the volume of the soil specimen is small. The effect of (2) may be minimised by increasing the relative humidity of the air space around the specimen.

Situation 2:

In a test system where both air and water are present, the air can dissolve in water. The volume of air that can dissolve in a unit volume of water is called volumetric coefficient of solubility, and the rate at which air dissolves in water is called diffusivity. The amount of dissolved air in water is governed by Henry's law, which states that the amount of dissolved gas in a liquid is proportional to its partial pressure above the liquid. The higher the amount of dissolved air in water, the higher the compressibility of the water becomes, thus affecting the response behaviour of the pore-water pressure measurement in the test system. This is well known in saturated soil tests where the use of de-aired water is highly recommended. In unsaturated soil tests where air pressure remains elevated above atmospheric pressure for a long duration, the water in the system and the soil will become more and more saturated with dissolved air. However, to date, no effect of dissolved air in the pore-water of unsaturated soil on its engineering properties has been reported.

Situation 3:

Air can diffuse through water and through the water phase of soils and other porous materials. The diffusion process is governed by Fick's first law and the mass rate of transfer across a unit area of water J_m depends on the coefficient of diffusion D and the concentration gradient ∇u_a (Equation 2.12).

$$J_m = -D\nabla u_a \tag{2.12}$$

Coefficients of diffusion for air through the water phase of soils and porous elements measured by Sides and Barden (1970) are summarised in Table 2.1. The implication of Table 2.1 is that air will diffuse through the rubber membrane, soil specimen and porous elements during an unsaturated soil test in the laboratory. Hence, caution needs to be exercised when the diffused air presents a measurement error.

It is common in laboratory unsaturated soil tests, where a high air-entry ceramic disk is used to separate the air and water phases, to track the volume of water flowing out from the unsaturated soil specimen. In such tests, the diffused air will accumulate below the high air-entry ceramic disk and interfere with water flowing out of the soil specimen and, hence, its volume measurement. Padilla et al. (2006) measured the average diffusion rate of air through a 15-bar ceramic disk as shown in Figure 2.3. The average diffusion rate increases as the applied air pressure increases. Padilla et al. (2006) recommended that the tests involved be flushed according to the frequency shown in Table 2.2.

Table 2.1 Coefficients of diffusion for air through different materials

Material	Coefficient of Diffusion, D $(m^2/s) \times 10^{-10}$
Free water	22.0
Latex rubber (0.5 mm thick)	1.1
Coarse porous stone (6 mm thick)	25.0
High air-entry ceramic disk (6 mm thick)	1.6
Kaolin (saturated, $w = 75\%$)	6.2
Dement clay (saturated, $w = 53\%$)	4.7
Jackson's-bentonite clay mix (saturated, $w = 39\%$)	0.1
Compacted west water clay ($w = 16.5\%$)	0.38

(data from Sides and Barden (1970)).

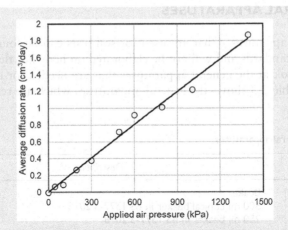

Figure 2.3 Average diffusion rate of air through saturated 15-bar high air-entry ceramic disk with applied air pressure (data from Padilla et al. (2006)).

Table 2.2 Frequency of flushing of high air-entry ceramic disks recommended by Padilla et al. (2006)

Ceramic disk	Frequency of flushing
1-bar	Once in three days
3-bar	Once in three days
5-bar	Once in two days
15-bar	Once a day for applied air pressures < 750 kPa
	Twice or more times a day for applied air pressure > 750 kPa

2.4 TEST ENVIRONMENT

Control of temperature and relative humidity of the test environment is needed when required, especially in the testing of unsaturated soils. When required, tests shall be conducted in an environment where the temperature and relative humidity can be controlled to within acceptable limits. For temperature, British Standards (BS1377, 1990) typically specify that the temperature should be kept to within ±4°C, whereas American Society for Testing and Materials (ASTM) typically specifies ±5°C. Commonly adopted standard temperature is taken as 20°C, but this may be difficult to achieve in tropical countries laboratories, and usually more attention is placed on recording the daily maximum and minimum temperatures and observing that the temperature fluctuation range is kept within ±4°C or ±5°C during the test. For relative humidity, there is no specific mention except for the storage of samples. Relative humidity may play a bigger role in testing unsaturated soils, as temperature and relative humidity affect the rate of drying of soil specimens during the test where such specimens are exposed to the ambient condition.

2.5 GENERAL APPARATUSES

The general apparatuses that are used for soil tests can be found in many books and standards and will not be elaborated on here. For this book, we shall only look at the general apparatuses that are needed for unsaturated soil tests. Table 2.3 shows the general apparatuses which are referred to in this book.

Table 2.3 General apparatuses

Apparatus	Description
	Beaker 500 ml conical beaker in BS 1377-3:2018 250 ml beaker in BS 1377-3:2018
	Oven Temperature of 105°C to 110°C to be maintained for drying soil samples in BS. Temperature of 45°C to 50°C are to be used for soil samples that experience physical changes at the higher drying temperature of 105°C to 110°C Temperature of 110 ± 5°C to be maintained for drying soil samples in BS
	Burette 50 ml Burette in BS 1377-3:2018
	Desiccator Used to cool specimen after oven dry. Can be either using vacuum pump or using silica gel
	Electronic weight scales Accuracy depends on the test 0.1 g for particle density test using gas jar 0.01 g for typical water content measurement 0.001 g for particle density test using small pycnometer
	Evaporating dish 150 mm diameter – BS 1377-2 (1990)
	Round bottom flask

(Continued)

Table 2.3 (Continued)

Apparatus	Description
	Flat bottom flask
	BS1377-2 Small pycnometer – 50 mL Large pycnometer – 1 L
	Funnel BS1377-2 (1990) Glass filter funnel – 100 mm diameter Buchner funnel – 150 mm diameter
	Gas jar 1 L in capacity – BS1377-2 (1990)
	Hand vacuum pump
	Burette/pipette pump
	Hydrometer Apparatus used to conduct sedimentation test by hydrometer method according to BS 1377-2 (1990)
	Measuring cylinder 25 mL, 50 mL, 100 mL, 500 mL, 1 L measuring cylinder BS1377-2 (1990)

(Continued)

Table 2.3 (Continued)

Apparatus	Description
	Pestle and mortar Used to pulverise material in Atterberg Limit (D4318-2017)
	Tube/tubing Used to provide water or air pressure
	Retort stand with clamp Typically used to support burette
	Stirrer Usually made of glass
12:00	Stopwatch Used to record time
	Tripod Used to place non-flat bottom apparatus such as Tempe cell or round bottom flask
	U tube Used as manometer
	Weight Used to apply small seating pressure to a specimen

REFERENCES

Huang, J. (2018). A simple accurate formula for calculating saturation vapor pressure of water and ice. *Journal of Applied Meteorology and Climatology*, 57(6): 1265–1272. Retrieved May 12, 2022, from https://journals.ametsoc.org/view/journals/apme/57/6/jamc-d-17-0334.1.xml

Padilla, J. M., Perera, Y. Y., Houston, W. N., Perez, P. N., & Fredlund, D. G. (2006). Quantification of air diffusion through high air-entry ceramic disks. In *Proceedings of the Fourth International Conference on Unsaturated Soils, ASCE, Carefree, AZ*, Vol. 2, pp. 1852–1863.

Sides, G. R. & Barden, L. (1970). The times required for the attainment of air-water equilibrium in clay soils. *Journal of Soil Science*, 21: 50–62. https://doi.org/10.1111/j.1365-2389.1970.tb01151.xj

Chapter 3

Sampling, storage and sample preparation

3.1 BACKGROUND

Reliable soil properties cannot be obtained without an appropriate sampling technique, proper storage and careful sample preparation. In testing saturated soils, the soil specimen is commonly re-conditioned by saturating the soil specimen. However, unsaturated soil commonly requires the soil to be tested at its natural (as is) condition of soil suction and water content. Thus, it is not an understatement to say that unsaturated soil testing requires a much higher sample quality compared to saturated soil testing.

In this chapter, sampling, storage and sample preparation will be described with emphasis on unsaturated soils. This chapter aims to guide engineers to

1. select the best sampling method depending on the test that is desired,
2. specify proper storage conditions for soil samples intended to be used for unsaturated soil testing, and
3. select the appropriate specimen preparation method for the intended test.

3.2 RELATED STANDARDS

1. ASTM D1452/D1452M (2016) Standard Practice for Soil Exploration and Sampling by Auger Borings.
2. ASTM D1586 (2011) Standard Test Method for Standard Penetration Test (SPT) and Split-Barrel Sampling of Soils.
3. ASTM D1587/D1587M (2015) Standard Practice for Thin-Walled Tube Sampling of Fine-Grained Soils for Geotechnical Purposes.
4. ASTM D3550/D3550M (2017) Standard Practice for Thick Wall, Ring-Lined, Split Barrel, Drive Sampling of Soils.
5. ASTM D4220/D4220M (2014) Standard Practices for Preserving and Transporting Soil Samples.
6. ASTM D4700 (2015) Standard Guide for Soil Sampling from the Vadose Zone.

7. ASTM D5299/D5299M (2017) Standard Guide for Decommissioning of Groundwater Wells, Vadose Zone Monitoring Devices, Boreholes, and Other Devices for Environmental Activities.
8. ASTM D5782 (2018) Standard Guide for Use of Direct Air-Rotary Drilling for Geoenvironmental Exploration and the Installation of Subsurface Water-Quality Monitoring Devices.
9. ASTM D5783 (2018) Standard Guide for Use of Direct Rotary Drilling with Water-Based Drilling Fluid for Geoenvironmental Exploration and the Installation of Subsurface Water-Quality Monitoring Devices.
10. ASTM D5784/D5784M (2013) Standard Guide for Use of Hollow-Stem Augers for Geoenvironmental Exploration and the Installation of Subsurface Water-Quality Monitoring Devices.
11. ASTM D5876/D5876M (2017) Standard Guide for Use of Direct Rotary Wireline Casing Advancement Drilling Methods for Geoenvironmental Exploration and Installation of Subsurface Water-Quality Monitoring Devices.
12. ASTM D6151/D6151M (2015) Standard Practice for Using Hollow-Stem Augers for Geotechnical Exploration and Soil Sampling.
13. D6169/D6169M (2013) Standard Guide for Selection of Soil and Rock Sampling Devices Used With Drill Rigs for Environmental Investigations.
14. ASTM D6232 (2016) Standard Guide for Selection of Sampling Equipment for Waste and Contaminated Media Data Collection Activities.
15. ASTM D6282/D6282M (2014) Standard Guide for Direct Push Soil Sampling for Environmental Site Characterizations.
16. ASTM D6286 (2012) Standard Guide for Selection of Drilling Methods for Environmental Site Characterization.
17. ASTM D6519 (2015) Standard Practice for Sampling of Soil Using the Hydraulically Operated Stationary Piston Sampler.
18. ASTM D6640 (2015) Standard Practice for Collection and Handling of Soils Obtained in Core Barrel Samplers for Environmental Investigations.
19. ASTM D7015 (2013) Standard Practices for Obtaining Intact Block (Cubical and Cylindrical) Samples of Soil.
20. BS 5930 (2015) Code of Practice for Ground Investigations.
21. BS EN 1997-2 (2007) Eurocode 7 – Geotechnical Design – Part 2: Ground Investigation and Testing.
22. BS EN ISO 22475-1 (2006) Geotechnical Investigation and Testing – Sampling Methods and Groundwater Measurement – Part 1: Technical Principles for Execution.
23. BS 22475-2 (2011) Geotechnical Investigation and Testing – Sampling Methods and Groundwater Measurements – Part 2: Qualification Criteria for Enterprises and Personnel.

24. BS 22475-3 (2011) Geotechnical Investigation and Testing. Sampling Methods and Groundwater Measurements. Conformity Assessment of Enterprises and Personnel by Third Party.

3.3 SAMPLING CATEGORY AND SAMPLE QUALITY

Sampling is commonly carried out either by drilling and sampling (making a borehole through drilling and then using a sampler), or by excavating and retrieving the soil through block sampling (BS EN ISO 22475-1 2006; Monnet 2015). The selection of sampling method is affected by several factors such as (BS EN ISO 22475-1 2006)

1. type of soil being sampled (sandy soil, clayey soil, etc.),
2. consistency of soil (soft, stiff, etc.),
3. depth of soil sample,
4. location of groundwater table, or
5. required sample quality.

Sample quality is divided into five classes where class 1 represents the best sample quality and class 5 the worst (BS EN ISO 22475-1 2006). The sample quality class will affect the type of geotechnical engineering properties that can be determined (BS EN 1997-2 2007; ASTM D4220/D4220M 2014) as shown in Table 3.1.

3.3.1 BS EN ISO 22475-1 (2006)

Sampling can be categorised into A, B and C depending on the drilling method, soil type and sampler used. Category A sampling can be used to obtain samples of quality class from 1 to 5 where the sampling process introduces very minimal disturbance to the soil structure, water content, void ratio and chemical composition. Geotechnical engineering parameters related to shear strength, compressibility and permeability must be obtained from samples obtained by category A sampling.

Category B sampling can be used to obtain sample quality class from 1 to 5 where natural water content can be retained but the structure of the soil has been disturbed and thus affecting its engineering properties. Category B sampling is commonly used to obtain samples for index tests such as grain size distribution, water content, Atterberg limits, density and chemical test.

Category C sampling can only obtain samples of quality class 5 where the soil has been completely remoulded and thus does not retain its natural structure and water content. Tables 3.1 and 3.2 show several sampling methods, along with their potential sample quality and requirement. Samples obtained using category C sampling are only useful for visual examination of soil type and determining the sequence of layers (profile).

Table 3.1 Sampling by drilling in soils without flushing medium based on BS EN ISO 22475-1 (2006)

Cutting technique	Designation	Tool	Preferred soil type	Sampling categories*	Sample class*
Rotary drilling	Rotary dry core drilling	Single-tube core barrel	Clay, silt, fine sand, silt	B (A)	4 (2–3)
	Auger drilling	Hollow-stem auger	Clay, silt, sand, organic soils	B (A)	3 (1–2)
	Auger drilling	Drill rods with shell or flight auger; hollow-stem auger	All soils above water surface, all cohesive soils below water surface	B	4 (3)
	Auger drilling with light equipment	Shell auger or spiral flight auger	Clay to medium gravel above water surface, cohesive soils below water surface	C (B)	5
Hammer driving	Percussive core drilling	Percussion clay cutter with cutting edge inside; also with sleeve (or hollow-stem auger)	Clay, silt and soils with particle size up to $D_e/3$	A	2 (1)
	Percussive drilling	Percussive clay cutter with cutting edge outside	Non-cohesive soil	B (A)	3 (2)
			Gravel and soils with a particle size up to $D_e/3$	B	4
	Small diameter hammer driving	Hammer driving linkage with tube sampler	Soils with a particle size up to $D_e/5$	C	5
Vibration drilling with an optional slow rotation	Resonance drilling	Thick wall sampler or single-tube core barrel with optional plastic lining tube	Cohesive soil	B	4
			Non-cohesive soil	C	5
Percussion	Cable percussion drilling	Cable with percussion shell auger	Clay and silt above water surface, clay below water surface	C (B)	4 (3)
		Cable with valve auger	Gravel and sand in water	C (B)	5 (4)

Pneumatic/continuous thrust	Small diameter pneumatic/continuous thrust drilling	Pneumatic/continuous thrust linkage, with tube sampler	Clay, silt, fine sand	C	5
Grabbing	Grab drilling	Cable with grab	Gravel, boulders of size less than $D_e/2$, cobbles (above water surface)	B	4
			Gravel, boulders of size less than $D_e/2$, cobbles (below water surface)	C	5

D_e is internal diameter of sampling tool.
* The sampling categories and quality classes in parentheses can be achieved provided sampling is carried in favourable conditions.

Table 3.2 Sampling by drilling in soils with flushing medium based on BS EN ISO 22475-1 (2006)

Cutting technique	Designation	Need flushing medium	Tool	Preferred soil type	Sampling categories*	Sample class*
Rotary drilling	Rotary core drilling	Yes	Single-tube core barrel	Clay, clayey and cemented composite soils, boulders	B (A)	3 (1–2)
			Double-tube core barrel		B (A)	4 (2–3)
			Triple-tube core barrel		A	3 (1–2)
			Double/triple-tube core barrel with extended inner tube	Clay, silt	A	2 (1)
	Reverse circulation drilling	Yes	Drill rods with hollow chisel	All soils	C (B)	5 (4)
Rotary hammer driving	Rotary percussive drilling	yes	Single- or double-tube core barrel	Clay, silt (cohesive soil)	A	2 (1)
				Fine sand (Non-cohesive soil)	B	4 (3)

D_e is the internal diameter of the sampling tool.
*The sampling categories and quality classes in parentheses can be achieved provided sampling is carried in favourable conditions.

Table 3.3 Sampling method using sampler which is able to produce sample class 1 based on BS EN ISO 22475-1 (2006)

Sampler type	Abbreviation	Penetration method	Suitable soil types	Sampling category*	Sample class*
Thin-walled open-tube	OS-T/W	static or dynamic driving	Cohesive or organic soils of soft or stiff consistency	A	1
			Cohesive or organic soils of stiff consistency	A	2 (1)
Thin-walled piston tube	PS-T/W	Static driving	Cohesive or organic soils of pasty or stiff consistency and sensitive soils	A	1
Thick-walled piston tube	PS-TK/W	Static driving	Cohesive or organic soils of soft to stiff consistency and sensitive soils	B (A)	2 (1)
Cylinder (large sampler)	LS	Static rotating	Clay, silt	A	1

*The sampling categories and quality classes in parentheses can be achieved provided sampling is carried in favourable conditions.

Table 3.4 Class requirement and of geotechnical engineering properties (BS EN 1997-2 2007)

Soil properties	(BS EN 1997-2 2007) Minimum class requirement	(BS EN 1997-2 2007) Minimum sampling category	ASTM D4220/ D4220M (2014) Group
Compressibility	1	A	C or D
Shear strength	1	A	C or D
Permeability	2	A	C or D
Density, density index, porosity	2	A	C or D
Boundary of strata – fine	2	A	C or D
Water content	3	B	B
Boundary of strata – broad	4	B	B
Particle size	4	B	B
Atterberg limits, particle density, organic content, water content	4	B	B
Sequence of layers	5	C	A

3.3.2 ASTM D4220/D4220M (2014)

The sampling classification listed in ASTM D4220/D4220M (2014) differs slightly from BS EN ISO 22475-1 (2006). The classification in ASTM D4220/D4220M (2014) is divided into four groups A, B, C and D as follows:

1. Group A is only used for visual identification or profiling (Equivalent to Category C of BS EN ISO 22475-1 (2006)).
2. Group B is used for water content determination, classification, compaction and laboratory test which requires bulk sample (Equivalent to Category B of BS EN ISO 22475-1 (2006)).
3. Groups C and D give the best sample quality (Equivalent to Category A of BS EN ISO 22475-1 (2006)). However, ASTM D4220/D4220M (2014) places more emphasis on group D, which is for highly sensitive or fragile samples. The difference between groups C and D samples is that the soil sample obtained using group D must be stored in the same orientation as it is sampled due to its sensitivity.

3.4 SAMPLING IN UNSATURATED SOIL

As mentioned previously, sampling for unsaturated soil testing demands a higher sample quality than for saturated soil testing. Unsaturated soil is often encountered near the ground surface above the groundwater table (Fredlund et al. 2012), and the zone is usually referred to as vadose zone

(ASTM D4700 2015). Obtaining undisturbed unsaturated soil is more difficult due to the following:

1. Restriction in using water or drilling fluid during the sampling process (ASTM D4546 2008; ASTM D4700 2015). In unsaturated soil testing, the natural water content and void ratio are very important as the soil properties of unsaturated soil are affected by its water content (which affects its suction). The use of drilling fluid may cause the soil to become saturated due to seepage of the fluid into the soil. Thus, a sampling method using water or drilling fluid is not suitable due to the possibility of changing the natural water content of the soil.
2. A sampler which depends on suction (i.e., piston sampler) to retrieve and hold the sample may not work in unsaturated coarse-grained sands and gravels due to the high air permeability of the soil that prevents the creation of suction (ASTM D4700 2015).

Drilling methods for unsaturated soil sampling should use air only as the flushing fluid to preserve the water content of the soil. Suitable drilling methods using air as the flushing fluid are listed in Table 3.1 and only rotary dry core drilling with a hollow-stem auger or percussive core drilling with cutting edge inside and a sleeve can be used to achieve the desired sample quality. If it is not possible to use any of the methods listed in Table 3.1, drilling with flushing fluid other than water may be used (Table 3.2) provided that the water content profile of the soil is known. For such soil samples, it may be possible to bring the soil specimen back to its in situ water content before the test.

Block sampling is the best method to obtain large amounts of high-quality undisturbed unsaturated soil. It is categorised as sampling category A (BS EN ISO 22475-1 2006) and the sample quality is class 1 (Monnet 2015). Block sampling procedures are described in ASTM D7015 (2013) and in BS EN ISO 22475-1 (2006). The procedures are schematically shown in Figure 3.1 and summarised as follows:

1. Prepare a box with side lengths around 0.30 m (the side length is arbitrary, but it must be at least 10 to 15 mm larger than the sample size).
2. Remove topsoil containing roots and other organic matters which are not desired.
3. Level and smooth the ground surface at which the soil sample is going to be sampled.
4. Excavate a trench surrounding the soil sample (without inducing disturbance to the soil sample) such that there is sufficient working space for a person to go in and trim the sample. The width and height of the pedestal are about 0.90 m and 1.20 m, respectively, and it is unnecessary for the trench to be all around the soil sample.

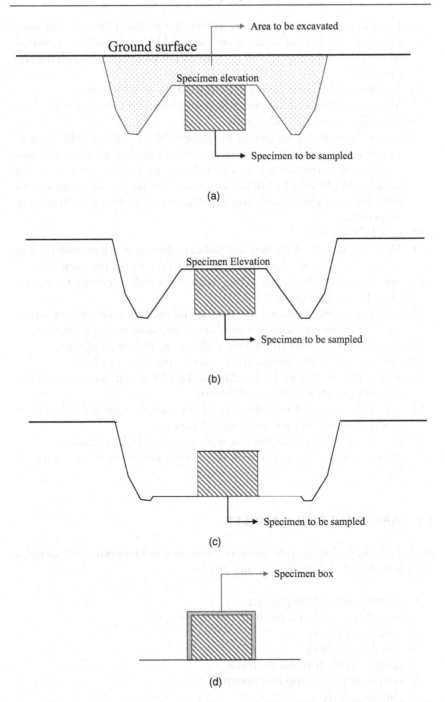

Figure 3.1 Block sampling process. (a) Stage 1 to 3. (b) Stage 4. (c) Stage 5 to 8. (d) Stage 9 to 19.

5. Use a chain saw or hand tools to trim the pedestal into the soil sample with the desired dimensions. Remoulded soil shall be removed from the sampling spot, and no water shall be in contact with the soil sample.
6. Measure and record the top elevation of the block sample.
7. Measure and record the groundwater elevation if encountered.
8. Visually classify the soil sample.
9. Cover the exposed surface of the sample block. This should be done as soon as possible if the sample is to be tested at its natural water content, as evaporation can affect the natural water content of the sample. ASTM D7015 (2013) suggests covering the soil sample with three layers of cheesecloth and painting it with melted wax to prevent evaporation.
10. Label the sample.
11. Mark the sample at the top and indicate the north orientation such as by using a piece of paper that is waxed to the top of the sample.
12. For easily disturbed soil, a box with both ends removed should be placed over the sample.
13. Fill up the space between the sample and the box with packing material to cushion the sample from disturbance during transportation.
14. Attach the top cover to the box with screws or bolts and nuts.
15. Cut the base of the sample using a wire saw, shovel or knife.
16. Slowly tilt the box and add additional packing materials to minimise gaps between the sample and the box.
17. Cover the exposed bottom part of the sample with three layers of cheesecloth and paint it with melted wax.
18. Attach a cover to close the box with screws or bolts and nuts.
19. Place a copy of the paper with the sample information on the top of the box.

3.5 LABELLING SOIL SAMPLES

In general, the following information is required in identifying soil samples (BS EN ISO 22475-1 2006; Monnet 2015):

1. identification of the project;
2. identification of trial pit, borehole number, etc.;
3. sample reference;
4. sampling method;
5. sample depth (top and bottom);
6. sample direction (top and bottom);
7. sampling date;
8. sample recovery; and
9. possible disturbance during the sampling process.

Due to climatic and groundwater conditions, the water content of unsaturated soil may vary greatly with depth (Fredlund et al. 2012). Thus, it is important to know the direction and depth of each specimen which is going to be tested. For example, air-entry value of sandy soils can be less than 10 kPa, which is equivalent to 1m depth of water. Thus, the water contents of a 1m long sample at the top and at the bottom of the sample may vary greatly. If the sample recovery is less than 100%, judgement is required to determine the actual depth of the tested specimen.

Sampling date is important to determine the possibility of the sample drying due to the long duration between sampling and testing. Similar with saturated soils, it is recommended that unsaturated soil testing should be carried out as soon as possible after sampling to avoid sample disturbance.

3.6 STORAGE OF UNSATURATED SOIL SAMPLES

Storage of the soil sample starts from the time when the soil sample is taken until the time that the soil sample is used for testing. Procedures for handling unsaturated soil samples are given in BS EN ISO 22475-1 (2006) for sampling category A or (ASTM D4220/D4220M 2014) for group D. Any disturbance which might be induced during the handling process (from site to the laboratory for testing) must be clearly indicated in the test as it might affect the test result.

Proper storage procedures for tube samples are given below:

1. BS EN ISO 22475-1 (2006) specifies that the ends of the sample tube shall be sealed with either plastic expandable packers or by a soil filling.
2. Wax must be applied on both ends of the soil sample to avoid evaporation. BS EN ISO 22475-1 (2006) specifies that for long-term sealing, microcrystalline wax up to 15% beeswax, paraffin or resin can be used. ASTM D4546 (2008) specifies that the temperature of the wax shall be 8°C to 14°C above the melting point, as very hot wax might penetrate pores and cracks, and dry the sample. Cheesecloth or aluminium foil may be placed as an interface between soil sample and the wax to avoid wax intrusion.
3. A rubber cap is placed at the top and bottom of the sample tube to protect the sample tube from impact and vibration and to keep the sample from falling out.
4. At the site, the sample tubes must be kept using a sample holder such as shown in Figure 3.2 to avoid the possibility of any accidental damage to the sample.
5. At the site, the sample tubes must be placed in a shaded area to avoid direct exposure of the soil sample to sunlight.

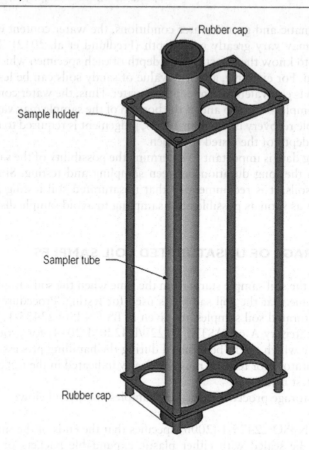

Figure 3.2 Storage of soil sample on site.

During transportation, BS EN ISO 22475-1 (2006) specifies that soil samples must be protected from vibration, shock and extreme temperatures. Samples shall be transported in a solid box or container such that the samples fit snugly to prevent bumping, rolling or dropping. Additional cushioning material such as sawdust, rubber, polystyrene, urethane foam or any other resilient material can be used to protect the samples. Cushioning between samples and the walls of the container shall have a minimum thickness of 25 mm, while the thickness of the cushion between samples and the bottom of the container shall be 50 mm.

At the laboratory, the sample tubes should be kept in a sample storage room that has a high relative humidity and constant temperature to preserve the water content of the soil samples (Monnet 2015). BS EN ISO 22475-1 (2006) specifies that the storage room temperature must be the same as the ground temperature (+6°C to +12°C) with relative humidity from 85% to 100%.

3.7 SAMPLE PREPARATIONS

Three types of soil specimens are used for unsaturated soil testing (ASTM D5333 2003; ASTM D6836-02 2008; ASTM D4829 2011):

1. undisturbed soil specimen,
2. re-constituted soil specimen, and
3. compacted soil specimen.

3.7.1 Undisturbed soil specimen

Undisturbed soil specimens can be prepared with or without a retaining ring (ASTM D6836-02 2008). A retaining ring is commonly used to maintain the shape of the specimen or to simulate k_0 condition. However, a retaining ring may cause friction, or the soil specimen may shrink within the retaining ring due to drying.

3.7.1.1 Specimen prepared from undisturbed samples with retaining ring

Preparation of undisturbed soil specimen with a retaining ring can be summarised as follows:

1. Place the retaining ring on top of the soil sample.
2. Use a trimming tool to slowly cut away the soil which is protruding beyond the edge of the retaining ring until the soil specimen slowly slides into the ring.
3. Trim the top and bottom surfaces of the soil specimen such that it is levelled with the ring.
4. Weigh the soil specimen in the retaining ring to the nearest 0.01 g.
5. Determine the initial water content from the left-over material.

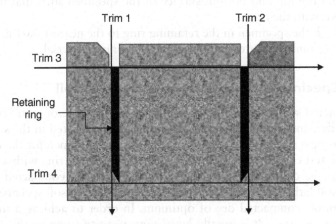

Figure 3.3 Specimen preparation using retaining ring.

3.7.1.2 Specimen prepared from undisturbed samples without retaining ring

Preparation of undisturbed soil specimen without retaining ring is as follows:

1. Trim the specimen to the desired dimensions using a trimmer. A cylindrical soil specimen is preferred for the ease of volume calculation as the soil may shrink and volume change measurement may be required.
2. Measure the dimensions (diameter/side length/height) at three locations (or four locations in BS1377: 1990) for each dimension to calculate the volume of the soil specimen.
3. Determine the average of each dimension to the nearest 0.1 mm.
4. Weigh the soil specimen to the nearest 0.01 g.

3.7.2 Specimen prepared from re-constituted soil

In some cases, it is not possible to prepare undisturbed soil specimens. Therefore, re-constitution of the soil specimen might be the only choice. Re-constituting the specimen must be done such that the specimen's condition can meet certain criteria such as duplicating the site condition. Preparation for the disturbed specimen is summarised as follows:

1. Air dry and crush the representative samples for an amount up to two times the specimen required to fill the specimen retaining ring.
2. Remove all particles with a diameter exceeding one-half the height of the retaining ring.
3. The sample may be moistened to the specified water content if required.
4. Place the moistened sample into the retaining ring or mould.
5. Tap or compact the soil to the specified density.
6. Trim the top and bottom surface of the specimen such that it is levelled with the ring.
7. Weigh the specimen in the retaining ring to the nearest 0.01 g.
8. Determine the water content of the remaining material.

3.7.3 Specimen prepared from compacted soil

A compacted soil specimen can be obtained by dynamic or static compaction. In dynamic compaction, the soil sample is compacted in the standard mould using a specific compaction effort. The soil specimen for the unsaturated soil test can then be obtained by pushing a tube or ring with a cutting edge into the compacted sample or trimmed from the compacted sample using a soil lathe. However, it is challenging to obtain a soil specimen from a soil sample compacted dry of optimum. In order to achieve a uniform/consistent specimen. It is usually more convenient to compact the soil into the soil specimen that is required for the test (i.e., water content and dry density) by using static compaction (Goh 2012; Priono et al. 2017).

REFERENCES

ASTM D1452/D1452m (2016). *Standard practice for soil exploration and sampling by auger borings.* West Conshohocken, PA: ASTM International.

ASTM D1586 (2011). *Standard test method for standard penetration test (spt) and split-barrel sampling of soils.* West Conshohocken, PA: ASTM International.

ASTM D1587/D1587m (2015). *Standard practice for thin-walled tube sampling of fine-grained soils for geotechnical purposes.* West Conshohocken, PA: ASTM International.

ASTM D3550/D3550m (2017). *Standard practice for thick wall, ring-lined, split barrel, drive sampling of soils.* West Conshohocken, PA: ASTM International.

ASTM D4220/D4220m (2014). *Standard practices for preserving and transporting soil samples.* West Conshohocken, PA: ASTM International.

ASTM D4546 (2008). *Standard test methods for one-dimensional swell or collapse of cohesive soils.* West Conshohocken, PA: ASTM International.

ASTM D4700 (2015). *Standard guide for soil sampling from the Vadose Zone.* West Conshohocken, PA: ASTM International.

ASTM D4829 (2011). *Standard test method for expansion index for soils.* West Conshohocken, PA: ASTM Standard.

ASTM D5299/D5299m (2017). *Standard guide for decommissioning of groundwater wells, vadose zone monitoring devices, boreholes, and other devices for environmental activities.* West Conshohocken, PA: ASTM International.

ASTM D5333 (2003). *Standard test method for measurement of collapse potential of soils.* West Conshohocken, PA: ASTM International.

ASTM D5782 (2018). *Standard guide for use of direct air-rotary drilling for geoenvironmental exploration and the installation of subsurface water-quality monitoring devices.* West Conshohocken, PA: ASTM International.

ASTM D5783 (2018). *Standard guide for use of direct rotary drilling with water-based drilling fluid for geoenvironmental exploration and the installation of subsurface water-quality monitoring devices.* West Conshohocken, PA: ASTM International.

ASTM D5784/D5784m (2013). *Standard guide for use of hollow-stem augers for geoenvironmental exploration and the installation of subsurface water-quality monitoring devices.* West Conshohocken, PA: ASTM International.

ASTM D5876/D5876m (2017). *Standard guide for use of direct rotary wireline casing advancement drilling methods for geoenvironmental exploration and installation of subsurface water-quality monitoring devices.* West Conshohocken, PA: ASTM International.

ASTM D6151/D6151m (2015). *Standard practice for using hollow-stem augers for geotechnical exploration and soil sampling.* West Conshohocken, PA: ASTM International.

D6169/D6169m (2013). *Standard guide for selection of soil and rock sampling devices used with drill rigs for environmental investigations.* West Conshohocken, PA: ASTM International.

ASTM D6232 (2016). *Standard guide for selection of sampling equipment for waste and contaminated media data collection activities.* West Conshohocken, PA: ASTM International.

ASTM D6282/D6282m (2014). *Standard guide for direct push soil sampling for environmental site characterizations.* West Conshohocken, PA: ASTM International.

ASTM D6286 (2012). *Standard guide for selection of drilling methods for environmental site characterization.* West Conshohocken, PA: ASTM International.

ASTM D6519 (2015). *Standard practice for sampling of soil using the hydraulically operated stationary piston sampler.* West Conshohocken, PA: ASTM International.

ASTM D6640 (2015). *Standard practice for collection and handling of soils obtained in core barrel samplers for environmental investigations.* West Conshohocken, PA: ASTM International.

ASTM D6836-02 (2008). *Standard test method for determination of soil water characteristic curve for desorption using hanging column, pressure extractor, chilled mirror hygrometer, or centrifuge.* West Conshohocken, PA: ASTM International.

ASTM D7015 (2013). *Standard practices for obtaining intact block (cubical and cylindrical) samples of soil.* West Conshohocken, PA: ASTM International.

BS EN 1997-2 (2007). *Eurocode 7 – geotechnical design – part 2: Ground investigation and testing.*

BS 5930 (2015). *Code of practice for ground investigations.* UK: BSI.

BS EN ISO 22475-1 (2006). *Geotechnical investigation and testing – sampling methods and groundwater measurement – part 1: Technical principles for execution.* UK: BSI.

BS 22475-2 (2011). *Geotechnical investigation and testing – sampling methods and ground water measurements – part 2: Qualification criteria for enterprises and personnel.* UK: BSI.

BS 22475-3 (2011). *Geotechnical investigation and testing. Sampling methods and groundwater measurements. Conformity assessment of enterprises and personnel by third party.* UK: BSI.

Fredlund, D. G., Rahardjo, H., & Fredlund, M. D. (2012). *Unsaturated soil mechanics in engineering practice.* Hoboken, New Jersey: John Wiley & Sons, Inc.

Goh, S. G. (2012). Hysteresis effects on mechanical behaviour of unsaturated soils. In *School of civil and environmental engineering.* Nanyang Technological University, Singapore, vol. Ph.D.

Monnet, J. (2015). *In situ tests in geotechnical engineering.* New Jersey: John Wiley & Sons, Inc., ISTE Ltd, UK.

Priono Rahardjo, H., Chatterjea, K., & Leong, E.-C. (2017). Laboratory investigation on hydraulic anisotropy behavior of unsaturated soil. *Canadian Geotechnical Journal* 54(7): 1034–1046.

FURTHER READING

Head, K. H. (2006). *Manual of soil laboratory testing, volume I, soil classification and compaction testing,* 3rd ed. Caithness,UK: Whittles Publishing.

Monnet, J. (2015). *In situ tests in geotechnical engineering.* New Jersy: John Wiley & Sons, Inc.

Chapter 4

Grain-size distribution, and specific gravity

4.1 BACKGROUND

Soil is an assemblage of discrete grains of various shapes and sizes (Head 2006). Thus, classification of soil is commonly based on the cumulative proportions of grains (by mass) under different ranges of sizes obtained by conducting grain-size analysis. There are two common methods to determine the grain-size distribution of soil, which are mechanical sieving and sedimentation test. Mechanical sieving is commonly carried out to determine the proportions of coarse-grained soils, while a sedimentation test (i.e., pipette and hydrometer tests) is carried out to determine the proportions of fine-grained soils.

In unsaturated soil mechanics, grain-size distribution or its parameters are commonly used to obtain properties such as soil-water characteristic curve (SWCC), hydraulic conductivity and shear strength. Such relationships are commonly known as pedo-transfer function. Considering the long test duration to obtain SWCC, pedo-transfer functions for SWCC are the most common, as they can provide a quick preliminary estimate until more accurate SWCC can be made available from tests.

The specific gravity of soil solid is used to calculate the void ratio and degree of saturation (ASTM D854-14 2014). Both parameters are commonly used in unsaturated soil mechanics, and their determinations are already well covered in test standards and classical soil mechanics textbooks. Hence, in this chapter, the emphasis will be on the determination of the grain-size distribution and specific gravity. Their roles in determining the SWCC will also be covered.

4.2 RELATED STANDARDS

1. BS EN ISO 17892-3 (2015) Geotechnical Investigation and Testing. Laboratory Testing of Soil. Part 3: Determination of Particle Density.
2. BS EN ISO 17892-4 (2016) Geotechnical Investigation and Testing. Laboratory Testing of Soil. Part 4: Determination of Particle Size Distribution.

DOI: 10.1201/b22304-4

3. BS 1377-2 (1990) Methods of Test for Soils for Civil Engineering Purposes. Part 2: Classification Tests.
4. ASTM D854-14 (2014) Standard Test Methods for Specific Gravity of Soil Solids by Water Pycnometer.
5. ASTM D421-85 (2007) Standard Practice for Dry Preparation of Soil Samples for Particle Size Analysis and Determination of Soil Constants.
6. ASTM D422-63 (2007) Standard Test Method for Particle-Size Analysis of Soils.
7. ASTM D2487-17 (2017) Standard Practice for Classification of Soils for Engineering Purposes (Unified Soil Classification System).
8. ASTM D6913/D6913M (2017) Standard Test Methods for Particle-Size Distribution (Gradation) of Soils Using Sieve Analysis.
9. ASTM D7928 (2017) Standard Test Method for Particle-Size Distribution (Gradation) of Fine-Grained Soils Using the Sedimentation (Hydrometer) Analysis.

4.3 SOIL CLASSIFICATION BASED ON GRAIN SIZE

Grain-size distribution is used to

1. classify soil according to predominant soil types such as gravel, sand, silt or clay and
2. estimate engineering properties.

In unsaturated soil mechanics, grain-size distribution is particularly useful in estimating the SWCC, which can then be used to estimate other unsaturated soil properties.

Unfortunately, different soil classification systems as shown in Table 4.1 and Figure 4.1 are in use. In some soil classifications, such as Unified Soil

Table 4.1 Soil classification based on different standards

	Minimum grain size (mm)			
Soil type	BS 1377:1990	ASTM D422-63	BS EN ISO 14688-1 2018	USDA
Large Boulders			>630	
Boulders	>200		200–630	
Cobbles	60–200		63–200	
Gravel	2–60	75–4.75	2–63	>2
Sand	0.06–2	4.75–0.075	0.063–2	0.05–2
Silt	0.002–0.06	0.075–0.005	0.002–0.064	0.002–0.05
Clay	<0.002	<0.005	<0.002	<0.002

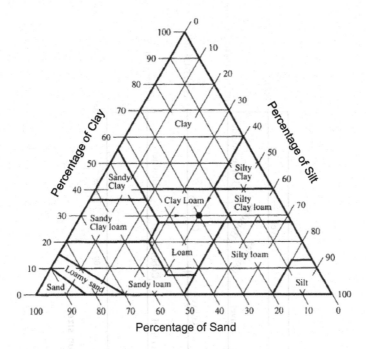

Figure 4.1 U.S. Department of Agriculture (USDA) soil classification.

Classification System (USCS) and BS 5930 (2015), additional tests are needed to determine whether the soil should be referred to as silt or clay. Comparison between different soil classification is shown in Figure 4.2.

4.4 CURVE-FITTING, GRAIN-SIZE DISTRIBUTION

Mechanical sieving provides discrete data points but does not differentiate the shape of the grains. The sedimentation test, on the other hand, based on Stokes' law assumes all grains to be an equivalent sphere. Sometimes there is a discontinuity between the overlapping data from the two tests because of the difference in assumptions. When this happens, a smooth best-fit curve is drawn through this overlapping zone which may not pass through the data points from both tests in the overlapping zone. A common criticism of the grain-size distribution curve is that the exact proportions of the grains within a size range are not known, and the grain-size D_x where x is the percent passing (e.g., D_{10}, D_{30}, D_{60}) is read from the curve assuming equal distribution of grains within the size range. As the grain-size distribution curve is used in unsaturated soil mechanics to estimate other engineering properties, it is desirable to represent the grain-size distribution curve using

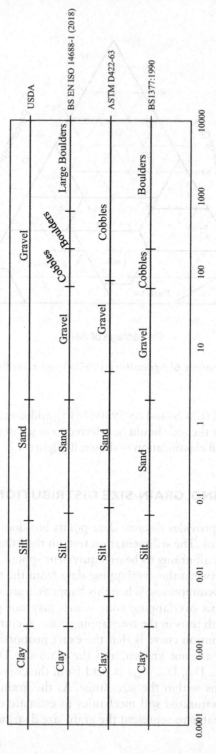

Figure 4.2 Comparison between soil classification system based on grain size.

an equation. It is possible to use an interpolation function such as a cubic spline to represent the grain-size distribution, but there are advantages in using an equation over an interpolation function:

1. The entire grain-size distribution can be represented by a set of parameters in the equation.
2. Parameters of the equation can be used for correlation purposes.

Table 4.2 shows some of the equations that can be used to fit the grain-size distribution curve. Most of the equations are adapted from equations to fit the SWCC, as both SWCC and grain-size distribution have a similar shape.

4.5 EFFECT OF GRAIN SIZE AND PACKING CONFIGURATION ON THE SWCC (CONTACTING SPHERES MODEL)

The effect of grain size on the SWCC can be described by using the contacting spheres model (Lu and Likos 2004). Figure 4.3. shows a schematic of the air-water interface between two identical spherical soil grains with radius R, where r_1 describes the curvature that concaves towards the water phase, while r_2 describes the curvature that convex from the water phase (toroidal approximation), and θ is filling angle. According to the contacting spheres model, r_1 causes a pressure decrease in the water phase while r_2 causes a pressure increase in the water phase.

There are three forces in the free body diagram:

1. Surface tension T_s along the interface described by r_1 (positive horizontal direction) F_1
2. Surface tension along the interface described by r_2 (negative horizontal direction) F_2
3. Air and water pressure applied on either side of the interface F_3

F_1, F_2 and F_3 are given as follows:

$$F_1 = 2(T_s \sin\alpha)(2r_3) = 4r_3 T_s \sin\alpha \tag{4.1}$$

$$F_2 = -4T_s r_1 \sin\alpha \tag{4.2}$$

$$F_3 = (u_a - u_w)(2r_1 \sin\alpha)2r_2 = 4r_1 r_2 (u_a - u_w)\sin\alpha \tag{4.3}$$

By balancing all three forces, it is possible to describe matric suction $(u_a - u_w)$ in terms of T_s, r_1 and r_2 as follows:

$$u_a = u_w - T_s\left(\frac{1}{r_1} - \frac{1}{r_2}\right) \tag{4.4}$$

Table 4.2 Curve-fitting, grain-size distribution

Author	Equation	Parameters
Zou (2018)	**Multimodal form:** $$P = 100 - m_1(D_{max} - D) + \sum_{i=2}^{n}(m_i - m_{i-1})R_i(D, D_{max}, c_i)$$ $$R_i(D, D_{max}, c_i) = \frac{1}{2}\left\langle (D_{max} - D) + \frac{1}{c_i}\ln\left\{\frac{\cosh\left[c_i(D_i - D)\right]}{\cosh\left[c_i(D_{max} - D_i)\right]}\right\}\right\rangle$$ $$c_i = \frac{2}{\log\left(\dfrac{d_i^-}{d_i^+}\right)}; D = \log d$$	P = percent passing d = grain diameter m_i = slope of linear segment d_{max} = maximum grain diameter d_i = grain diameter at the intersection between linear segment i and linear segment i-1 d_i^- = convergence point between the curve and linear segment i-1 d_i^+ = convergence point between the curve and linear segment i+1
Fredlund et al. (2000)	**Unimodal form:** $$P = \frac{1}{\left\{\ln\left[\exp(1) + \left(\dfrac{a_{gr}}{d}\right)^{m_{gr}}\right]\right\}}\left[1 - \left(\frac{\ln\left(1 + \dfrac{d_{r,gr}}{d}\right)}{\ln\left(1 + \dfrac{d_{r,gr}}{d_{min}}\right)}\right)^{7}\right]$$ **Bimodal form:** $$P = w\left\{\frac{1}{\ln\left[\exp(1) + \left(\dfrac{a_{gr,1}}{d}\right)^{n_{gr,1}}\right]^{m_{gr,1}}}\right\} + (1-w)\left\{\frac{1}{\ln\left[\exp(1) + \left(\dfrac{a_{gr,2}}{d}\right)^{n_{gr,2}}\right]^{m_{gr,2}}}\left[1 - \left(\frac{\ln\left(1 + \dfrac{d_{r,gr}}{d}\right)}{\ln\left(1 + \dfrac{d_{r,gr}}{d_{min}}\right)}\right)^{7}\right]\right\}$$	$a_{gr}, n_{gr}, m_{gr}, d_{r,gr}$ = curve-fitting parameters of Fredlund et al. (2000) for unimodal form $a_{gr,l}, n_{gr,l}, m_{gr,l}, d_{r,gr}$ = curve-fitting parameters of Fredlund et al. (2000) for bimodal form with i = l is an index which can be either 1 or 2 d_{hp}, n_{hp}, m_{hp} = curve-fitting parameters of Haverkamp and Parlange (1986) d_m = mean grain size s_d = standard deviation of the grain size $d_{m,1}$ = mean grain size of the coarser grain $d_{m,2}$ = mean grain size of the finer grain $s_{d,1}$ = standard deviation of the coarse grain size $s_{d,2}$ = standard deviation of the finer grain size

Haverkamp and Parlange (1986)

Unimodal form:

$$P = \cfrac{1}{\left[1 + \left(\cfrac{d_{hp}}{d}\right)^{n_{hp}}\right]^{m_{hp}}} \qquad m_{hp} = 1 - \cfrac{1}{n_{hp}}$$

Satyanaga et al. (2013)

Unimodal form:

$$P = \left\{1 - \left[0.15\ln\left(1 + \cfrac{0.075}{d}\right)\right]\right\}\left[erfc\left[x\left(d_{max}, d_m, s_d\right)\right]\right]$$

Bimodal form:

$$P = \left\{1 - \left[0.15\ln\left(1 + \cfrac{0.075}{d}\right)\right]\right\}\left[\begin{array}{l} w \times erfc\left[x\left(d_{max,1}, d_{m,1}, s_{d,1}\right)\right] \\ + (1-w)\,erfc\left[x\left(d_{max,2}, d_{m,2}, s_{d,2}\right)\right]\end{array}\right]$$

$$x\left(d_{max}, d_m, s_d\right) = \left(\cfrac{\ln\left(\cfrac{d_{max} - d_m}{d_{max} - d}\right)}{s_d}\right); \quad erfc(x) = \int_{-\infty}^{x}\cfrac{1}{\sqrt{2\pi}}\exp\left(-\cfrac{x^2}{2}\right)dx$$

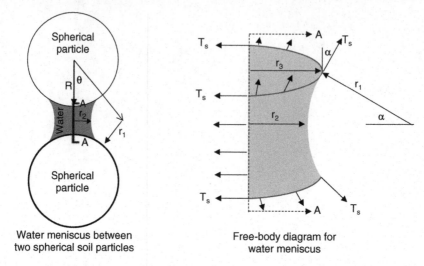

Water meniscus between   Free-body diagram for
two spherical soil particles   water meniscus

Figure 4.3 Idealised air-water interface geometry in unsaturated soil re-drawn
from Lu and Likos (2004).

Relationships between r_1, r_2, c, and R are given as follows:

$$r_1 = R\left(\frac{1}{\cos\theta} - 1\right) \tag{4.5a}$$

$$r_2 = R\tan\theta - r_1 \tag{4.5b}$$

Substituting Equation 4.5 into Equation 4.4, $(u_a - u_w)$ can be described
using R and θ as follows:

$$u_a - u_w = \frac{T_s}{R} s^*; s^* = \frac{\cos\theta(\sin\theta + 2\cos\theta - 2)}{(1 - \cos\theta)(\sin\theta + \cos\theta - 1)} \tag{4.6}$$

where s^* is normalised suction as a function of filling angle θ. Equation 4.6
shows that $(u_a - u_w)$ is directly a function of R. When R decreases, $(u_a - u_w)$
will increase. Thus, under the same filling angle θ, fine-grained soil will have
much larger $(u_a - u_w)$ compared to coarse-grained soil.

The packing configuration of identical spherical grains can be represented
by simple cubic (SC) packing for loose condition giving a void ratio of 0.91
and tetrahedral (TH) packing for dense condition giving a void ratio of 0.34
(Lu and Likos 2004), as shown in Figure 4.4. Dallavalle (1943) shows that
the volume of water lenses normalised by the volume of one sphere for SC
packing, and its water content can be represented as:

$$\frac{V_l}{V_s} = \frac{3}{2}\left(\frac{1}{\cos\theta} - 1\right)^2 \left[1 - (0.5\pi - \theta)\tan\theta\right] \tag{4.7a}$$

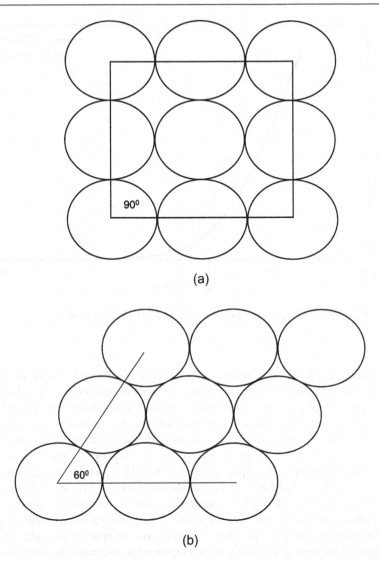

(a)

(b)

Figure 4.4 Packing order of spherical grains redrawn from Lu and Likos (2004).
(a) Simple cubic packing which represent the loosest packing order (e = 0.91).
(b) Tetrahedral packing which represent the densest packing order (e = 0.34).

$$w = \frac{3}{G_s}\frac{V_l}{V_s} = \frac{9}{2G_s}\left(\frac{1}{\cos\theta}-1\right)^2 \left[1-(0.5\pi-\theta)\tan\theta\right] \qquad (4.7b)$$

The water content for tetrahedral (TH) packing is given as:

$$w = \frac{6}{G_s}\frac{V_l}{V_s} = \frac{9}{G_s}\left(\frac{1}{\cos\theta}-1\right)^2 \left[1-(0.5\pi-\theta)\tan\theta\right] \qquad (4.8)$$

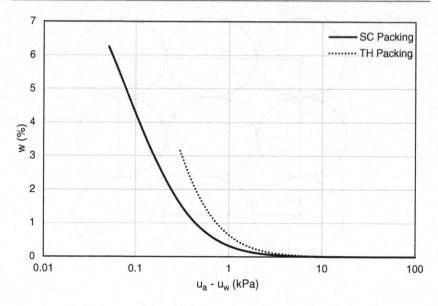

Figure 4.5 SWCC comparison between SC and TH packing order.

However, water lenses may overlap when θ is larger than 45° for SC packing and 30° for TH packing, thus violating the assumption in deriving the volume of water lenses. Figure 4.5 shows the comparison between TH and SC packing configurations for solid grains with R = 1 mm and G_s = 2.65. For SC packing, the maximum water content with respect to θ = 45° is 6.25%, while saturated water content (e = 0.91) is 34.3%. For TH packing, maximum water content with respect to θ = 30° is 3.21%, while saturated water content (e = 0.34) is 12.8%. It is also shown that the SWCC of TH packing is located above SC packing, which is not in accord with experimental observation on SWCC-w (Salager et al. 2010; Wijaya and Leong 2017). Despite these limitations, packing configurations of spheres provide a good description on how grain size can be employed to estimate the pore size to estimate the SWCC of soil.

4.6 TEST METHODS FOR DETERMINING GRAIN-SIZE DISTRIBUTION

Test method for grain-size distribution depends on the adopted standard. In this section, the general principles from several standards are outlined. Inspection of the soil is important to determine the set of procedures required to obtain the grain-size distribution. Head (2006) recommends the following to be considered:

1. Size of the largest grains present to determine the amount of required soil sample
2. Range of grain sizes to determine the testing methods
3. Soil characteristics and stability of soil grains to determine the complexity of the test and the required preparation methods

There are three types of commonly employed tests that will be discussed in this chapter:

1. Wet sieving (ASTM D1140-00 2000; BS 1377-2 1990; BS EN ISO 17892-4 2016)
2. Dry sieving by using a mechanical shaker (BS EN ISO 17892-4 2016; BS 1377-2 1990; ASTM D6913/D6913M 2017)
3. Sedimentation test by using hydrometer (ASTM D7928 2017; BS 1377-2 1990; BS EN ISO 17892-4 2016)

Wet sieving is carried out to obtain an accurate separation between fine- and coarse-grained soils. Dry sieving is conducted to obtain the grain-size distribution for coarse-grained soils, while a sedimentation test is conducted to obtain the grain-size distribution of fine-grained soils.

In sieve analysis, it is important to ensure that the sieve is not overloaded. The maximum mass to be retained on each sieve following BS 1377-2 (1990) is given in Table 4.3.

The recommended test according to soil type is given as follows:

1. Clean sands and gravels: dry sieving
2. Soils containing silts or clay: wet sieving for separation and dry sieving for the retained materials
3. Predominantly clayey and silty soils: wet sieving for separation and material passing the 63 μm sieve is collected for sedimentation test while material retained on the 63 μm sieve is for dry sieving

4.6.1 Sample preparations

Table 4.4 shows the minimum quantity of soil required based on the maximum size of material present in substantial proportion. Figure 4.6 shows the recommended minimum quantity of soil required for sieving based on the largest significant particle size according to BS 1377-2 (1990) and ASTM D422-63.

For some residual soils, especially volcanic soils, the result of the grain-size distribution test may be affected by the preparation method. Usually, oven drying of such soils is avoided and instead the soil is air dried to a water content necessary to carry out the test (Wesley (2010)). Difficulties are also encountered for soils with a high proportion of allophane as it tends to flocculate. Hence, a dispersing agent is important for such soils when conducting sedimentation tests to avoid flocculation.

Table 4.3 Maximum mass of material to be retained on each test sieve at the completion of sieving (BS 1377-2 1990)

Test sieve aperture size	Maximum mass on sieve of diameter		
	450 mm	300 mm	200 mm
Mm	kg	kg	g
50	10	4.5	—
37.5	8	3.5	—
28	6	2.5	—
20	4	2	—
14	3	1.5	—
10	2	1	—
6.3	1.5	0.75	—
5	1	0.5	—
3.35	—	—	300
2	—	—	200
1.18	—	—	100
Mm			
600	—	—	75
425	—	—	75
300	—	—	50
212	—	—	50
150	—	—	40
63	—	—	25

Table 4.4 Mass of soil sample for sieving determined based on the maximum size of material present in substantial proportion (more than 10%)

Test sieve aperture (mm)	BS 1377-2 (1990)	BS EN ISO 17892-4 (2016)
	Minimum mass of sample (kg)	
63	50	40
50	35	
37.5	15	14
28	6	
20	2	2
14	1	
10	0.5	0.5
6.3	0.2	0.3
5	0.2	
3.35	0.15	
2 or smaller	0.1	0.1

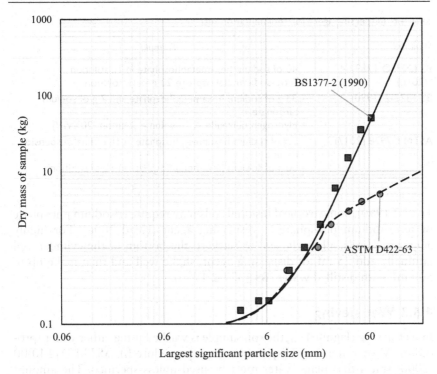

Figure 4.6 Minimum mass of sample required for grain size test (BS 1377-2, 1990 and ASTM D422-63).

4.6.2 Preparation of dispersing agent

Silty clay materials tend to clump together. Thus, a dispersing agent is required to separate the soil grains. Table 4.5 and Table 4.6 show the composition of the dispersing agent for wet sieving and sedimentation test, respectively, for the various standards.

A dispersing agent may not work on some soils, which can be observed by the formation of relatively large crumbs or flocs which fall rapidly through the water leaving a sharply defined, clear layer above the suspension (BS

Table 4.5 Dispersing agent for wet sieving

Standard	Formula
BS EN ISO 17892-4 (2016)	2 g of sodium hexametaphosphate in 1L solution. Hydrogen peroxide (20%V/V) to remove organic material. Hydrochloric acid (0.2M) to remove carbonate.
BS 1377-2 (1990)	2 g of sodium hexametaphosphate in 1L solution.
ASTM D1140-00 (2000)	40 g of sodium hexametaphosphate in 1L solution.

Table 4.6 Dispersing agent for sedimentation test

Standard	Formula
BS EN ISO 17892-4 (2016)	- 40 of sodium hexametaphosphate in I L solution - Tetra-sodium diphosphate 20 g in I L solution
BS 1377-2 (1990)	- 33 g of sodium hexametaphosphate and 7 g of sodium carbonate - Hydrogen peroxide. A 20-volume solution (20%V/V)
ASTM D7928 (2017)	- 5 g of sodium hexametaphosphate in I L of final suspension volume - 50 g of sodium hexametaphosphate in I L of solution

1377-2 1990). Some tropical residual soils may require trisodium phosphate or tetra-sodium phosphate as dispersing agent (Head 2006). For highly aggregated soils, allow the soil to settle to the bottom of the cylinder and gradually add 25 mL of dispersing agent, shake well and then make up to 500 mL with distilled water (BS 1377-2 1990).

4.6.3 Wet sieving

In wet sieving (Figure 4.7), the soil sample is washed using either water (procedure A) or water with dispersing agent (procedure B). ASTM D1140-00 (2000) states that plain water must be used unless specified. The amount of fine materials are determined by the amount of soil that passes through sieve with openings of either 75 μm (#200) for ASTM D1140-00 (2000) or 63 μm for BS 1377-2 (1990). In general, the test procedures for wet sieving are as follows:

1. Oven dry (ASTM D1140-00 2000) or air dry the soil sample and determine its mass.
2. Place the sample in a container.
3. Pour water to cover the soil sample (procedure A) and optionally add dispersing agent (procedure B) to the water. The dispersing agent can be either 2 g/L sodium hexametaphosphate (BS 1377-2 1990) or 50 ml of standard dispersant solution such as sodium hexametaphosphate per litre (Head 2006).
4. Agitate the soil to separate the fine grains from the coarse grains (ASTM D1140-00 2000). If a dispersing agent is used, BS 1377-2 (1990) recommends that the soil mixture be stirred continuously for 1 hour.
5. Pour the water with the suspended fine grains into the nested sieves (coarser sieve at the top). BS 1377-2 (1990) specified that the sieve size is 2 mm at the top and 63 μm at the bottom (see Table 4.7).
6. Repeat steps 3 to 5 several times until the water becomes clear.
7. Put back all the soil into the container.

Figure 4.7 Wet sieving.

8. Oven dry the soil and determine the leftover soil mass.
9. The differences between soil mass before and after washing is the mass of the fine-grained soils, which is reported to the nearest 0.1% (BS 1377-2 1990; ASTM D1140-00 2000).
10. Report the wet sieving result using record sheet such as that shown in Figure 4.8.

4.6.4 Dry sieving

Dry sieving is best carried out for soil that has negligible fine materials or clean granular materials (Head 2006). Otherwise, wet sieving must be carried out first to precisely determine the percentage of coarse-grained and fine-grained soils.

In dry sieving (Figure 4.9), a set of sieves is stacked with the largest aperture sieve at the top and the smallest aperture sieve at the bottom. A pan is placed below the smallest aperture sieve. Recommended sieve sizes according to different standards are shown in Table 4.7. It is not necessary to use all of the sieve sizes (BS EN ISO 17892-4. 2016; BS 1377-2 1990; ASTM D6913/D6913M. 2017) and additional sieve sizes outside the standard set shown in Table 4.7 can be added but the largest aperture sieve size must allow 100% passing of the soil sample (ASTM D6913/D6913M 2017).

Table 4.7 Sieve sizes according to different standards

ASTM 6913-17 (Alternative)	ASTM 6913-17 (Standard)	BS 1377-2 (1990) (Complete)	ISO 3310-1 (2016)	ISO 3310-2 (2016)
Lid	Lid	Lid	Lid	Lid
3 in.	75 mm	75 mm	125 mm	125 mm
2 in.	50 mm	63 mm	90 mm	90 mm
1-1/2 in.	37.5 mm	50 mm	63 mm	63 mm
1 in.	25.0 mm	37.5 mm	45 mm	45 mm
3/4 in.	19.0 mm	28 mm	31.5 mm	31.5 mm
3/8 in.	9.5 mm	20 mm	22.4 mm	22.4 mm
No. 4	4.75 mm	14 mm	16 mm	16 mm
No. 10	2.00 mm	10 mm	11.2 mm	11.2 mm
No. 20	850 μm	6.3 mm	8 mm	8 mm
No. 40	425 μm	5 mm	5.6 mm	5.6 mm
No. 60	250 μm	3.35 mm	4 mm	4 mm
No. 100	150 μm	2 mm	2.8 mm	2.8 mm
No. 140	106 μm	1.18 mm	2 mm	2 mm
No. 200	75 μm	600 μm	1.4 mm	1.4 mm
Pan	Pan	425 μm	1 mm	1 mm
		300 μm	710 μm	710 μm
		212 μm	500 μm	500 μm
		150 μm	355 μm	355 μm
		63 μm	250 μm	250 μm
		Pan	180 μm	180 μm
			125 μm	125 μm
			90 μm	90 μm
			63 μm	63 μm
			45 μm	45 μm
			32 μm	32 μm
			25 μm	25 μm
			20 μm	20 μm

1. Oven dry the soil such that the amount of soil for the sieve analysis is in accord with Figure 4.6 or Table 4.4.
2. Prepare the sieve sizes in accord with Table 4.7. Additional sieves between the sizes shown may be added.
3. Weigh each of the sieves.
4. Place the soil into the topmost sieve. The sample can be divided into several portions as well.
5. Secure the sieve set in a mechanical shaker and turn it on for around 10 to 20 minutes. A very long shaking duration does not guarantee a

Grain-Size Distribution – Finer Materials Separation

Project details

Project ID: BK-01
Project Name: Example of Grain size distribution test
Contract No: CN-01
Company: Geotech Consultant
Client Name: SI Contractor
Technician: Martin Wijaya
Testing date: 15-10-19
Project Engineer: Martin Wijaya

Sample details

BH No.:   BH-01
Sample reference: TW-1
Sample depth top: 1.00   m
Sample depth bottom: 2.00   m

Specimen details

Specimen reference: Sieve-1
Specimen depth top: 1.00   m
Specimen depth bot (m): 2.00   m

Test details	Notation
Standard: *ASTM D1140 Method A*	M_{sieve}: *Mass of sieve*
Oven dried specimens	$M_{sieve+soil}$: *Mass of sieve + mass of soil*
Container ID: 203	M_{soil}: *Mass of soil*
$M_{container}$: 200 gram	$M_{container}$: *Mass of container*
M_{c+ws}: 1455 gram	M_{c+ws}: *Mass of container + mass of wet soils*
M_{c+DS}: 1144.3 gram	M_{c+DS}: *Mass of container + mass of oven dried soils*

Wet sieving

Sieve size	Container	$M_{container}$	M_{c+ws}	M_{c+DS}	M_{soil}
mm	ID	gram	gram	gram	gram
0.425	M1	428.31	685	588.65	160.34
0.075	M2	294.71	532.09	389.93	95.22

Wet sieving summary

Water content of sieved soils: 32.90 %
Dissolving agent: None
Weight of dissolving agent: 0.00 gram
Total mass of soils: 344.57 gram
Collected fined soil using dry sieving: 8.57 gram
Mass of collected fine grained soils: 97.58 gram
Mass of collected coarse grained soils: 246.99 gram
%Fine grained soils: 28.3 %
%Coarse grained soils: 71.7 %

Remark

Figure 4.8 Wet sieving record sheet.

better result, as the soil grains might be able to pass through an opening that is slightly oversized (Head 2006).

6. Repeat steps 4 and 5 until all the soils have been sieved.
7. Record the weight of each sieve with soil. The mass of soil on each sieve should not exceed the allowed maximum mass shown in Table 4.4.

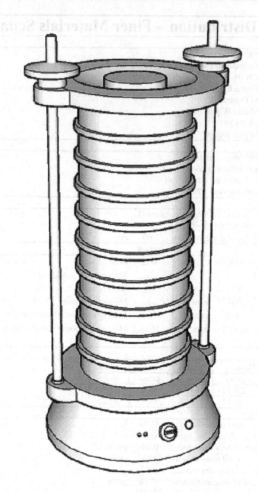

Figure 4.9 Sieve shaker.

8. Calculate the percentage passing of soil for each sieve.
9. Plot the grain-size distribution curve and determine all the required parameters as shown in Figure 4.10.

4.6.5 Sedimentation test based on hydrometer

A sedimentation test is conducted to determine the size of soil grains based on Stoke's law which is defined in Equation 4.9.

$$v = \frac{D^2 g (\rho_s - \rho_L)}{18\eta} \tag{4.9}$$

Grain Size Distribution – Sieve Analysis

Project details

Project ID: BK-01
Project Name: Example of Grain size distribution test
Contract No: CN-01   Technician: Martin Wijaya
Company: Geotech Consultant   Testing date: 15-10-19
Client Name: SI Contractor   Project Engineer: Martin Wijaya

Sample details

BH No.:   BH-01
Sample reference: TW-1
Sample depth top: 1.00   m   Sample depth bottom: 2.00   m

Specimen details

Specimen reference: Sieve-1
Specimen depth top: 1.00   m   Specimen depth bot (m): 2.00   m

Initial mass of soil prior to sieve	*Soil mass after sieve*		Wet sieving: Yes	
$M_{Container}$: 0.00 gram	M_{pan}: 270.00 gram	$M_{soil+Pan}$:	367.58 gram	
$M_{Container+Soil}$: 344.57 gram	Total soil mass after sieve:		344.44 gram	
M_{Soil}: 344.57 gram	Soil mass difference:		−0.13 gram	

Sieve analysis

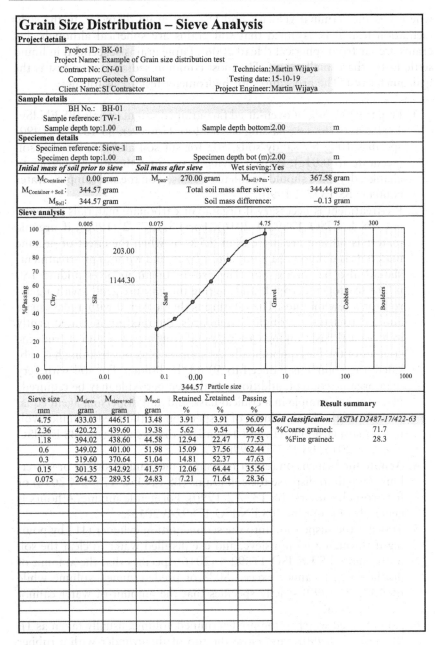

Sieve size	M_{sieve}	$M_{sieve+soil}$	M_{soil}	Retained	Σretained	Passing	Result summary	
mm	gram	gram	gram	%	%	%		
4.75	433.03	446.51	13.48	3.91	3.91	96.09	*Soil classification:*	*ASTM D2487-17/422-63*
2.36	420.22	439.60	19.38	5.62	9.54	90.46	%Coarse grained:	71.7
1.18	394.02	438.60	44.58	12.94	22.47	77.53	%Fine grained:	28.3
0.6	349.02	401.00	51.98	15.09	37.56	62.44		
0.3	319.60	370.64	51.04	14.81	52.37	47.63		
0.15	301.35	342.92	41.57	12.06	64.44	35.56		
0.075	264.52	289.35	24.83	7.21	71.64	28.36		

Figure 4.10 Dry sieving record sheet.

where v is the terminal velocity, and D is the grain diameter. In the sedimentation test, it is assumed that soil grains are spherical although clay grains are far from spherical (Head 2006). Large grains suspended in liquid settle faster than small grains. The most common sedimentation test is the hydrometer test. The procedures for hydrometer test are as follows:

1. Prepare 20g–30g of oven-dried fine-grained soil materials (BS EN ISO 17892-4 2016), i.e., passing sieve #200. However, BS 1377-2 (1990) specifies 100g for sandy soils, 50g for silt soil and 30g for clay soil (accurate to 0.01g). If oven drying may cause the soil properties to change, dry mass should be calculated based on the soil sample's water content (BS 1377-2 1990).

2. Pre-treatment is necessary if organic or carbonate materials are present (BS EN ISO 17892-4 2016). For BS 1377-2 (1990), if organic materials present are more than 0.5% or when the effect of organic materials is unknown, a parallel test (with and without pre-treatment) should also be carried out. The pre-treatment involves adding 150 mL of hydrogen peroxide and stirring the mixture gently with a glass rod for a few minutes. The mixture is then covered with a glass plate and left to stand overnight. Afterwards, the mixture is heated gently in a conical flask, taking care to avoid frothing over, and stirring the content of the flask periodically. As soon as vigorous frothing has subsided, the volume is reduced to 50 mL by boiling. For soil with very high organic content, additional hydrogen peroxide may be required.

3. Place the soil sample in a conical flask and add dispersing agent equal to 10% of the final volume of suspension. If the predetermined volume of suspension is 1L, add 100 mL of dispersing agent (BS EN ISO 17892-3 2015; BS 1377-2 1990).

4. Agitate the suspension (either by using a mechanical shaker or stirrer) long enough to disperse the soil grains (BS EN ISO 17892-4 2016) or for around 4 hours (BS 1377-2 1990). For gentle mixing, 4 hours is generally appropriate (BS EN ISO 17892-3 2015).

5. Transfer the suspension into the sedimentation cylinder (1L) by pouring it through a 63 μm sieve, and use distilled water to clear the soil in the sieve. BS EN ISO 17892-4 (2016) specifies that the amount of distilled water cannot exceed 80% of predetermined volume while BS 1377-2 (1990) strictly specifies that the amount of water cannot exceed 500 mL.

6. Add distilled water to the predetermined volume, usually taken as 1L (BS 1377-2 (1990)), and close the top of the cylinder with a rubber bung. If too much distilled water is added, remove the excess water through evaporation (BS EN ISO 17892-4 2016).

7. Place the cylinder into a constant temperature bath ($25°C+/-0.5°C$) for at least 1 hour (BS 1377-2 1990).

8. Prepare a sedimentation cylinder with a reference solution consisting of the same volume of dispersing agent solution and add distilled water such that the reference cylinder has the same volume usually 1L (BS EN ISO 17892-4 2016).

9. Place the hydrometer in the reference solution. Once the hydrometer is stable, take the hydrometer reading R'_0 at the upper rim of the meniscus to the nearest 0.0005 g/mL (nearest 0.5 reading), and then rinse the hydrometer with distilled water (BS EN ISO 17892-3 2015; BS 1377-2 1990).

10. Agitate the soil suspension by turning the sedimentation cylinder end to end for 60 times in 2 minutes (BS EN ISO 17892-4 2016; BS 1377-2 1990).

11. Start the timer at the end of agitation stage, remove the rubber bung and place the sedimentation cylinder in the testing place (BS EN ISO 17892-4 2016).

12. Disperse the froth on the surface of the suspension by adding two drops of methylated spirit (BS 1377-2 1990).

13. Place the hydrometer in the suspension and ensure it floats freely for around 15 seconds (BS EN ISO 17892-3 2015; BS 1377-2 1990).

14. Take the hydrometer reading (R'_h) at the upper rim of the meniscus at short intervals of time. The first three readings must be taken within 5 minutes (i.e., 0.5 minute, 1 minute and 2 minutes). Record the reading to the nearest 0.0005 g/ml (nearest 0.5) (BS EN ISO 17892-3 2015). BS 1377-2 (1990) specifies four readings which are taken at 0.5 minute, 1 minute, 2 minutes and 4 minutes.

15. After the first three readings, remove the hydrometer and rinse it with distilled water (BS EN ISO 17892-4 2016).

16. Record the temperature of the suspension once in the first 15 minutes and then for every hydrometer reading to the nearest 0.1°C (BS EN ISO 17892-3 2015).

17. Repeat steps 13 and 14 for at least three times until the proportion finer than 2 μm has been determined (BS EN ISO 17892-3 2015; BS 1377-2 1990). BS 1377-2 (1990) specifies that the reading must be taken at 8 minutes, 30 minutes, 2 hours, 8 hours, 24 hours and twice in the following day until the proportion finer than 2μm is obtained.

4.7 TEST METHODS FOR DETERMINING SPECIFIC GRAVITY

Specific gravity test assumes that soil solid is not soluble in water (ASTM D854-14 2014), and there is no enclosed void. If an enclosed void is present, the soil must be crushed (BS EN ISO 17892-3 2015). Accurate determination of specific gravity (G_s) is especially important for unsaturated soil analysis, as it is used to determine the void ratio and the degree of saturation. Inaccuracy in determining G_s may cause the degree of saturation to be calculated to be higher than 100% and will affect the shrinkage curve and the saturation-based SWCC. Different test methods may be employed for different ranges of grain size as described next:

1. Small fluid pycnometer is suitable for soils that pass through 4.75 mm sieve size (ASTM D854-14 2014), 4 mm sieve size (BS EN ISO 17892-3 2015) or for soils consisting of clay, silt and sand sizes (less than 2 mm) grains (BS 1377-2 1990). For soil grains larger than the recommended size, BS 1377-2 (1990) and BS EN ISO 17892-3 (2015) specify that the large grains must be grounded to grain size smaller than the recommended size.
2. Large fluid pycnometer is suitable for non-cohesive soil finer than 20 mm (coarser grain shall be broken down to be less than this size) (BS 1377-2 1990).

Some other methods such as the following will not be discussed in detail in this book:

1. Gas jar or gas pycnometer method is suitable for soils of any grain size, especially when 10% of the soil grains are retained on 37.5 mm sieve (BS 1377-2 1990).
2. Gas pycnometer (BS EN ISO 17892-3 2015).

While different grain sizes within a sample may have different G_s values, it is sufficient to determine the average value or conduct different tests for each grain-size fraction (BS 1377-2 1990; ASTM D854-14 2014).

BS 1377-2 (1990) and BS EN ISO 17892-3 (2015) prefer to express specific gravity in terms of solid density (and thus, has the unit of density). In this book, the term "specific gravity" (G_s) is preferred, as it is unitless and is commonly used.

4.7.1 Specific gravity test based on small fluid pycnometer

For a specific gravity test based on a small pycnometer, the balance should be accurate to 0.01g according to ASTM D854-14 (2014) and BS EN ISO 17892-3 (2015) or 0.001g according to BS 1377-2 (1990). The pycnometer can be either a stoppered flask, a stoppered iodine flask or a volumetric flask with 250 mL minimum capacity. Stoppered flask and stoppered iodine flask are preferred, as both mechanically set the volume. The test procedures are as follows:

1. Determine the mass of pycnometer (and stopper if a stopper is used) (M_1) (ASTM D854-14 2014; BS 1377-2 1990).
2. Determine the dry mass of the specimen required. ASTM D854-14 (2014) recommends different specimen dry masses according to soil type, as shown in Table 4.8. BS 1377-2 (1990) specifies that at least two tests (5g to 10g for each test) must be conducted, whereas BS EN ISO 17892-3 (2015) specifies that the specimen must be at least 10g. The specimen can be either moist or oven dried.

Table 4.8 Recommended mass of test specimen for specific gravity test using small pycnometer (ASTM D854-14 2014)

Soil type	Specimen dry mass (g) When using 250 mL pycnometer	Specimen dry mass (g) When using 500 mL pycnometer
SP, SP-SM	60±10	100±10
SP-SC, SM, SC	45±10	75±10
Silt or Clay	35±5	50±10

For moist specimen, determine the water content of the specimen (ASTM D854-14 2014). Put 100 mL of water into a mixing container or blender, add the soil and blend (ASTM D854-14 2014). Pour the slurry into the pycnometer by using a funnel and rinse the soil that remains in the funnel by using a wash/spray squirt bottle (ASTM D854-14 2014). Based on the water content of the specimen, determine the dry mass of soil (M_s) to calculate the weight of dry pycnometer, stopper and dry mass of soil (M_2). Another alternative is to oven dry the specimen at the end of the test to obtain the dry mass of soil (BS EN ISO 17892-3 2015).

For oven-dried specimen, the specimen shall be dried in an oven set at 105°C to 110°C (BS 1377-2 1990; ASTM D854-14 2014; BS EN ISO 17892-1 2014), but if the specific gravity may change due to the high temperature, the specimen must be oven dried at a lower temperature (BS EN ISO 17892-1 2014; BS 1377-2 1990). BS 1377-2 (1990) specifies that the temperature must not exceed 80°C for such soils. To determine the specific gravity of an oven-dried specimen, break up the soil clods using a mortar and pestle. Place the funnel into the pycnometer where the stem of the funnel goes beyond the calibration mark, spoon the soil solids directly into the funnel and then rinse the funnel by using a wash/spray squirt bottle (ASTM D854-14 2014). Record the weight of the pycnometer, stopper and soil (M_2).

3. Remove the stopper and add de-aired distilled water to cover the soil in the bottle in order to make a slurry (BS 1377-2 1990) approximately 10 to 20 mm depth (BS EN ISO 17892-3 2015). De-air the soil slurry by either mechanical agitation, heating (hot plate or Bunsen burner), vacuum (vacuum desiccator) or a combination of heat and vacuum. Mechanical agitation is sufficient for coarse specimens while other methods can be used for both coarse and fine specimens (BS EN ISO 17892-3 2015). ASTM D854-14 (2014) recommends two hours duration for heating and vacuum methods as the soil is agitated, while for a combination of heat and vacuum, the temperature must not be more than 40°C and the vacuum is applied for at least 1 hour.

Whenever heating is involved, slurry should be agitated to prevent the slurry from drying and sticking to the pycnometer. BS 1377-2 (1990) recommends using a vacuum desiccator with a suction of about 20 mm of mercury (2.7 kPa) to avoid the air inside the soil to bubble too violently for at least 1 hour until no further loss of air is apparent. Stir the soil in the bottle with a chattaway spatula or vibrate the bottle and then wash off any soil stuck to the spatula with de-aired water and repeat this step until no more air comes out from the soil (BS 1377-2 1990). This step is very important, as most errors in this test come from a failure to completely remove the entrapped air.

4. Filled the pycnometer with de-aired water. It is important to ensure that a clear water layer forms at the top surface prior to placing the stopper to avoid soil grains coming out from the pycnometer. Placed the stopper carefully to avoid any trapped air. Clean the outside of the pycnometer and let the pycnometer's temperature come to equilibrium with the room's temperature. Record the weight of the pycnometer with water and soil solids (M_3) (ASTM D854-14 2014; BS 1377-2 1990, BS EN ISO 17892-3 2015). Record the temperature of the soil slurry inside the pycnometer by inserting the sensing portion of the thermometric device to the appropriate depth of immersion (ASTM D854-14 2014).

5. Clean out the bottle and fill it with de-aired water and placed the stopper while ensuring no air is trapped inside the pycnometer and then record the weight of the pycnometer, water and stopper (M_4). This step can also be done prior to the test (BS EN ISO 17892-3 2015). Calculate G_s as follows (BS 1377-2 1990; BS EN ISO 17892-3 2015):

$$G_s = \frac{M_2 - M_1}{(M_4 - M_1) - (M_3 - M_2)} \tag{4.10}$$

6. If M_4 is measured at different times (referred to as pycnometer calibration) during the test, there may be a significant difference in the temperature between calibration and testing stages. It is thus required to measure the temperature during the calibration stage and to apply a correction factor to the testing stage (ASTM D854-14 2014) or, use a water bath to control the temperature (BS EN ISO 17892-3 2015). Density of water (ρ_w) for both pycnometer calibration stage ($\rho_{w,c}$) and testing stage ($\rho_{w,t}$) can be obtained as follows (ASTM D854-14 2014):

$$\rho_w = 1.00034038 - T(7.77 \times 10^{-6}) - T^2(4.9 \times 10^{-6}) \tag{4.11}$$

where T is temperature in °C. Alternatively, it can be obtained as follows (BS EN ISO 17892-3 2015):

$$\rho_w = \frac{1}{\left\{1+\left[\left(2.31T-2\right)^2-182\right]10^{-6}\right\}}$$ (4.12)

7. Calculate the volume of the pycnometer (V_p) and specific gravity of the soil solids (G_s) as follows (ASTM D854-14 2014):

$$V_p = \frac{M_4 - M_1}{\rho_{w,c}}$$ (4.13)

$$G_s = \frac{M_2 - M_1}{\left(M_1 + V_p\rho_{w,t}\right)-\left(M_3 - M_2\right)}$$ (4.14)

where $\rho_{w,t}$ is the density of water during the testing stage.
8. If G_s at 20°C is required, apply the correction factor to the calculated G_s as follows (ASTM D854-14 2014):

$$G_{s,20^0C} = G_s\left(\frac{\rho_w}{0.9982063}\right)$$ (4.15)

9. If soil grains have R percent of soil retained on the 4.75 mm sieve with $G_{s,20°C}$ equals to G_1, and P percent of soil pass through 4.75 mm sieve with $G_{s,20°C}$ equals to G_2, the average specific gravity at 20°C ($G_{ave,20°C}$) is calculated as follows (ASTM D854-14 2014):

$$G_{ave,20^\circ C} = \frac{1}{\dfrac{R}{100G_1}+\dfrac{P}{100G_2}}$$ (4.16)

4.7.2 Specific gravity test based on large fluid pycnometer

Large pycnometer (approximately 1L capacity) test is less accurate than the gas jar method and is suitable for clay soils (BS 1377-2 1990). The test must be carried out using a balance with 0.5g accuracy (BS 1377-2 1990). The test procedures are as follows:

1. Prepare two sets of 400g of oven-dried specimen (BS 1377-2 1990).
2. Clean the pycnometer along with the screw cap assembly and record its weight (M_1) (BS 1377-2 1990).
3. Place the specimen inside the pycnometer along with the screw cap assembly and record the weight of pycnometer and the soil (M_2) (BS 1377-2 1990).
4. Add water to pycnometer until half full and stir the mixture thoroughly with a glass rod to remove the air trapped in the soil (BS 1377-2 1990).

5. Fit the screw cap assembly and tighten to the reference point and fill the pycnometer with water (BS 1377-2 1990).
6. Agitate the pycnometer by shaking or rolling while covering the opening of the conical top with a finger. Allow air to escape and the froth to disappear (BS 1377-2 1990).
7. Top up the pycnometer with de-aired water so that the water level is flushed with the hole in the conical stopper and ensure neither air bubble nor froth is trapped inside (BS 1377-2 1990).
8. Clean the outside of the pycnometer and record the weight of pycnometer, water and soil (M_3) (BS 1377-2 1990).
9. Clean the pycnometer and then fill the pycnometer with de-aired water. Ensure that no air is trapped inside the pycnometer and the water surface is flush with the hole. Record the weight of the pycnometer with water (M_4) (BS 1377-2 1990).
10. Calculate the specific gravity as follows (BS 1377-2 1990):

$$G_s = \frac{M_2 - M_1}{(M_4 - M_1) - (M_3 - M_2)} \tag{4.17}$$

11. Repeat the test with the second specimen and take the average (BS 1377-2 1990).

4.7.3 Specific gravity test based on gas jar method

The gas jar method is suitable for soil containing up to 10% of grains retained on 37.5 mm test sieve and, any grains retained on a 50 mm test sieve shall be broken down to less than 37.5 mm size (BS 1377-2 1990). The test procedure is given as follows:

1. Prepare 200 g for fine-grained soils and 400 g for coarse-grained soils for a test and prepare for at least two tests (BS 1377-2 1990).
2. Oven dry the specimen at 105°C to 110°C. If it is suspected that the high temperature may affect the specific gravity, the temperature must not be higher than 80°C (BS 1377-2 1990).
3. Prepare a clean dry 1L gas jar and record its weight (M_1) to the nearest 0.2g (BS 1377-2 1990).
4. Transfer the soil into the gas jar, and record the weight (M_2) to the nearest 0.2g (BS 1377-2 1990).
5. Add approximately 500 mL of water at room temperature (±2°C), placed the rubber stopper, and leave it for 4 h (BS 1377-2 1990).
6. Agitate the gas jar by hand until the soil grains are in suspension (BS 1377-2 1990).
7. Place the gas jar in the shaking apparatus and shake for 20 to 30 minutes.

8. Remove the stopper and wash any soil grains with distilled water back into the jar. Disperse any froth with spray distilled water and add distilled water to the gas jar within 2 mm of the top, and allow the soil to settle for a few minutes.

9. Fill the gas jar to the brim with more distilled water, place a glass plate on the top of the jar and ensure that there is no air trapped under the plate. Record the weight of the gas jar and glass plate with soil and water (M_3).

10. Dry the gas jar, wash it out thoroughly and fill it to the brim with water at room temperature ±2°C. Record the weight of the gas jar with water (M_4).

11. Calculate the specific gravity as follows:

$$G_s = \frac{M_2 - M_1}{(M_4 - M_1) - (M_3 - M_2)} \tag{4.18}$$

12. Repeat steps 2 to 10 for the second set of specimens and take the average if the difference between the two measurements is less than 0.03. Express the average value to the nearest 0.01 (BS 1377-2 1990)

4.7.4 Specific gravity test based on gas pycnometer method

Helium is preferred as the measurement gas. Alternatively, any gas that gives good diffusion into soil pores may be used. The principle behind the gas pycnometer method is to compare the difference between gas pressure when the sample is present and when the sample is absent. The pressure difference is converted to a volume by using Boyle's law. The test procedure is given as follows (BS EN ISO 17892-3 2015):

1. Prepare 10g of oven-dried specimen. If the specimen has coarse-grained material, it should be crushed prior to the test.

2. Determine the volume of the sample chamber (V_c) by using appropriate pressure readings in accordance with the equipment manufacturer's instructions.

3. Determine the volume of the expansion chamber (V_r) in accordance with the equipment manufacturer's instructions.

4. Calculate the volume of the specimen V_s as follow (BS EN ISO 17892-3 2015):

$$V_s = V_c + \frac{V_r}{1 - \left[\dfrac{p_1 - p_0}{p_2 - p_0}\right]} \tag{4.19}$$

$$\rho_s = \frac{m_s}{V_s} \qquad\qquad (4.20)$$

$$G_s = \frac{\rho_s}{\rho_w} \qquad\qquad (4.21)$$

where p_0 is the initial pressure in the equipment, which can be atmospheric pressure, p_1 is the pressure in the system before opening the isolation valve, p_2 is the pressure in the system after opening the isolation valve, ρ_s is the solid density.

REFERENCES

ASTM D1140-00 (2000). *Standard test method for amount of materials in soils finer than the no. 200 (75-mm) sieve.* West Conshohocken, PA: ASTM.

ASTM D2487-17 (2017). *Standard practice for classification of soils for engineering purposes (unified soil classification system).* West Conshohocken, PA: ASTM International.

ASTM D6913/D6913m (2017). *Standard test methods for particle-size distribution (gradation) of soils using sieve analysis.* West Conshohocken, PA: ASTM International.

ASTM D7928 (2017). *Standard test method for particle-size distribution (gradation) of fine-grained soils using the sedimentation (hydrometer) analysis..* West Conshohocken, PA: ASTM International.

ASTM D854-14 (2014). *Standard test methods for specific gravity of soil solids by water pycnometer.* West Conshohocken, PA: ASTM International.

BS 1377-2 (1990). *Methods of test for soils for civil engineering purposes classification tests.*

BS 5930 (2015). *Code of practice for ground investigations.* UK: BSI.

BS EN ISO 17892-1 (2014). *Geotechnical investigation and testing – laboratory testing of soils part 1: Determination of water content.* UK: BSI.

BS EN ISO 17892-3 (2015). *Geotechnical investigation and testing. Laboratory testing of soil. Part 3: Determination of particle density.* UK: BSI.

BS EN ISO 17892-4 (2016). *Geotechnical investigation and testing. Laboratory testing of soil. Part 4: Determination of particle size distribution..* UK: BSI.

Dallavalle, J. M. (1943). *Micrometrics.* London: Pitman.

Fredlund, M. D., Fredlund, D. G., & Wilson, G. W. (2000). An equation to represent grain-size distribution. *Canadian Geotechnical Journal*, 37(4): 817–827.

Haverkamp, R. & Parlange, J. Y. (1986). Predicting the water-retention curve from particle-size distribution: 1. Sandy soils without organic matter. *Soil Science*, 142(6): 325–339.

Head, K. H. (2006). *Manual of soil laboratory testing, volume I, soil classification and compaction testing*, 3rd ed. Caithness, UK: Whittles Publishing.

Lu, N. & Likos, W. J. (2004). *Unsaturated soil mechanics.* Hoboken, NJ: J. Wiley, c2004.

Salager, S., El Youssoufi, M. S., & Saix, C. (2010). Definition and experimental deter-mination of a soil-water retention surface. *Canadian Geotechnical Journal, 47*(6): 609–622.

Satyanaga, A., Rahardjo, H., Leong, E.-C., & Wang, J.-Y. (2013). Water characteris-tic curve of soil with bimodal grain-size distribution. *Computers and Geotechnics, 48*(0): 51–61.

Wesley, L. D. (2010). *Geotechnical engineering in residual soils*. Hoboken, New Jersey: John Wiley & Sons, Inc.

Wijaya, M. & Leong, E. C. (2017). Modelling the effect of density on the unimodal soil-water characteristic curve. *Géotechnique, 67*(7): 637–645.

Zou, L. (2018). Effects of grain-size distribution and hysteresis on soil-water char-acteristic curve (SWCC). In *School of civil and environmetal engineering* (vol. Ph.D.). Singapore: Nanyang Technological University.

FURTHER READING

Head, K. H. (2006). *Manual of soil laboratory testing, volume I, Soil classification and compaction testing*, 3rd ed. North America, Whittles Publishing,

Lu, N. & Likos, W. J. (2004). *Unsaturated soil mechanics*. Hoboken, NJ: J. Wiley.

Chapter 5

Atterberg limits and shrinkage test

5.1 BACKGROUND

Behaviour of fine-grained soils is highly affected by water content. The soil becomes a slurry at high water content and becomes hard and brittle as the soil dries. A simple method to distinguish a fine-grained soil behaviour is through its Atterberg limits (Atterberg 1911).

The Atterberg limits (shrinkage limit, plastic limit and liquid limit) separates fine-grained soils into four physical states: solid, semi-solid, plastic and liquid. In unsaturated soil mechanics, the shrinkage test provides not only the shrinkage limit but also the volume change with water content, which is commonly referred to as the shrinkage curve.

5.2 RELATED STANDARDS

1. BS 1377-1 (2016) Methods of Test for Soils for Civil Engineering Purposes Part 1: General Requirements and Sample Preparation.
2. BS 1377-2 (1990) Methods of Test for Soils for Civil Engineering Purposes Classification Tests.
3. BS 5930 (2015) Code of Practice for Ground Investigations, BSI.
4. BS EN ISO 17892-1 (2014) Geotechnical Investigation and Testing – Laboratory Testing of Soils Part 1: Determination of Water Content.
5. ASTM D427-04 (2004) (Withdrawn 2008) Standard Test Method for Shrinkage Factors of Soils by the Mercury Method, ASTM International.
6. ASTM D2216 (2010) Standard Test Methods for Laboratory Determination of Water (Moisture) Content of Soil and Rock by Mass, ASTM International.
7. ASTM D2487-17 (2017) Standard Practice for Classification of Soils for Engineering Purposes (Unified Soil Classification System). West Conshohocken, PA 19428-2959, ASTM International.

DOI: 10.1201/b22304-5

5.3 THEORY

5.3.1 Fine-grained soil classification

Fine-grained soil is usually classified as either silts or clays. The classification of a fine-grained soil into either silt or clay is usually based on its plasticity. Silt is a fine-grained soil which is non-plastic, making its behaviour similar to coarse-grained soil (BS 5930 2015) except it has much smaller soil particles, while clay is a fine-grained soil which is plastic (Holtz et al. 2011). Thus, classifying fine-grained soil solely on its particle size is inappropriate (BS 5930 2015) as clay and silt cannot be distinguished. According to BS 5930 (2015) most fine-grained soils are always a mixture of clay- and silt-size particles. Thus, fine-grained soils should be described as either silt or clay depending on the plastic properties and secondary descriptor for fine-grained soils (i.e., silty clay, or clayey silt) is allowed only for field distinction when differentiation is difficult. However, ASTM D2487-17 (2017) allows the use of secondary descriptors to describe fine-grained soils.

5.3.2 Water in soils

Water in soils may exist in different forms (Head 2006):

1. Water of hydration within the solid particle (unremovable through oven drying).
2. Adsorbed water, which exists in solid state at the particle surface (unremovable through oven drying at 110°C).
3. Hygroscopic water, which is held by van der Waals forces and not tightly held (removable through oven drying).
4. Capillary water, which is held by surface tension (removable through oven drying).
5. Gravitational water (removable through drainage and oven drying).

As a water content test is commonly conducted by oven drying the soil specimen at 105°C–110°C, only hygroscopic water, capillary water and gravitational water are removed. Thus, the so-called dry condition does not necessarily mean that there is no more water in the soil but that no more water can be removed at a temperature not exceeding 110°C (BS 1377-2 1990).

The amount of water in a soil has a strong influence on the behaviour of soil, especially for fine-grained soils. Thus, measurement of water content is very important in geotechnical engineering, especially for unsaturated soils. Gravimetric water content (w) is the most commonly used in geotechnical engineering to express the water content of soil and is calculated using Equation 5.1.

$$w = \frac{M_w}{M_s} 100\%$$ (5.1)

where M_w is the mass of water, and M_s is the mass of soil solid. Volumetric water content (θ) is often used in unsaturated soil mechanics to express soil water content and is calculated using Equation 5.2.

$$\theta = \frac{V_w}{V_t}$$ (5.2)

where V_w is the volume of water and V_t is the total volume of the specimen. When the prefix "gravimetric" or "volumetric" is not mentioned in geotechnical engineering, it usually referred to gravimetric water content.

Description of the soil's physical states based on its water content can be easily described by using an idealised shrinkage curve which is shown in Figure 5.1. The shrinkage curve is a curve which describes the relation between water content and the void ratio (Wijaya et al. 2015; Fredlund et al. 2002). In an idealised shrinkage curve, the soil starts drying from the slurry condition (with degree of saturation equals 100%). The soil dries along the saturation line, and the relationship between water content and the void ratio is linear and is given by Equation 5.3.

$$e = wG_s$$ (5.3)

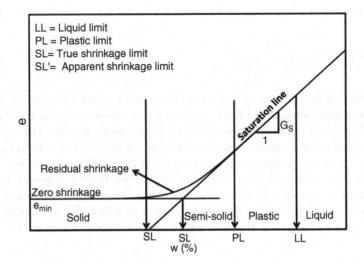

Figure 5.1 Idealised shrinkage curve.

On the saturation line, soil suction has the same effect as the net normal stress, i.e., volume and shear strength change due either to the change in net normal stress or change in pore-water pressure.

A soil with water content above its liquid limit behaves as a viscous fluid. As such a soil dries below its liquid limit, the soil reaches a plastic state. The soil shrinks and the shear strength increases (Fredlund et al. 2012). The undrained shear strength of soil at the liquid limit is approximately 1.7 kPa (Aitkinson 2007), and its average effective stress is approximately 8 kPa (Wood 1990). Once the soil's water content reaches the plastic limit, the soil starts to desaturate (Fredlund et al. 2002, 2012) and the undrained shear strength may increase by up to two orders in magnitude (Aitkinson 2007). It is still open for discussion if the plastic limit is the water content at the air-entry value. The difference between liquid limit (LL) and plastic limit (PL) is defined as plasticity index (PI) and is used in the plasticity chart for soil classification, as well as a number of empirical relations for many saturated soil properties. Another three indices that are commonly used are relative consistency (C_r), liquidity index (I_L) and activity which are defined according to Equations 5.4, Equations 5.5, and Equations 5.6, respectively.

$$C_r = \frac{LL - w}{PI} \tag{5.4}$$

$$LI = \frac{w - PL}{PI} \tag{5.5a}$$

Alternatively,

$$LI = 1 - C_r \tag{5.5b}$$

$$\text{Activity} = \frac{PI}{\%\text{Clay}(< 2\,\mu\text{m})} \tag{5.6}$$

As the soil desaturates further from PL, the soil reaches a semi-solid phase. The associated shrinkage phase is referred to as residual shrinkage (Leong and Wijaya 2015; Peng and Horn 2013). At this point, the effect of soil suction on the volume change and shear strength of the soil is less than the net normal stress as the soil becomes unsaturated ($S_r < 100\%$). On further drying, the soil reaches the zero-shrinkage phase (no further volume changes due to a decrease in water content), and the corresponding minimum void ratio is referred to as e_{min}. The water content at which the soil reaches the zero-shrinkage phase is the true shrinkage limit (SL). Contrast the true shrinkage limit to the shrinkage limit defined at the intersection between the zero-shrinkage line and the saturation line (BS 1377-2 1990; Fredlund et al. 2002). Leong and Wijaya (2015) differentiate the two shrinkage limits as follows:

1. Apparent *SL'*, which is given by the intersection between the zero-shrinkage line and the saturation line, is commonly used in geotechnical engineering.
2. True SL is the water content at which the soil experiences no further change in volume as its water content decreases further and is commonly used in soil science.

True shrinkage limit is more difficult to determine, as the shrinkage curve for some soils does not asymptote to a horizontal line near zero water content. The apparent shrinkage limit is easier to determine as it only requires the weight and volume of soil before and after oven-dried condition to obtain e_{min} and assume a horizontal line and *SL'* can then be determined from e_{min} using Equation 5.7.

$$SL' = \left(\frac{e_{min}}{G_s}\right)100\% \qquad (5.7)$$

5.3.3 Shrinkage and swelling curves of soils

The role of the shrinkage curve is not limited to describing the shrinkage limit of soils from the slurry condition. The shrinkage curve can also be used to estimate the volume change of undisturbed or compacted soils induced by drying. Conversely, the swelling curve of soils can be obtained by wetting the soil instead of drying the soil. The swelling curve can be used to estimate the wetting-induced volume change of soils.

The swelling curve is rarely determined as it is more difficult to wet the soil while measuring its weight and volume. However, some studies show that there is negligible hysteresis between shrinkage and swelling curves (Fleureau et al. 2002; Gould et al. 2011). Thus, Wijaya (2017) suggests that the shrinkage curve can be a proxy for the swelling curve. The limit of the swelling curve is defined by the swelling potential of soil, which is defined in Chapter 13.

5.3.4 Classification of soil shrinkage curve

The basic soil shrinkage curve consists of two linear segments (saturation line and zero-shrinkage line) as shown in Figure 5.1. Natural and compacted soil shrinkage curves may not follow the basic shrinkage curve for the following reasons:

1. presence of occluded air may cause the degree of saturation of the soil to never reach 100%;
2. presence of macropore that may cause structural shrinkage (Cornelis et al. 2006; Peng and Horn 2013); or
3. presence of peat, organic matter or bentonite (montmorillonite) may cause the soil to not have a true shrinkage limit or no zero-shrinkage line (Peng and Horn 2013; Wijaya et al. 2019).

To simplify the description of the soil shrinkage curve, Leong and Wijaya (2015) use a number of linear segments (two, three and four linear segments) to describe the shrinkage curve as shown in Figure 5.2. Figure 5.2a, c and e show the schematic illustration and the parameters describing the shrinkage curve while Figure 5.2b, d and e show examples of shrinkage curve data from literature.

The two linear segments shrinkage curve in Figure 5.2a is for the basic shrinkage curve as shown in Figure 5.1 except the first linear segment is referred to as loading line instead of as saturation line. The saturation line is for a soil at 100% degree of saturation and is strictly applicable to soils starting from the slurry condition. However, the soil is not necessarily fully saturated for undisturbed or compacted soils as occluded air might exist in the soil. Thus, the slope of loading line (m_2) is given by Equation 5.8.

$$m_2 = \frac{G_s}{S_0} \tag{5.8}$$

where S_0 is the initial degree of saturation. For the zero-shrinkage line, the slope (m_1) is commonly taken as 0 unless peat, organic matter or montmorillonite is present. Table 5.1 lists the equations that can be used to represent the shrinkage curve with two linear segments. In constructing the shrinkage curve, specific gravity (G_s), minimum void ratio (e_{min}) and initial degree of saturation (S_0) must be obtained while the curvature parameter can be estimated (Wijaya and Leong 2015; Leong and Wijaya 2015; Fredlund et al. 2002).

The three linear segments shrinkage curve in Figure 5.2b is commonly encountered for peats and organic soils (Leong and Wijaya 2015; Peng and Horn 2013), while the four linear segments shrinkage curve in Figure 5.2c is applicable when macropores are present. The difficulty in curve fitting three and four linear segments shrinkage curve using a single equation increases and a more complex equation is required as shown in Table 5.2.

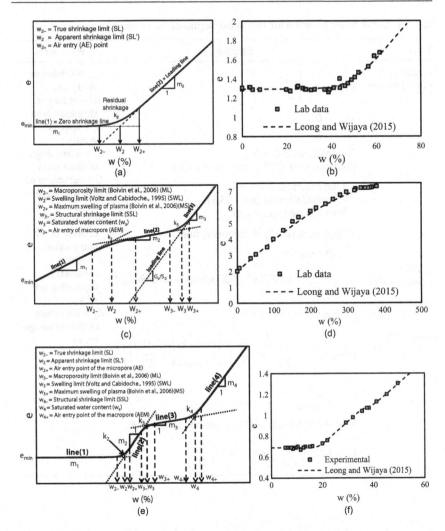

Figure 5.2 Shrinkage curve classification based on the number of linear segments (Leong and Wijaya 2015; Wijaya et al., 2019). (a) 2 linear segments shrinkage curve. (b) Kaolin (Wijaya, 2017). (c) Three linear segments shrinkage curve. (d) Histosol undisturbed soil core (Peng et al., 2007). (e) Four linear segments shrinkage curve. (f) Lixisol A (Cornelis et al., 2006).

Table 5.1 Curve fitting equation for two linear segments shrinkage curve

Model	Equation	Parameters
Frelund et al. (2002) model	$e(w) = e_{min}\left[\dfrac{w^C}{SL'^C} + 1\right]^{-1/C}$ $$\dfrac{e_{min}}{SL'} = \dfrac{G_s}{S_0}$$ Recommended C value Fredlund et al. (2002) recommendation: 1. undisturbed soil = 9.57 2. initially slurried soil = 25.31 3. compacted spo = 8.47 Wijaya et al. (2015) recommendation: 1. all soils = 9	e_{min} = minimum void ratio SL' = apparent shrinkage limit G_s = specific gravity S_0 = initial degree of saturation C = curvature of the shrinkage curve w_{AEV} = water content at air-entry value
Giraldez et al. (1983) model	$e = \dfrac{G_s}{S_0 \cdot w_{AEV}^2}\left(0.0267w^3 + 0.23w^2 w_{AEV} + 0.7249w_{AEV}^3\right)$	β = slope parameter depending on AEV
Kim et al. (1992) model	$e = e_{min}\exp(-\beta wG_s) + w\dfrac{G_s}{S_0}$	K_2 = curvature of the shrinkage curve
Leong and Wijaya (2015)	$e(w) = e_{min} + \dfrac{G_s}{2S_0}\left\{w + \dfrac{1}{k_2}\ln\left\{\dfrac{\cosh[k_2(w - SL')]}{\cosh(k_2 \cdot SL')}\right\}\right\}$ $$k_2 = \dfrac{2}{SL - w_{AE}}$$ Wijaya and Leong (2015) recommendation on k_2: $$k_2 = 4.32\left(\dfrac{G_s}{S_0 e_{min}}\right) \text{ or } k_2 = 4.32\left(\dfrac{G_s}{SL'}\right)$$	

5.3.5 Effect of stress history on soil shrinkage curve

Wijaya (2017) investigated the effect of stress history on the shrinkage curve. Figure 5.3a shows a kaolin specimen which was loaded to 1600 kPa and then unloaded. Figure 5.3b shows the comparison between pre- and post-compression shrinkage test results. Figure 5.3b shows that compression will affect the shrinkage curve of soil. Thus, the shrinkage test must be conducted by considering the stress history of the soil.

Table 5.2 Curve fitting equation for multi-linear segments shrinkage curve

Reference	Equation	Parameters
Leong and Wijaya (2015)	$$e(w) = e_{min} + m_l w + \sum_{i=2}^{n} T_i(w, w_i, m_i, m_{i-1}, k_i)$$ $$T_i(w, w_i, m_i, m_{i-1}, k_i) = \frac{(m_i - m_{i-1})}{2}\left\langle w + \frac{1}{k_i}\ln\left[\frac{\cosh[k_i(w - w_i)]}{\cosh[k_i w_i]}\right]\right\rangle$$ k_i can be either curve fitted or obtained from: $$k_i = \frac{2}{w_{i+} - w_{i-}};\ k_i = \frac{1}{w_i - w_{i-}};\ k_i = \frac{1}{w_{i+} - w_i}$$	e_{min} = minimum void ratio w = gravimetric water content m_l = slope of linear segment i w_i = gravimetric water content between segment i and segment i-1 k_i = curvature parameter between segment i and segment i-1 SL' = apparent shrinkage limit G_s = specific gravity S_0 = initial degree of saturation
Chertkov (2000, 2003)	$$e = \begin{cases} e_{min}; & \upsilon < \upsilon_{SL} \\ e_{min} + \mu(\upsilon - \upsilon_{SL})^2 \dfrac{\rho_w^2}{\rho_s}; & \upsilon_{SL} \le \upsilon \le \upsilon_{AE} \\ \upsilon; & \upsilon > \upsilon_{AE} \end{cases}$$	υ = moisture ratio = $w.G_s$ υ_{AE} = moisture ratio at air-entry value m_2 = slope of the loading line β = slope parameter depending on AEV μ = model parameter
Cornelis et al. (2006) or modified Chertkov (2000, 2003) equation (ModC)	$$e = \begin{cases} e_{min}; & \upsilon < \upsilon_{SL} \\ e_{min} + \mu(\upsilon - \upsilon_{SL})^2 \dfrac{\rho_w^2}{\rho_s}; & \upsilon_{SL} \le \upsilon \le \upsilon_{AE} \\ 2\mu(\upsilon - \upsilon_{SL})\upsilon + e_{min} \\ \quad -\mu(\upsilon_{AE} - \upsilon_{SL})(\upsilon_{AE} + \upsilon_{SL}); & \upsilon_{AE} < \upsilon < \upsilon_{LL} \end{cases}$$	υ_{SL} = moisture ratio at true shrinkage limit υ_{LL} = moisture ratio at liquid limit γ, ζ, ξ = model parameter $\upsilon_{SL'}$ = moisture ratio at apparent shrinkage limit υ_{inf} = moisture ratio at inflection point e_s = saturated void ratio β = model parameter $a_0, a_G, b_G, e_r, m_G, \phi_G, \alpha_G, \alpha_{er}, e_{r0}$ = model parameters

(Continued)

Table 5.2 (Continued)

Reference	Equation	Parameters
Cornelis et al. (2006) or modified Groenevelt and Grant (2001, 2002) (ModGG)	$$e = e_{min} + \gamma\left[\exp\left(\frac{-\xi}{\upsilon^\varsigma}\right)\right]$$	
McGarry and Malafant (1987)	$$e = \begin{cases} e_{min} + \dfrac{\upsilon}{\upsilon_{SL'}}\left(\upsilon_{SL'} - e_{min} - e_n\right) & 0 < \upsilon < \upsilon_{SL'} \\[2mm] e_n + \upsilon & \upsilon_{SL'} \le \upsilon \le \upsilon_{SWL} \\[2mm] e_s + \dfrac{\upsilon}{\upsilon_{SWL}}\left(\upsilon_{SWL} - e_s + e_n\right) & \upsilon_{SWL} < \upsilon \le \upsilon_S \end{cases}$$ $$e = e_{min} + \frac{e_s}{1 + \exp\left[-\beta(\upsilon - \upsilon_{inf})\right]}$$	
Gould et al. (2011)	$$e = f(w, a_G) - f(w, b_G) - f(0, a_G) + f(0, b_G) + e_r$$ $$f(x,y) = -\frac{m}{\phi\pi}\left\langle\phi(x-y)\tan^{-1}\left[\phi(x-y)\right] = 0.5\ln\left\{1 + \left[\phi(x-y)\right]^2\right\}\right\rangle$$ to account for net normal stress: $$a_G = \alpha_G \ln\left(1 + \frac{\sigma}{\sigma_0}\right) + a_0$$ $$e_r = \alpha_{er} \ln\left(1 + \frac{\sigma}{\sigma_0}\right) + e_{r0}$$	

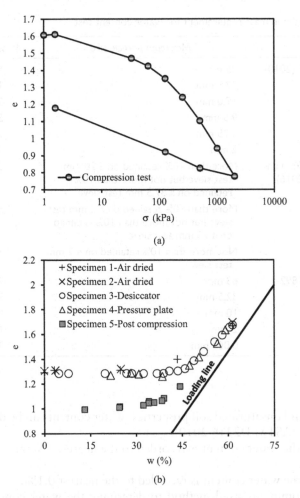

Figure 5.3 Post-compression shrinkage curve. (a) Compression test on Specimen 5. (b) Pre-compression shrinkage curve (Wijaya et al., 2015) and Post-compression shrinkage curve (Specimen 5).

5.4 TEST METHODS

5.4.1 Gravimetric water content test

Gravimetric water content test (or water content test) is one of the most common tests in geotechnical engineering, as the water content of the specimen is always required in determining the Atterberg limits, saturated soil

Table 5.3 Specimen mass requirement for water content test

Standard	Maximum particle size	Mass requirement
ASTM D2216 (2010)	75 mm	5 kg
	37.5 mm	I kg
	19.0 mm	250 g
	9.5 mm	50 g
	4.75 mm	20 g
	2 mm	20 g
BS 1377-2 (1990) and BS 1377-1 (2016)	More than 10% retained on a 20 mm test sieve but not more than 10% retained on a 37.5 mm test sieve	3 kg
	More than 10% retained on a 2 mm test sieve but not more than 10% retained on a 20 mm test sieve	300 g
	Not more than 10% retained on a 2 mm test sieve	30 g
BS EN ISO 17892-1 (2014)	63 mm	21,000 g
	32.5 mm	3,000
	10 mm	500
	2 mm	100
	0.063 mm	30

properties and unsaturated soil properties. Water content can be determined using either (ASTM D2216 2010):

Method A: the water content is recorded to the nearest 1% (referee method).

Method B: the water content is recorded to the nearest 0.1%.

The most commonly used method to determine the water content of the specimen is oven drying, which will be explained in this section. The procedures to determine water content by oven drying are as follows:

1. Prepare a specimen with mass according to Table 5.3.
2. Prepare a clean moisture can or tray (including lid if it is used) and record its weight (M_c) (BS 1377-2 1990; ASTM D2216 2010; BS EN ISO 17892-1 2014).
3. Place the specimen into the can/tray and record the weight of both specimen and can/tray ($M_{c,ws}$) (BS 1377-2 1990; ASTM D2216 2010; BS EN ISO 17892-1 2014).
4. Place the specimen in the oven and use a drying temperature of either (105°C – 110°C) (BS 1377-2 1990) or 110 ± 5°(ASTM D2487-17 2017). For certain soil types such as peats or highly organic soils, 60°C

is preferred (Head 2006). For material that contains gypsum, the temperature cannot exceed 80°C (BS 1377-2, 1990).

5. Check the change in the sample weight at 4-hour intervals until the difference between each 4-hour interval does not exceed 1% for method A or 0.1% for method B of the original mass of the sample (BS 1377-2 1990; ASTM D2216 2010). In most cases, 12 (ASTM D2487-17 2017), 16 (BS 1377-2 1990; ASTM D2216 2010; BS EN ISO 17892-1 2014) to 24 hours (BS 1377-2 1990) are sufficient to dry a soil.

6. Take out the specimen and place it in a desiccator to cool without it absorbing water from the air (BS 1377-2 1990; ASTM D2216 2010; BS EN ISO 17892-1 2014).

7. Record the weight of the can/tray with the dry specimen ($M_{c,ds}$) (BS 1377-2 1990; ASTM D2216 2010; BS EN ISO 17892-1 2014).

8. Calculate the gravimetric water content of the specimen using Equation 5.9 (BS 1377-2 1990; ASTM D2216. 2010; BS EN ISO 17892-1 2014):

$$w = \frac{M_{c,ws} - M_{c,ds}}{M_{c,ds} - M_c} 100\% \tag{5.9}$$

9. If the water content is to be related to the Atterberg limits, it is necessary to do a correction such that the water content used for the correlation (i.e., liquidity index and relative consistency) is only representing the particle size used for the Atterberg limits (i.e., 425 μm for BS standard). The correction can be done using Equation 5.10 (BS 1377-2 1990):

$$w_a = w\left(\frac{100}{P_a}\right) - w_r\left(\frac{100 - P_a}{P_a}\right) \tag{5.10}$$

where w_a is the corrected water content for Atterberg limit correlation, P_a is the percentage by dry mass of the portion of the soil sample passing the 425 μm, w_r is the water content of the fraction retained on the 425 μm sieve (for BS standard). The water content w_r is necessary when the particles retained on the 425 μm sieve (for BS standard) are porous and absorbing water; otherwise, w_r is set to 0.

5.4.2 Specimen preparations for liquid limit and plastic limit tests

The liquid limit test can be done on a natural soil or a sieved sample. A natural sample can be used when there is little to no material retained on a 425 μm sieve. The procedures for natural material are as follows:

1. Prepare at least 500 g of soil in its natural state and remove any observable particles retained on a 425 μm sieve.
2. Determine the mass of all coarse particles that is present in the soil in its natural state (BS 1377-2 1990).
3. Transfer the soil to a flat glass plate (BS 1377-2 1990).
4. Add distilled water to the soil and mix thoroughly with palette knives until the soil and water become a homogeneous paste (BS 1377-2 1990).
5. Place the paste in an airtight container for 24 hours (BS 1377-2 1990).

The procedures for a sieved sample are as follows:

1. Prepare at least 300g of soil that passed through the 425 μm sieve (BS 1377-2 1990). The use of a dispersing agent is not allowed.
2. Take a representative specimen and determine its water content (w_n).
3. Record the weight of the specimen (M_w).
4. Place the soil in a container and then submerged under distilled water until it becomes a slurry (BS 1377-2 1990).
5. Conduct wet sieving of the slurry with a 425 μm sieve and collect the filtrate using a receiver pan. Use as little distilled water as possible. Wash the fines in the 425 μm sieve.
6. Oven dry the soil particles retained on the 425 μm sieve and record their weight (M_r).
7. Allow the soil particles in the receiver pan to settle and decant any clear water above the suspension.
8. Dry the suspension using warm air, vacuum or pressure until it forms a stiff paste.
9. Calculate the dry mass of the soil using Equation 5.11.

$$M_s = \frac{100}{100 + w_n} M_w \qquad (5.11)$$

10. Calculate the percentage of the soil passing through the 425 μm using 5.12.

$$P_a = \frac{M_s - M_r}{M_s} \qquad (5.12)$$

5.4.3 Liquid limit

There are two commonly used methods for the determination of liquid limit:

1. Casagrande liquid limit test
2. Fall cone penetrometer test

Casagrande test is the earliest method used to obtain liquid limit. However, the fall cone penetrometer test is easier to conduct, gives more consistent results and is less prone to error, as it is a static test which depends on soil shear strength (BS 1377-2 1990).

5.4.3.1 Liquid limit based on Casagrande apparatus

Procedures for the liquid limit test based on the Casagrande apparatus are as follows:

1. Prepare 300 g of soil paste and place it on the glass plate.
2. Mix the paste for around 10 minutes and add more distilled water if necessary.
3. Place a portion of the soil paste into the Casagrande cup without entrapping air and level off the soil paste surface parallel to the base.
4. Hold the grooving tool normal to the surface of the cup with the chamfered edge facing the direction of movement and split the soil into two equal parts.
5. Turn the crank handle at the rate of two revolutions per second so the cup is lifted and dropped until the two parts of the soil come into contact at the bottom of the groove along a distance of 13 mm and record the number of bumps at which this occurs.
6. Repeat steps 3 to 5 until two consecutive consistent numbers of bumps can be achieved. If the soil paste slides on the surface of the cup instead of within the soil, the test is discarded and steps 2 to 5 are repeated.
7. If two consistent consecutive numbers of bumps can be achieved, take about 10 g of soil with a spatula from the portions of the sample that have flowed together and determine the water content of the soil.
8. Remove the soil from the cup. Clean, wash and dry the Casagrande cup and the grooving tool. Add more water to the paste.
9. Repeat steps 2 to 8 at least three times. Plot the number of bumps with their respective water contents.
10. Make a straight (regression) relationship and determine the liquid limit from the regression line corresponding to the water content at 25 bumps. Present the liquid limit as a whole number.

5.4.3.2 Liquid limit based on fall cone test

Procedures for a liquid limit test based on a fall cone penetrometer are as follows:

1. Prepare 300 g of soil paste and place it on the glass plate.

2. Mix the soil paste for around 10 minutes and add more distilled water if necessary.
3. Place a portion of the soil paste into the cup using the palette knife and push it with the palette knife without entrapping air. Repeat until the cup is full.
4. Use a straightedge to level the soil surface at the cup level.
5. Adjust the fall cone position such that the tip of the cone just touches the surface of the soil. The correct position is indicated by moving the cup slightly and the cone tip should mark the soil surface.
6. Record the initial dial gauge reading to the nearest 0.1 mm.
7. Release the cone for 5 ± 1s.
8. Record the final dial gauge reading to the nearest 0.1 mm and calculate the penetration, which is the difference between final and initial dial readings.
9. Lift up the cone and clean it.
10. Add soil paste to patch the soil in the cup and repeat steps 3 to 9.
11. Record the average of the two penetrations if the difference between first and second penetration readings is not more than 0.5 mm. If the difference is more than 0.5 mm and less than 1 mm, patch the soil in the cup and redo steps 3 to 9 to get the third penetration. If the overall range is not more than 1 mm, record the average of the three penetrations, otherwise, remove all the soil from the cup and repeat steps 2 to 11.
12. Take 10 g of the soil paste around the penetrated area to determine its water content.
13. Remove the soil paste from the cup, wash and clean the cup.
14. Add more distilled water to the soil paste and repeat steps 2 to 12 for at least three more readings such that the range of penetration from 15 mm to 25 mm is obtained.
15. Plot water content versus cone penetration and plot a best-fit straight line through the data points (i.e., regression line), and determine the liquid limit which is the water content on the straight line corresponding to a cone penetration of 20 mm.
16. Present the liquid limit as a whole number and report the percentage of material passing the 425 μm sieve.

5.4.4 Plastic limit test

Procedures for the plastic limit test are as follows:

1. Prepare 20 g of the soil paste and place it on a scratch-free glass plate.
2. Let the soil dry on the plate such that it can be shaped into a ball.
3. Mould the ball with fingers and roll it between palms such that there is a slight crack on its surface.
4. Divide the sample into two 10 g sub-samples.

5. Divide each sub-sample into four equal parts.
6. Mould each part with fingers to form a soil thread of 6 mm in diameter.
7. Roll the thread between the fingers, from the fingertip to the second finger joint of one hand, and the surface of the glass plate.
8. Use small pressure to reduce the thread diameter to about 3 mm by using forward and back movement of the hand.
9. Pick up the soil, mould it between the fingers to dry it further.
10. Repeat steps 7 to 9 until the thread shears both longitudinal and transversely when it has been rolled to about 3 mm diameter. Usually, a short metal rod of 3 mm diameter is used as a guide for the required soil thread. Water content of the soil at this point is taken as the plastic limit.
11. Gather the portions of the crumbled soil thread and placed it in a moisture content can.
12. Repeat steps 6 to 11 for the other three equal parts of the sub-sample and place all four crumbled parts together in the same moisture content can and determine the water content.
13. Repeat steps 5 to 12 for the other sub-samples.
14. Take the average water content of the two samples as the plastic limit and present it to the nearest whole number.

5.4.5 Shrinkage limit and shrinkage curve test

Shrinkage tests can be conducted on a cylindrical specimen (volumetric shrinkage) or on a soil bar (linear shrinkage). The soil bar is formed from soil paste prepared in the same manner as for the liquid limit test, except only 150g of the fraction passing sieve 425 µm is required. Whichever specimen is used, the shrinkage limit of the soil, the shrinkage ratio and the volumetric shrinkage for a given change of moisture content can be determined.

5.4.5.1 Volumetric shrinkage

Shrinkage limit can be conducted on either an intact sample (definitive method) or from a sieved sample which is made into a slurry paste (subsidiary method) (BS 1377-2 1990) . For the definitive method, the ideal specimen size is 38 mm to 51 mm in diameter and a length of one to two times the diameter (BS 1377-2 1990).

The principle behind the shrinkage limit or the shrinkage curve is to record the volume and weight of the specimen as the specimen dries. For shrinkage limit determination (ASTM D427-04 2004), only the initial and final mass and volume of the specimen need to be recorded (the final mass and volume refer to the oven-dry condition) and the apparent SL' can be obtained. But shrinkage curve determination (BS 1377-2 1990) requires multiple readings to be taken between initial wet to dry conditions. It is possible to estimate the shrinkage curve by using SL' (Wijaya et al. 2015; Wijaya and Leong 2015; Fredlund et al. 2002) when the shrinkage curve only has two linear segments, and the zero-shrinkage line is present. Thus, it is always recommended to determine the shrinkage curve when dealing with unsaturated soils.

The challenge in determining the shrinkage limit and the shrinkage curve of soil is in determining the volume of the specimen as the soil specimen may not shrink homogeneously and uniformly. Shrinkage may distort the specimen's shape, making volume measurement challenging. Several methods that can be employed in measuring the specimen's volume are listed in Table 5.4. Specimen volume measurement using mercury described in BS1377-2 (1990) is intentionally omitted.

Table 5.4 Volume measurement method for shrinkage test updated from Wijaya et al. (2015)

Method	Destructive	Description
Direct measurement method	Non-destructive	Using measurement devices such as the Vernier caliper to obtain the dimensions of the specimen. It is assumed that the specimen has a regular geometry such as a cylinder so that the volume can be easily calculated. In order to account for the non-homogeneous geometry, several measurements are taken to obtain the average dimension. The simplicity, speed and non-destructive are the advantages of this method. This method may not work on irregular-shaped specimen.
Volume displacement method	Destructive /Non-destructive	The soil specimen is coated with wax or saran resin dissolved in ethyl ketone (Brasher et al. 1966) and then submerged into a liquid. The volume of the liquid displaced equals to the total volume of the specimen. It can be used to measure the volume of the irregular-shaped specimen. The disadvantages of this method are destructive and when the clods have large pores, more viscous saran resins are needed (Guillermo et al. 2001) or alternatively by using mercury (ASTM D427-04 2004; BS 1377-2 1990)

(Continued)

Table 5.4 (Continued)

Method	Destructive	Description
Two non-mixing liquid method	Destructive	Two types of liquids are used to produce different buoyancy forces on the soil specimen. The difference in the buoyancy force is then used to calculate the density of the specimen and hence the total volume of the specimen. This method is similar to the volume displacement method.
Gamma-ray attenuation dual-energy method	Non-Destructive	Dual-energy gamma beam with different mass absorption coefficients is applied to obtain different mass attenuation coefficients and a number of gamma-ray quanta produced and hence bulk density and total volume. This method requires special apparatus and is the most capital-expensive method compared to the others; however, it can be used to measure irregular-shaped specimen and it is not destructive.
Photogrammetry method	Non-destructive	Soil specimen pictures are taken at different views to construct 3D specimen model.
Laser scanner method	Non-destructive	Laser scanner is used to obtain a very dense point cloud to construct a 3D specimen model.

As volume measurement depends on the technique employed, the shrinkage limit test can be briefly described as follows:

1. Determine the initial water content of the specimen when possible (i.e., using the left-over sample) to estimate the water content and void ratio of the specimen during the test.
2. Record the initial weight and volume of the specimen and estimate the water content of the specimen using Equation 5.13.

$$w_i \left[\left(\frac{M_i}{M_{\mathrm{ref}}} \right) \left(\frac{w_{\mathrm{ref}}}{100} + 1 \right) \right] 100\% \qquad (5.13)$$

where w_i is the gravimetric water content at measurement i, M_i is the mass of soil at measurement i, M_{ref} is the reference mass, while w_{ref} is the reference gravimetric water content. As M_{ref} and w_{ref} are obtained at the same time, they can be obtained either before or after the shrinkage test. If M_{ref} and w_{ref} are obtained prior to the shrinkage test, M_{ref} is the mass of the specimen at the beginning of the test, while w_{ref} can only be obtained by using other specimens or left-over sample.

3. Place the specimen on a watch glass and exposed it to air at room temperature for 2h (BS 1377-2 1990) or place it in the pressure plate and apply suction to the specimen.
4. Record the weight and volume of the specimen.
5. Repeat steps 3 and 4 until there is no more change in the weight of the specimen.
6. For a very expansive clay (i.e., bentonite), there may be no more change in weight while the water content is still relatively high. For such a specimen, place it in the oven for 5 to 20 minutes to accelerate the drying process by using a temperature lower than 105°C (BS1377-1 recommends oven temperature from 45°C to 50°C while ASTM D4318-17 recommends oven temperature less than 60°C).
7. Record the weight and volume of the specimen.
8. Repeat steps 6 and 7 until there is very little water inside the specimen and record the final mass and volume of the specimen.
9. Oven dry the specimen to a constant mass to determine its water content.
10. Use the final mass as M_{ref} and final water content as w_{ref} and determine the actual water content of the specimen by using Equation 5.9 and calculate the void ratio of the specimen from each measurement using Equation 5.14.

$$e_i = V_i G_s \frac{\left(1 + w_{ref}/100\right)}{M_{ref}} - 1 \qquad (5.14)$$

where e_i is the void ratio at measurement I, V_i is the volume of the specimen at measurement i.

The shrinkage ratio, R_s, of the specimen is given by Equation 5.15:

$$R_s = \frac{m_d}{V_d} \qquad (5.15)$$

where m_d is the mass and V_d is the volume of the specimen after oven drying. The shrinkage ratio provides a convenient way to calculate the volumetric shrinkage of the soil, V_s, from an initial water content w, as shown in Equation 5.16.

$$V_s = \frac{w - SL'}{R_s} \tag{5.16}$$

5.4.5.2 Linear shrinkage

The linear shrinkage test is performed on a soil bar formed from soil paste using a mould, as shown in Figure 5.4. The test procedures are as follows:

1. Use about 150 g of the dried soil and make it into a paste at its liquid limit. The paste is prepared in the same manner as the liquid limit test.
2. Fill the mould with the soil paste using a palette knife ensuring that there is no air trapped within.

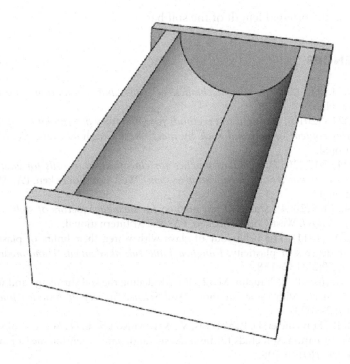

Figure 5.4 Mould for linear shrinkage test.

3. Level the soil along the top of the mould with a palette knife and clean the surrounding of the mould of any adhering soil with a damp cloth.
4. Place the mould at a location where the soil specimen can air dry slowly until the soil has shrunk away from the walls of the mould. A closed chamber with low humidity (60%–70%) may be suitable.
5. Then dry it in the oven at a temperature not exceeding 65°C until shrinkage has largely stopped and complete the drying at 105°C to 110°C.
6. Remove the mould and soil from the oven to cool.
7. Measure the mean length of the soil bar, L_d. Some soils may result in a curved soil bar during drying. In this case, remove the soil carefully from the mould and measure the lengths of the top and bottom surfaces, and record L_d as the mean of these two lengths.

The linear shrinkage is given by Equation 5.17.

$$\text{Linear shrinkage} = \left(1 - \frac{L_d}{L_0}\right) \times 100\% \qquad (5.17)$$

where L_0 is the original length of the soil bar.

REFERENCES

Aitkinson, J. H. (2007). *The mechanics of soils and foundations*. London: McGraw-Hill.

ASTM D2216 (2010). *Standard test methods for laboratory determination of water (moisture) content of soil and rock by mass*. West Conshohocken, PA: ASTM International.

ASTM D2487-17 (2017). *Standard practice for classification of soils for engineering purposes (Unified Soil Classification System)*. West Conshohocken, PA: ASTM International.

ASTM D427-04 (2004). *Standard test method for shrinkage factors of soils by the mercury method*. West Conshohocken, PA: ASTM International.

Atterberg, A. (1911). The behaviour of clays with water, their limits of plasticity and their degrees of plasticity. *Kungliga Lantbruksakademiens Handlingar och Tidskrift*, 50(2): 132–158.

Boivin, P., Garnier, P., & Vauclin, M. (2006). Modelling the soil shrinkage and water retention curves with same equations. *Soil Science Society of America Journal*, 70(4): 1082–1093.

Brasher, B. R., Franzmeier, D. P., Valassis, V., & Davidson, S. E. (1966). Use of saran resin to coat natural soil clods for bulk density and water retention measurement. *Soil Science*, 101(2): 108.

BS 1377-1 (2016). *Methods of test for soils for civil engineering purposes Part 1: General requirements and sample preparation*. UK: BSI.

BS 1377-2 (1990). *Methods of test for soils for civil engineering purposes Classification tests*.

BS 5930 (2015). *Code of practice for ground investigations*. UK: BSI.

BS EN ISO 17892-1 (2014). *Geotechnical investigation and testing – Laboratory testing of soils Part 1: Determination of water content*. UK: BSI.

Cornelis, W. M., Corluy, J., Medina, H., Diaz, J., Hartmann, R., Van Meirvenne, M., & Ruiz, M. E. (2006). Measuring and modelling the soil shrinkage characteristic curve. *Geoderma*, 137(1–2): 179–191.

Fleureau, J. M., Verbrugge, J. C., Huergo, P. J., Correia, A. G., & Kheirbek-Saoud, S. (2002). Aspects of the behaviour of compacted clayey soils on drying and wetting paths. *Canadian Geotechnical Journal of Agricultural Science*, 39(6): 1341–1357.

Fredlund, D. G., Rahardjo, H., & Fredlund, M. D. (2012). *Unsaturated soil mechanics in Engineering practice*. Hoboken, New Jersey: John Wiley & Sons, Inc..

Fredlund, M. D., Wilson, G. W., & Fredlund, D. G. (2002). Representation and estimation of the shrinkage curve. In *Proceedings of 3rd International Conference on Unsaturated Soils, UNSAT 2002*, pp. 145–149.

Gould, S. J. F., Kodikara, J., Rajeev, P., Zhao, X. L., & Burn, S. (2011). A void ratio – water content – net stress model for environmentally stabilized expansive soils. *Canadian Geotechnical Journal*, 48(6): 867–877.

Guillermo, O. S., Roberto, R. F., & Gimenez, D. (2001). Measurement of soil aggregate density by volume displacement in two non-mixing liquids. *Soil Sciience Society of America Journal*, 65(5): 1400–1403.

Head, K. H. (2006). *Manual of soil laboratory testing, volume I, soil classification and compaction testing*, 3rd ed. Caithness, UK: Whittles Publishing.

Holtz, R. D., Kovacs, W. D., & Sheahan, T. C. (2011). *An introduction to geotechnical engineering*, 2nd ed. Upper Saddle River: Pearson Education, Inc.

Leong, E. C. & Wijaya, M. (2015). Universal soil shrinkage curve equation. *Geoderma*, 237: 78–87.

Peng, X. & Horn, R. (2013). Identifying six types of soil shrinkage curves from a large set of experimental data. *Soil Science Society of America Journal*, 77(2): 372–381.

Peng, X., Horn, R., & Smucker, A. (2007). Pore shrinkage dependency of inorganic and organic soils on wetting and drying cycles. *Soil Science Society of America Journal*, 71(4): 1095–1104.

Wijaya, M. (2017). Compression, Shrinkage and Wetting-Induced Volume Change of Unsaturated soils. In *Civil and Environmental Engineering (Ph.D.)* Nanyang Technological University: Singapore.

Wijaya, M. & Leong, E. C. (2015). Estimation of Soil Shrinkage Curve. In *Proceedings of Unsaturated soils: Research & Applications. CRC Press/Balkema*, vol. 1, pp. 785–789.

Wijaya, M., Leong, E. C., & Abuel-Naga, H. (2019). Shrinkage curves for powder and granular bentonites. *E3S Web of Conferences*, 92: 07009.

Wijaya, M., Leong, E. C., & Rahardjo, H. (2015). Effect of shrinkage on air-entry value of soils. *Soils and Foundations*, 55(1): 166–180.

Wood, D. M. (1990). *Soil behaviour and critical state soil mechanics*. London: Cambridge University Press.

FURTHER READING

Leong, E. C. & Wijaya, M. (2015). Universal soil shrinkage curve equation. *Geoderma*, 237: 78–87.

Wijaya, M., Leong, E. C., & Rahardjo, H. (2015). Effect of shrinkage on air-entry value of soils. *Soils and Foundations*, 55(1): 166–180.

Chapter 6

Compaction

6.1 BACKGROUND

Compacted soils are used in a number of civil engineering projects, especially as fill material. Fill material is compacted either by rolling, ramming or vibrating. Compaction reduces the air voids in the soil. The quality of compaction is related to the dry density of the compacted materials. Therefore, it is important to specify the desired dry density that must be achieved by compaction in civil engineering projects. Compaction is the only topic in classical soil mechanics textbooks where the soils are in an unsaturated state. Compaction is also used to prepare soil specimens for saturated and unsaturated soil tests. In this chapter, Proctor, vibrating hammer and static compaction test procedures will be laid out.

6.2 RELATED STANDARDS

1. BS 1377-1 (2016) Methods of Test for Soils for Civil Engineering Purposes – Part 1: General Requirements and Sample Preparation.
2. BS 1377-4 (1990) Methods of Test for Soil for Civil Engineering Purposes – Part 4: Compaction-Related Tests, BSI.
3. ASTM D698-12 (2012) Standard Test Methods for Laboratory Compaction Characteristics of Soil Using Standard Effort (12,400 ft-lbf/ft3 (600 kN-m/m3)).
4. ASTM D1557-12 (2012) Standard Test Methods for Laboratory Compaction Characteristics of Soil Using Modified Effort (56,000 ft-lbf/ft3 (2,700 kN-m/m3)).
5. ASTM D7382-08 (2008) Standard Test Methods for Determination of Maximum Dry Unit Weight and Water Content Range for Effective Compaction of Granular Soils Using a Vibrating Hammer.

6.3 THEORY

There are two important factors that determine the dry density of soil through compaction:

1. compaction effort, and
2. water content.

Higher compaction effort will lead to higher soil dry density. However, it is not as straightforward for water content. At the beginning, when soil is compacted using the same compaction effort, higher water content will lead to higher density as the water makes the soil easier to compact until the water content reaches the optimum water content (w_{opt}) and the dry density reaches maximum dry density. However, when more water is added, excess pore-water pressure which is generated during the compaction will push the soil particles apart. Thus, under the same compaction effort, an increase in water content beyond w_{opt} will reduce the dry density. Relationship between water content and dry density using the same compaction effort is referred to as the compaction curve.

Several contour lines which represent the degree of saturation through the basic soil properties relationship is usually plotted together with the compaction curve. The contour lines can be plotted according to Equation 6.1.

$$\rho_d = \frac{G_s}{1 + \dfrac{w \cdot G_s}{S_r}}$$

(6.1)

The degree of saturation line which represents the fully saturated condition ($S_r = 100\%$) is commonly referred to as zero air void (ZAV) line. It is not possible for the compaction curve to cut the ZAV line due to the presence of occluded air bubbles that are generated when the soil is compacted. Thus, a compacted soil is always unsaturated.

Static compaction test on uniformly graded and coarse clean sands may not be reliable, and a vibrating hammer must be used instead (BS 1377-4 1990; ASTM D7382-08 2008).

6.4 TEST METHODS

Compaction test is commonly differentiated into

1. standard compaction test or standard Proctor test (ASTM D698-12 2012; BS 1377-4 1990),
2. modified compaction or modified Proctor test (ASTM D1557-12 2012; BS 1377-4 1990),
3. vibrating hammer (BS 1377-4 1990; ASTM D7382-08 2008), or
4. static compaction test (Priono et al. 2017; Goh 2012).

The differences between standard and modified Proctor tests are shown in Table 6.1. The selection of whether to conduct a standard or modified Proctor test is based on the compaction requirement of the site. Depending on the particle size of the specimen, ASTM recommends three testing methods for standard and modified Proctor compaction – namely, Methods A, B and C, while BS recommends five grading zone classifications, which are zone 1 to zone 5. Both ASTM and BS test procedures are summarised in Table 6.2. The vibrating hammer is mainly for granular soils satisfying the soil suitability shown in Table 6.3. The ASTM and BS test procedures for vibrating hammer compaction test are summarised in Table 6.4.

BS 1377-4 (1990) allows the same soil sample to be used multiple times after progressively increasing the water content, provided that the soil is not susceptible to crushing (and thus, requires a smaller sample size). If the soil contains soft granular material such as limestone or sandstone or if the soil is susceptible to crushing, the soil sample cannot be reused (BS 1377-1 2016; BS 1377-4 1990). ASTM D698-12 (2012) and ASTM D1557-12 (2012) strictly do not allow the soil sample to be reused.

If the expected dry density is known, the mass of water can be calculated by Equation 6.2 (ASTM D7382-08 2008):

$$M_w = M_s \left(\frac{\rho_w}{\rho_d} - \frac{1}{G_s} \right)$$

(6.2)

where M_w is the mass of water, M_s is the dry mass of the test specimen, ρ_d is the expected dry density and G_s is the specific gravity.

Table 6.1 Comparison between standard and modified Proctor compaction test (ASTM D1557-12 2012; ASTM D698-12 2012; BS 1377-4 1990)

ASTM D698-12 (2012)	ASTM D1557-12 (2012)
Standard compaction test	Modified compaction test
• rammer: 5.5 lbf (24.5 N) • layers: 3 • drop height: 12 in. (305 mm) • compaction effort: 12,400 ft-lbf/ft³ (600 kNm/m³)	• rammer: 10 lbf (44.48 N) • layers: 5 • drop height: 18 in. (457.2 mm) • compaction effort: 56,000 ft-lbf/ft³ (2,700 kNm/m³)

BS 1377-4 (1990)	
Standard compaction test	Modified compaction test
• rammer: 2.5 kg • layers: 3 • drop height: 300 mm	• rammer: 4.5 kg • layers: 3 to 5 • drop height: 450 mm

Table 6.2 Standard test procedures for Proctor compaction tests

Standards: ASTM D698-12 (2012) and ASTM D1557-12 (2012)

Method A	Method B	Method C
• Mould diameter: 4in. (101.6 mm)	• Mould diameter: 4in (101.6 mm)	• Mould diameter: 6in (152.4 mm)
• Material: passing No. 4 (4.75 mm) sieve	• Material: passing 3/8in. (9.5 mm) sieve	• Material: passing 3/4in. (19 mm) sieve
• Specimen dry mass: 16 kg	• Specimen dry mass: 16 kg	• Specimen dry mass: 45 kg
• Sub-specimen mass: 2.3 kg	• Sub-specimen mass: 2.3 kg	• Sub-specimen mass: 5.9 kg
• Blows per layer: 25	• Blows per layer: 25	• Blows per layer: 56
• Usage: may be used if 25% or less by mass of the material is retained on the No. 4 (4.75 mm) sieve	• Usage: may be used if 25% or less by mass of the material is retained on the 3/8in (9.5 mm) sieve	• Usage: may be used if 30% or less by mass of the material is retained on the 3/4in (19 mm) sieve

Standard: BS 1377-4 (1990)

Grading Zone	P_{20}	$P_{37.5}$	Minimum specimen mass (a)	Minimum specimen mass (b)	Type of mould
1	100	100	6 kg	15 kg	1 L
2	95	100	6 kg	15 kg	1 L
3	70	100	15 kg	40 kg	CBR
4	70	95	15 kg	40 kg	CBR
5	70	90	15 kg	40 kg	CBR
X	<70	<90	Not suitable for static compaction test		

P_{20} Minimum percentage passing test sieve size 20 mm
$P_{37.5}$ Minimum percentage passing test sieve size 37.5 mm
(a) Soil particles not susceptible to crushing during compaction
(b) Soil particles susceptible to crushing during compaction
1 L = one-litre compaction Mould
CBR = California bearing ratio
Grading zone x is not suitable for a static compaction test. The vibrating hammer can be considered.

6.4.1 Moist sample preparation

Compaction test can be carried out either by using a moist sample (which is the preferred method) or a dry sample. Preparation for moist sample can be done as follows (BS 1377-1 2016; ASTM D698-12 2012; ASTM D1557-12 2012):

1. Sieve the sample in accord with Table 6.4.
2. Determine the mass of both retained (oversized) and passing (test fraction) portions of the sample.
3. Oven dry the oversized fraction and determine its dry mass.
4. If more than 0.5% of the total dry mass of the sample is the oversized fraction, wash the oversized fraction, oven dried and record its oven-dry mass.

Table 6.3 Vibrating hammer compaction test (ASTM D7382-08 2008; BS 1377-4 1990)

Condition	ASTM D7382-08 (2008)	BS 1377-4 (1990)
Soil suitability	$P_{50} = 100\%$ $P_{19} > 70\%$ If $P_{0.425}$ is nonplastic, $P_{0.075} < 35\%$ If $P_{0.425}$ is plastic, $P_{0.075} < 15\%$	$P_{20} < 70\%$
Hammer	Frequency: 53 Hz to 58 Hz Weight: 53 N to 89 N Impact energy: 9.5 m.N to 12 m.N	Frequency: 25 Hz to 60 Hz Power consumption: 600 W to 800 W Voltage: 110V with earth leakage circuit breaker between the hammer and the mains supply Total applied force during vibration: 300 N to 400 N
Final height	Must be above the mould but not higher than 10 mm	Final height of the specimen must be between 127 mm to 133 mm

Table 6.4 Standard test procedures for vibrating hammer compaction test

ASTM D7382-08 (2008)

Method A	Method B
• Mould diameter: 6 in. (152.4 mm) • Material: passing 19.0 mm sieve • Layers: three • Dry mass: 7 kg • Wet mass: 9 kg • Vibration duration: 60 ± 5 s	• Mould diameter: 11 in. (279.4 mm) • Material: passing 50 mm sieve • Layers: three • Dry mass: 45 kg • Wet mass: 57 kg • Vibration duration: 52 ± 5 s

BS 1377-4 (1990)

• Mould: CBR mould
• Layers: three
• Vibration duration: 60 ± 2 s

5. Determine and record the water content of the test fraction by using small samples of the test fraction (BS 1377-1 2016; ASTM D698-12 2012; ASTM D1557-12 2012). Take at least two samples for representative value (BS 1377-1 2016).

6. Based on the water content determined in step 5, determine the dry mass of the test fraction.

7. Based on the dry mass of the oversized and test fractions, determine the percent of oversized fraction P_C and percent of test fraction P_F.

8. From the test fraction, select and prepare five sub-samples with the total mass of all sub-samples to be in accord with Table 6.2.

Table 6.5 Required standing time for the moist specimens (BS 1377-4 1990; ASTM D698-12 2012; ASTM D1557-12 2012)

ASTM D698-12 (2012) and ASTM D1557-12 (2012)	
Classification	Minimum standing time, h
GW, GP, SW, SP	No requirement
GM, SM	3
All other soils	16

BS 1377-4 (1990)	
Cohesive soil	24

9. Determine the water content required for the sub-samples. It is desired for the sub-samples water content to be around the optimum water content (preferably, two sub-samples are at the wet of optimum while two sub-samples are at the dry of optimum). The optimum water content for cohesive soil is typically less than the plastic limit (ASTM D698-12 2012). Thus for cohesive soils, water content of 8%–10% of plastic limit can be used as the starting water content while for sandy and gravelly soils, water content of 4%–6% is suitable as the starting water content (BS 1377-4 1990). In general, increment of 1%–2% are suitable for sandy and gravelly soils and 2%–4% are suitable for cohesive soils (BS 1377-4 1990).

10. Add/remove water to the sample to achieve the desired water content. Water can be added by using a spray bottle or removed by air or oven drying. Temperature for oven drying should not exceed 60°C. Mix the soil thoroughly and let the soil stand/cure in a closed container with duration in accord with Table 6.5.

6.4.2 Dry sample preparation

Dry sample preparation is not preferred. However, when the soil is too damp to be friable, dry sample preparation can be used. The procedure for dry sample preparation is as follows:

1. Dry the sample in air or by oven drying. Temperature for oven drying should not exceed 60°C.
2. Break up the sample thoroughly without breaking the individual soil particles (using a rubber mallet).
3. Sieve the sample in accord with either Method A, B or C (Table 6.2).
4. Determine the mass of both retained (oversized) and passing (test fraction) portions of the sample.

5. From the test fraction, select and prepare five sub-samples by using either mechanical splitting or quartering process to obtain the sub-samples with the mass for each sub-sample to be in accord with Table 6.2.

6. Determine the water content required for the five sub-samples. It is desirable for the water content of the five sub-samples to be around the optimum water content (preferably, two sub-samples are at the wet of optimum while two sub-samples are at the dry of optimum). The optimum water content for cohesive soil is typically less than the plastic limit (ASTM D698-12 2012). Thus, for cohesive soils, water content of 8%–10% of plastic limit can be used as starting water content while for sandy and gravelly soils, water content of 4%–6% is suitable as a starting water content (BS 1377-4 1990). In general, increments of 1%–2% are suitable for sandy and gravelly soils and 2%–4% are suitable for cohesive soils (BS 1377-4 1990).

7. Add/remove water to the sample to achieve the desired water content. Water can be added by using a spray bottle or removed by air or oven drying. Temperature for oven drying should not exceed 60°C. Thoroughly mix the soil and let the soil stand/cure in a closed container with a duration in accord with Table 6.5.

6.4.3 Compaction test procedures

6.4.3.1 Standard and modified compaction test

A typical compaction apparatus is shown in Figure 6.1, with test procedures for the standard and modified compaction tests as follows:

1. Choose either to conduct a standard or modified Proctor test (Table 6.1) in accord with the compaction requirement of the site.
2. Sieve the sample and select to do either Method A, B or C.
3. Prepare the sample under either moist or dry condition.
4. Record the mass of the mould and base plate M_m and the volume of the compaction mould V_m.
5. Assemble the compaction apparatus. The mould must rest on the base plate, and the extension collar must rest on the mould.
6. Place loose soil into the mould and spread it into a layer of uniform thickness. Lightly tamp the soil such that it is not in a loose state using the manual rammer.
7. Compact the soil layer in accord with Table 6.1 and the number of blows in accord with Table 6.2. In operating the manual rammer, take care to avoid lifting the guide sleeve during the rammer stroke, and the guide sleeve must be within 5° of vertical. The blow rate is 25 blow/min with the rammer pattern shown in Figure 6.2 to provide uniform

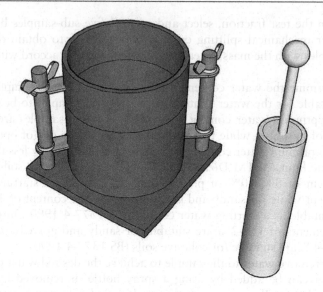

Figure 6.1 Compaction mould and hammer.

coverage of the soil surface. When the rammer pattern causes the surface to be uneven, judgement is required to adjust the rammer pattern. Every layer must have approximately the same thickness with the last layer extending into the collar. On each layer except the last layer, any soil that has not been compacted (sticking to the wall or extending above the compacted surface) shall be trimmed and discarded. The total amount of soil shall be such that the last compacted layer slightly extends into the collar but does not extend more than approximately 6 mm above the top of the mould. If the last layer extends beyond this limit or is below the top of the compaction mould, the compaction point shall be discarded.

8. Remove the collar and base plate from the mould by trimming and loosening the soil in the collar. Rotate the mould before removing the collar and base plate to reduce soil sticking to the collar or base plate.

9. Trim the compacted specimen by using a straightedge to level the specimen at the top and bottom of the mould and remove gravel-sized particles that are encountered. Patch any holes on the top surface with trimmed soil, pressing in with fingers.

10. Determine and record the mass of the specimen and mould $M_{s,m}$.

11. Determine bulk density ρ_b of the specimen using Equation 6.3:

$$\rho_b = \frac{M_{s,m} - M_m}{V_m} \tag{6.3}$$

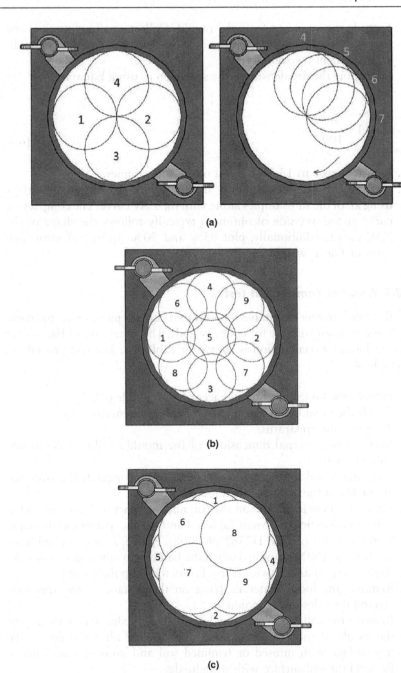

Figure 6.2 Rammer pattern (ASTM D698-12 2012; ASTM D1557-12 2012; ASTM D7382-08 2008). (a) Rammer pattern for 4in (101.6 mm) Mould. (b) Rammer pattern for 6in (152.4 mm) Mould. (c) Rammer pattern (Vibrating hammer).

12. Extrude the specimen from the mould. Collect small amounts of soil from the top, middle and bottom of the soil specimen to determine the water content by oven drying.
13. Determine the dry density of the specimen ρ_d using Equation 6.4.

$$\rho_d = \frac{\rho_b}{1 + \dfrac{w}{100}}$$

(6.4)

14. Repeat steps 5 to 13 for the rest of the sub-samples.
15. Plot the compaction curve (dry density versus the water content) by using all of the sub-samples along with the ZAV curve. The compaction curve at the wet side of optimum typically follows the shape of the ZAV curve. Additionally, plot 95% and 90% degrees of saturation contour line as well (BS 1377-4 1990).

6.4.3.2 Vibrating compaction test

The vibrating hammer is suitable for soils that contain particles up to coarse gravel, and it is not suitable for cohesive soils. The suitability of the soil for vibrating hammer compaction is shown in Table 6.3. The test procedures are as follows:

1. Prepare the sample and mould according to Table 6.4.
2. Weigh the mould with base plate and collar extension (M_d).
3. Assemble the apparatus.
4. Measure the internal dimensions of the mould to obtain the mould volume ($A_d h$).
5. Place moist soil in the mould such that when compacted, it occupies about 1/3 of the height of the mould body.
6. Place the circular tamper on the soil and compact it according to the vibration duration shown in Table 6.4 and rammer pattern as shown in Figure 6.1c for (ASTM D7382-08 2008). For BS, a total applied force of 300N to 400N is applied when the hammer is vibrated (Table 6.3).
7. Repeat steps 4 and 5 twice (three layers to fill up the mould).
8. Remove any loose materials lying on the surface of the specimen around the sides of the mould.
9. Remove the collar extension and use a straightedge across the top of the mould. Remove any oversized particles and fill any holes on the top surface with unused or trimmed soil and press in with fingers. Re-level the soil surface with straightedge.
10. Take four readings over the surface of the specimen to obtain the average height (h) of the specimen. The final height of the specimen must be in accord with Table 6.3.

11. Weigh the soil, mould and base plate (M_{sd}).
12. Extrude the specimen from the mould and determine the sample's water content (w).
13. Calculate the bulk density ρ_b of each compacted specimen using Equation 6.4.

$$\rho_b = \frac{M_{sd} - M_d}{A_d \cdot h}$$

(6.5)

14. Calculate the dry density ρ_d of each compacted specimen using Equation 6.6.

$$\rho_d = \frac{\rho_b}{1 + \dfrac{w}{100}}$$

(6.6)

15. Repeat steps 1 to 12 for the rest of the sub-samples
16. Plot the compaction curve (dry density versus the water content) by using all of the sub-samples together with the ZAV curve. The compaction curve at the wet side of optimum typically follows the shape of the ZAV curve. Additionally, plot the 95% and 90% degrees of saturation contour lines as well (BS 1377-4 1990).

6.4.3.3 Static compaction test

In static compaction, soil is compressed one-dimensionally in a mould using a compression machine until a specific dry density. To ensure uniformity of the soil specimen, the soil specimen is compacted in lifts. For dynamic compaction, usually, a constant compaction effort is used for each layer, while for static compaction, each soil layer is compressed to a specific dry density. For a tall specimen ($l/d = 2$), a static compaction apparatus similar to Goh (2012) can be used. The apparatus consists of multiple rings (base, split ring and cap ring), which can be assembled as each lift of the specimen is statically compacted, as shown in Figure 6.3. The height of each ring corresponds to the compacted lift height. The purpose of the multiple rings is to allow uniform compression of each lift by flipping the apparatus after each lift is statically compacted as indicated by the order of the compacted layers in Figure 6.3. When a lift is to be added, a split ring is added and soil is added and then pressed with a compression machine. The compacted soil specimen at a certain water content and dry density has a certain matric suction and prestress. Care must be exercised in interpreting the test data of a compacted soil specimen.

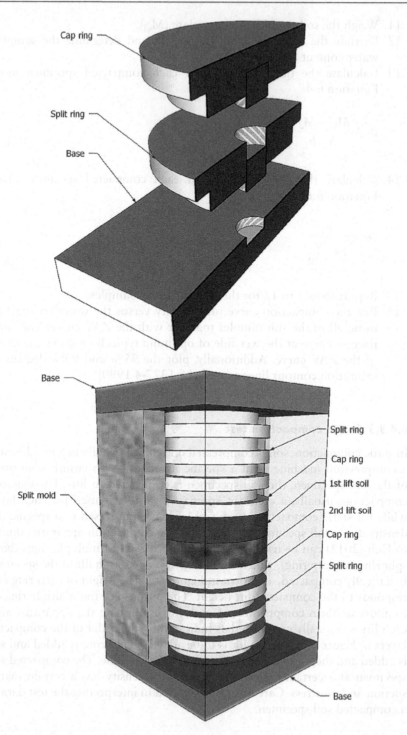

Figure 6.3 Static compaction apparatus for a tall specimen.

REFERENCES

ASTM D1557-12 (2012). *Standard test methods for laboratory compaction characteristics of soil using modified effort (56,000 ft-lbf/ft^3 (2,700 kN-m/m^3)).* West Conshohocken, PA: ASTM International.

ASTM D698-12 (2012). *Standard test methods for laboratory compaction characteristics of soil using standard effort (12,400 ft-lbf/ft^3 (600 kN-m/m^3)).* West Conshohocken, PA: ASTM International.

ASTM D7382-08 (2008). *Standard test methods for determination of maximum dry unit weight and water content range for effective compaction of granular soils using a vibrating hammer.* West Conshohocken, PA: ASTM International.

BS 1377-1 (2016). *Methods of test for soils for civil engineering purposes – Part 1: General requirements and sample preparation.* UK: BSI.

BS 1377-4 (1990). *Methods of test for soil for civil engineering purposes – Part 4: Compaction-related tests.* UK: BSI.

Goh, S. G. (2012). Hysteresis effects on mechanical behaviour of unsaturated soils. In *School of Civil and Environmental Engineering*, Singapore Nanyang Technological University, vol. Ph. D.

Priono Rahardjo, H., Chatterjea, K., & Leong, E.-C. (2017). Laboratory investigation on hydraulic anisotropy behavior of unsaturated soil. *Canadian Geotechnical Journal*, 54(7): 1034–1046.

FURTHER READINGS

Kodikara, J. (2012). New framework for volumetric constitutive behaviour of compacted unsaturated soil. *Canadian Geotechnical Journal*, 49: 1227–1243.

Chapter 7

Suction measurement

7.1 BACKGROUND

The energy state of soil water at a particular elevation and atmospheric pressure is referred to as total potential (Soil Science Society of America Glossary of Terms). Hence, the total potential of soil water is the sum of several energy components: gravitational potential, matric potential (or hydrostatic potential below the groundwater table), osmotic potential and velocity potential. When considering a soil element, gravitational potential is irrelevant and velocity potential is negligible as flow velocity in soil is small (and neglected) hence total potential is the sum of matric potential and osmotic potential. In engineering, unit of stress is preferred and hence suction is more commonly used in place of potential. Collectively, matric suction and osmotic suction are referred to as soil suction or total suction.

Soil suction or total suction, π, is an important stress variable for unsaturated soils. The term "soil suction" was first introduced in soil physics in relation to the soil-water-plant system (Buckingham 1907; Gardner and Widtsoe 1921; Richards 1928; Schofield 1935; Edlefsen and Anderson 1943; Childs and Collis-George 1948; Bolt and Miller 1958; Corey and Kemper 1961; Corey et al. 1967). Although the effect of soil suction on the mechanical properties of an unsaturated soil is dependent on the degree of saturation of the soil, it is measurable (Passioura 1980) unlike suction stress (Karube et al. 1996; Lu and Likos 2006).

Generally, soil suction can be divided into two major components: matric suction (u_a-u_w) and osmotic suction π (Krahn and Fredlund 1972). Matric suction is a stress-state variable but not osmotic suction. Matric suction is attributed to capillary and adsorptive forces (Stephens 1996; Or et al. 2003; Lu and Zhang 2019), whereas osmotic suction is attributed to the presence of salt in the pore water. More formal definitions of total, matric and osmotic suctions are given by Aitchison (1965):

Matric suction is the "equivalent suction derived from the measurement of the partial pressure of water vapour in equilibrium with the soil water, relative to the partial pressure of the water vapour in equilibrium with a solution identical in composition with the soil water".

DOI: 10.1201/b22304-7

Osmotic suction is the "equivalent suction derived from the measurement of partial pressure of the water vapour in equilibrium with a solution identical in composition with the soil water, relative to the partial pressure of water vapour in equilibrium with free pure water".

Total suction is the "equivalent suction derived from the measurement of partial pressure of the water vapour in equilibrium with the soil water, relative to the partial pressure of water vapour in equilibrium with free pure water".

It is generally accepted that total suction is the sum of matric and osmotic suctions (Krahn and Fredlund 1972; Edil and Motan 1984; Thyagaraj and Salini 2015).

7.2 MATRIC SUCTION

Although matric suction is attributed to capillary and adsorptive forces, it has long been conceptually associated with capillary forces alone. The capillary rise in a tube is often used as a simple model to associate matric suction with pore size (Fredlund et al. 2012). The capillary rise height, h_c, is given by Equation 7.1.

$$h_c = \frac{2T_s}{\rho_w g R_s} \tag{7.1}$$

where
T_s = surface tension of water (0.0728 N/m at 20°C)
ρ_w = density of water
g is gravitational constant (9.81 m/s^2)
R_s = radius of curvature of the meniscus of the air-water interface in a capillary tube of radius r (given by r/cos α)
α = contact angle between meniscus and capillary tube

There have been many attempts to associate the parameters of the grain-size distribution with R_s in Equation 7.1. Using Equation 7.1, the pore-water pressure u_w is given by Equation 7.2.

$$u_w = -\rho_w g h_c \tag{7.2}$$

In a soil system where both air and water phases exist in the voids, the air potential given by u_a (which is generally atmospheric, i.e., 0 kPa) should be included on the right-hand side of Equation 7.2, and ignoring adsorptive forces, Equation 7.3 is obtained.

$$u_a - u_w = \rho_w g h_c \tag{7.3}$$

To avoid confusion in this book, (u_a-u_w) is defined as matric suction as commonly used in the literature, and the stress due to adsorptive forces is called adsorption, i.e., matric potential is the summation of matric suction and adsorption. Adsorption is contributed by several components: cation hydration, inner-layer surface hydration, and particle surface hydration (Khorshidi et al. 2016). Cation hydration is attributed to the Coulomb electric force at the atomic scale and diminishes rapidly as hydration water increases beyond a few layers of water molecules. Cation hydration strongly bonds the water on the exchangeable cations and corresponds to suction on the order of a few hundred of MPa. Inner-layer surface hydration (expansive soil) and particle surface hydration are attributed to van der Waals force. The van der Waals force is much weaker than the Coulomb force by orders of magnitude (Mitchell and Soga 2005). The water held by the van der Waals force forms a moveable film on the particle surface and the suction associated with the moveable water film is below a few hundred MPa.

At present, no known measurement of adsorption is possible, and it has only been inferred (Lu 2016; Lu and Zhang 2019). In contrast, matric suction can be directly measured using tensiometers (Stannard 1992; Ridley and Burland 1993; Marinho and Chandler 1995; Guan and Fredlund 1997; He et al. 2006) and can be applied to test specimens in the laboratory using the axis-translation technique (Hilf 1956) by independently controlling pore-air and pore-water pressures to give the same magnitude of (u_a-u_w). Laboratory tests using the axis-translation principle have advanced the current understanding of unsaturated soil mechanics by ensuring the effect of a measurable stress-state variable on the engineering behaviour of soils.

7.3 OSMOTIC SUCTION

The role of osmotic suction in engineering properties of unsaturated soils is poorly understood. Osmotic suction or osmotic pressure of a salt solution is the pressure measured across a semi-permeable membrane where distilled water is present on the other side of the semi-permeable membrane. It is defined as the pressure that must be applied to the salt solution to prevent solvent (water) movement across the semi-permeable membrane due to the osmotic pressure gradient. It is unclear how the difference in salt concentration within a soil affects its engineering behaviour. The osmotic suction of a soil is usually measured indirectly by measuring the electrical conductivity of the pore water. To determine the osmotic pressure, the chart shown in Figure 7.1, relating osmotic pressure and electrical conductivity of salt solutions, is often used (USDA 1960). However, such a chart is meant to be used to estimate the salt concentration of the pore water that roots are exposed to in the soil. Water moves into the roots from the soil by osmosis and relies on the osmotic gradient. High salt concentration in the pore water will hinder or even reverse the direction of water movement across the root

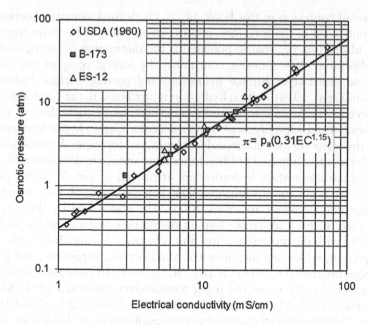

Figure 7.1 Relationship of electrical conductivity with osmotic suction for various salt concentrations(from Leong et al. 2007).

cellular semi-permeable membrane. Figure 7.1 can be more conveniently represented by Equation 7.4 (Leong et al. 2007).

$$\pi = P_a\left(0.31E_c^{1.15}\right) \tag{7.4}$$

where
p_a = atmospheric pressure (~101.325 kPa)
E_C = electrical conductivity in mS/cm

As osmotic suction has units of stress, it is often thought to have the same effect as other stresses in soils. The misunderstanding has led to some research to investigate the effect of osmotic suction on the mechanical response of unsaturated soils. For example, Ho and Pufahl (1987) and Peterson (1990) showed that soil strength increases as osmotic suction increases, whereas Yong and Warkentin (1975), Tang et al. (1997) and Mokni et al. (2014) found the reverse. The salt concentration of the pore water affects the soil structure through its influence on the adsorptive forces. The adsorptive forces have a greater effect on the soil structure of clays with high plasticity through their effect on the double diffuse layer of clay particles (Chattopadhyay 1972; Di Maio 1996; Tang et al. 1997; Arifin and Schanz 2009; Thyagaraj and Salini

2015). Soils with pore water of high salt concentration were observed to exhibit an aggregated structure and affect the matric suction (Thyagaraj and Salini 2015; Mokni et al. 2014). The effects were more evident in saturated soil tests (Peterson 1990; Rao et al. 2006; Wahid et al. 2011a, b; Witteveen et al. 2013; Mokni et al. 2014). Tests on soils with little or no double diffuse layer showed no changes in soil strength (Blight 1983; Katte and Blight 2012; Leong and Abuel-Naga 2018).

7.4 TOTAL SUCTION

Total suction or soil suction is the sum of matric and osmotic suction. In terms of measurement, total, matric and osmotic suctions in a soil can be illustrated conceptually using Figure 7.2. From Figure 7.2, total suction is given by the pressure difference across a semi-permeable membrane separating pure water from a soil whose pore water contains salts. In most geotechnical engineering problems, unsaturated soils are a common result of environmental changes that affect primarily the matric suction. Changes in osmotic suction are less significant (Fredlund and Rahardjo 1993) and normally these are inherently accounted for during the laboratory tests. Hence, for most geotechnical engineering problems that do not involve chemical contamination, a matric suction change is equivalent to a change in total suction, as osmotic suction is relatively constant.

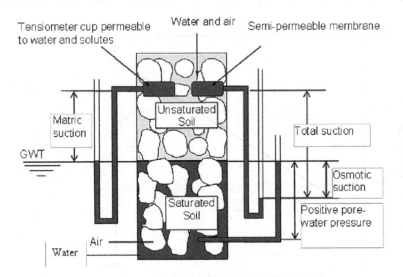

Figure 7.2 Illustration of suctions in soils (modified from He 1999).

Total suction is related to relative humidity through the thermodynamic relationship given in Equation 7.5.

$$\psi = \frac{RT}{v_{w0}\omega_v}\ln(R_h)$$ (7.5)

where
ψ = total suction or soil suction (kPa)
R = universal (molar) gas constant [i.e., 8.31432 J/(mol K)]
T = absolute temperature [i.e., T = (273.16 + t°C) (K)]
t = temperature in °C
v_{w0} = specific volume of water or the inverse of the density of water [i.e., $(1/\rho_w)$ (m³/kg)]
ρ_w = density of water (i.e., 998 kg/m³ at t = 20°C)
ω_v = molecular mass of water vapour (i.e., 18.016 kg/kmol)
R_h = relative humidity

Figure 7.3 shows a plot of total suction versus R_h for t = 20°C computed using Equation 7.2. There are many instruments that can be used to measure relative humidity (Leong et al. 2003). These instruments can be classified into primary standard, transfer standard and secondary instruments. Primary standard instruments which rely on fundamental principles and base units of measurement are of the highest accuracy and are used in national calibration laboratories. One such instrument is the gravimetric hygrometer (Weiderhold 1997) which weighs a certain amount of dry gas and compares it with the weight of the test gas at the same volume to

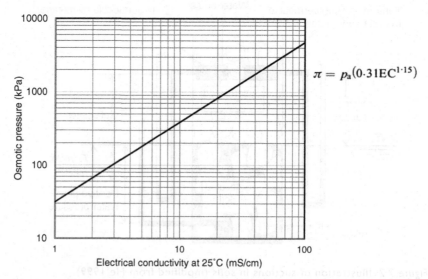

$$\pi = p_a(0.31EC^{1.15})$$

Figure 7.3 Relative humidity versus total suction relationship at t = 20°C.

calculate the amount of water present in the gas. Hence, the water vapour pressure can be calculated. Other base units of measurement besides mass are temperature and pressure. However, primary standard instruments are difficult to operate and not practical for normal usage. At the other end of the accuracy spectrum are the secondary instruments for relative humidity measurement. Secondary instruments are non-fundamental and must be calibrated against a transfer standard. To obtain accurate readings, secondary instruments need frequent recalibrations. Examples of secondary instruments are thin-film capacitive and resistive sensors. In between the primary and secondary standards are the transfer standard instruments which operate on fundamental principles and can provide good, stable and repeatable readings. However, precaution is needed when using these instruments, as improper usage can give incorrect results. Examples of transfer standard instruments are the chilled-mirror hygrometer, electrolytic hygrometer and psychrometer. Thermocouple psychrometers have been used for the measurement of total suction (Brown 1970; Krahn and Fredlund 1972; Richards 1974; van der Raadt et al. 1987; Hamilton et al. 1979; Lee and Wray 1992; Ridley 1993). The accuracy of psychrometer is known to be affected by many factors. Non-constant air temperature can cause fluctuations in relative humidity readings (e.g., Krahn and Fredlund 1972; Ridley 1993; He 1999). Krahn and Fredlund (1972) found that fluctuation in the temperature of the measurement environment of ±0.001°C can cause and an error of ±10 kPa in total suction measurement. The accuracy of the psychrometer is highly dependent on the operator's skill (Weiderhold 1997). Currently, the chilled-mirror hygrometer has seen an increase in usage for measuring the total suction of soils in the laboratory. All transfer and secondary standards instruments have a reduced accuracy above relative humidity above 95%. The chilled-mirror hygrometer is no exception. As seen in Figure 7.3, 95% relative humidity corresponds to suctions below about 7,000 kPa. It is very difficult to get accurate suction readings for relative humidity greater than 99% which corresponds to the suction of 1,000 kPa and below.

REFERENCES

Arifin, A. F. & Schanz, T. (2009). Osmotic suction of highly plastic clays. *Acta Geotechnica*, 4(3): 177–191.

Blight, G. E. (1983). Aspects of the capillary model for unsaturated soils. In Proceedings of the Seventh Asian Regional Conference. on Soil Mechanics and Foundation Engineering, Haifa, Israel, Vol. 1, pp. 3–7.

Bolt, G. H. & Miller, R. D. (1958). Calculation of total and component potentials of water in Soil. *Transactions, American Geophysical Union*, 39: 917–928. https://doi.org/10.1029/TR039i005p00917

Brown, R. W. (1970). Measurement of water potential with thermocouple psychrometers. Construction and Applications. U.S.D.A. Forest Service Res. Paper INT-80, 27pp.

Buckingham, E. (1907). *Studies on the movement of soil moisture.* Bulletin No. 38. Washington, DC: US Department of Agriculture, Bureau of Soils.

Chattopadhyay, P. K. (1972). Residual shear strength of some pure clay minerals [PhD thesis]. Canada: University of Alberta, Edmonton.

Childs, E. C. & George, N. C. (1948). Soil geometry and soil-water equilibria. *Faraday Society, 3:* 78–85.

Corey, A. T. & Kemper, W. D. (1961). Concept of total potential in water and its limitations. *Soil Science, 91:* 299–302.

Corey, A. T., Slatyer, R. O., & Kemper, W. D. (1967). Comparative terminologies for water in the soil-plant-atmosphere system. In R. M. Hagan, H. R. Haise, & T. W. Edminster (Eds.), *Irrigation of agricultural lands.* https://doi.org/10.2134/agronmonogr11.c23

Di Maio. (1996). Exposure of bentonite to salt solution: Osmotic and mechanical effects. *Géotechnique 1996, 46(4):* 695–707.

Edil, T. B. & Motan, S. E. (1984). Laboratory evaluation of soil suction components, *Geotechnical Testing Journal, 7(4):* 173–181.

Edlefsen, N & Anderson, A. (1943). Thermodynamics of soil moisture. *Hilgardia, 15(2):* 31–298. https://doi.org/10.3733/hilg.v15n02p031

Gardner, W. & Widtsoe, J. A. (1921). The movement of soil moisture. *Soil Science, 11:* 215–232.

Guan, Y. & Fredlund, D. G. (1997). Use of the tensile strength of water for the direct measurement of high soil suction. *Canadian Geotechnical Journal, 34:* 604–614.

Hamilton, J. M., Daniel, D. E., & Olson, R. E. (1979). Measurement of hydraulic conductivity of partially saturated soils. In T. F. Zimmie & C. O. Riggs (Eds.), *Permeability and groundwater contaminant transport, ASTM STP 746.* Philadelphia: ASTM, pp. 182–196.

He, L. (1999). Evaluation of instruments for measurement of suction in unsaturated soils. MEng Thesis, Nanyang Technological University, Singapore.

He, L., Leong, E. C., & Algamal, A. (2006). A miniature tensiometer for measurement of high matric suction. In G. A. Miller, C. E. Zapata, S. L. Houston, & D. G. Fredlund (Ed.), *Proceedings of 4th International Conference on Unsaturated Soils,* Carefree, Arizona, April 2–6, 2006, pp. 1897–1907.

Ho, Y. A. & Pufahl, D. E. (1987). The effects of brine contamination on the properties of fine grained soils. In *Geotech. Practice for Waste Disposal. ASCE Special Conference,* pp. 547–561.

Karube, D., Kato, S., Hamada, K., & Honda, M. (1996). The relationship between the mechanical behavior and the state of pore-water in unsaturated soil. *Geotechnical Engineering Journal, JSCE, 535*(III-34): 83–92.

Katte, V. & Blight, G. (2012). The roles of solute suction and surface tension in the strength of unsaturated soil. In C. Mancuso, C. Jommi, & F. D'Onza (Eds.), *Unsaturated soils: Research and applications.* Berlin, Heidelberg: Springer, pp 431–437.

Krahn, J. & Fredlund, D. G. (1972). On total, matric and osmotic suction. *Journal of Soil Science, 114(5):* 339–348.

Lee, H. C. & Wray, W. K. (1992). Evaluation of soil suction instruments. In *Proceedings of the 7th International Conference on Expansive Soils,* Dallas, Texas, August 3–5, Vol. 1, 231–238.

Leong, E.-C. & Abuel-Naga, H. M. (2018). Contribution of osmotic suction to shear strength of unsaturated high plasticity silty soil. *Geomechanics for Energy and the Environment, 15(9):* 65–73. https://doi.org/10.1016/j.gete.2017.11.002

Leong, E. C., Widiastuti, S., Lee, C. C., & Rahardjo, H. (2007). Accuracy of suction measurement. *Geotechnique*, *57*(6): 547–556.

Leong, E. C., Tripathy, S., & Rahardjo, H. (2003). Total suction measurement of unsaturated soils with a device using the chilled-mirror dewpoint technique, *Géotechnique*, *53*(2): 173–182.

Lu, N. (2016). Generalized soil water retention equation for adsorption and capillarity. *Journal of Geotechnical and Geoenvironmental Engineering*, *142*(10): 04016051. https://doi.org/10.1061/(ASCE)GT.1943-5606.0001524

Lu, N. & Zhang, C. (2019). Soil sorptive potential: Concept, theory, and verification. *Journal of Geotechnical and Geoenvironmental Engineering*, *145*(4): 04019006. https://doi.org/10.1061/(ASCE)GT.1943-5606.0002025

Lu, N. & Likos, W. J. (2006). Suction stress characteristic curve for unsaturated soil. *Journal of Geotechnical and Geoenvironmental Engineering*, *132*(2): 131–142. https://doi.org/10.1061/(ASCE)1090-0241(2006)132:2(131)

Marinho, F. A. M. & Chandler, R. J. (1995). Cavitation and the direct measurement of soil suction. In *Unsaturated Soils: Proceedings of the 1st International Conference On Unsaturated Soils*, Paris, France, pp. 623–629.

McGeorge, W. T. (1954). Diagnosis and improvement of Saline and Alkaline soils. *Soil Science Society of America Journal*, *18*(3): 348–348.

Mokni, N., Romero, E., & Olivella, S. (2014). Chemo-hydro-mechanical behaviour of compacted Boom Clay: Joint effects of osmotic and matric suctions. *Géotechnique*, *64*(9): 681–693.

Passioura, J. B. (1980). The meaning of matric potential. *Journal of Experimental Botany*, *31*(123): 1161–1169.

Peterson, R. W. (1990). The influence of soil suction on the shear strength of unsaturated soil. Misc. Paper No GL-90-17. Vicksburg, MS: U.S. Corps of Army Engineers, Waterways Experimental Station.

Rao, S. M., Thyagaraj, T., & Thomas, H. R. (2006). Swelling of compacted clay under osmotic gradients. *Géotechnique*, *56*(10): 707–713.

Richards, L. A. (1928). The usefulness of capillary potential to soil-moisture and plant investigations. *Journal of Agricultural Research*, *37*: 719–42. https://doi.org/10.1097/00010694-192103000-00003

Ridley, A. M. & Burland, J. B. (1993). A new instrument for the measurement of soil moisture suction, *Geotechnique*, *43*(2): 321–324.

Ridley, A. M. (1993). The Measurement of Soil Moisture Suction. PhD thesis, Imperial College of Science, Technology and Medicine, University of London.

Schofield, R. K. (1935). The pF of water in soil. *Transactions Third International Congress of Soil Science*, *2*: 37–48.

Tang, G. X., Graham, J., & Fredlund, D. G. (1997). Effects of osmotic suction on strength of unsaturated highly plastic clays. In *Proceedings of the 50th Canadian Geotechnical Conference*, Golden Jubilee, Ottawa, pp. 641–648.

Thyagaraj, T. & Salini, U. (2015). Effect of pore fluid osmotic suction on matric and total suctions of compacted clay. *Géotechnique*, *65*(11): 952–960.

van der Raadt, P., Fredlund, D. G., Clifton, A. W., Klassen, M. J., & Jubien, W. E. (1987). Soil suction measurement at several sites in Western Canada. Transportation Res. Rec. 1137. In Soil Mech. Considerations in Arid and Semi-Arid Areas. Washington, DC: Transportation Research Board, pp. 24–35.

Wahid, A. S., Gajo, A., & Di Maggio, R. (2011a). Chemo-mechanical effects in kaolinite. Part 1: prepared samples. *Géotechnique*, *61*(6): 439–447.

Wahid, A. S., Gajo, A., & Di Maggio, R. (2011b). Chemo-mechanical effects in kaolinite. Part 2: exposed samples and chemical and phase analyses. *Géotechnique*, 61(6): 449–457.

Weiderhold, P. R. (1997). *Water vapour measurement, methods and instrumentation*. New York, NY: Marcel Dekker.

Witteveen, P., Ferrari, A., & Laloui, L. (2013). An experimental and constitutive investigation on the chemo-mechanical behaviour of a clay. *Géotechnique*, 63(3): 244–255.

Yong, R. N. & Warkentin, B. P. (1975). *Soil properties and behaviour*. Amsterdam: Elservier Scientific Publishing Co.

FURTHER READING

ASTM E104-20a (2020). Standard practice for maintaining constant relative humidity by means of aqueous solutions. *1998 Annual Book of ASTM Standards*, 1103, Water and environmental technology, 781–783.

Camuffo, D. (2014). *Appendix 2 – summary of key equations to calculate humidity variables, microclimate for cultural heritage (second edition)*. Elsevier, pp. 507–509, https://doi.org/10.1016/B978-0-444-63296-8.15002-4.

Lang, A. R. G. (1967). Osmotic coefficients and water potentials of sodium chloride solutions from 0 to 40°C. *Australian Journal of Chemistry*, 20: 2017–2023.

Chapter 8

Matric suction measurement

Direct methods

8.1 INTRODUCTION

Matric suction can be determined by measuring the pore-water pressure and pore-air pressure (frequently pore-air pressure is at atmospheric condition, i.e., 0 kPa). Direct measurement is considered to be more reliable than indirect measurement, that involves the use of a proxy. However, direct methods of matric suction measurement have a limited range and a short stable period when measuring a high matric suction.

Direct measurement methods require a high-air entry (HAE) ceramic disk/cup to separate the air phase from the water phase. The HAE ceramic disk serves as an impermeable boundary for air but is permeable to water under air pressure below the air-entry value of the ceramic disk. For the ceramic disk to work properly, it must be saturated prior to use. The higher the air-entry value of the ceramic disk, the higher the water pressure and the longer the duration needed to properly saturate the disk. Thus, it is important to choose the appropriate ceramic disk according to the range of suction measurements that is of interest. The air-entry value (AEV) of the ceramic disk can be determined based on the maximum pore size of the ceramic disk and is given by Equation 8.1.

$$AEV = \frac{2T_s}{R_{max}} \tag{8.1}$$

Where T_s is the surface tension of water which is 72.75 mN/m at 20 °C, and R_{max} is the radius of the maximum pore size of the ceramic disk. Some typical HAE disks and their AEV together with their coefficient of permeability with respect to water (k_d) have been compiled and are shown in Tables 8.1–8.3.

DOI: 10.1201/b22304-8

Table 8.1 List of HAE disks used at imperial college (Blight 1966)

Type of disks	Porosity n	Coefficient of permeability with respect to water k_d	Air-entry value AEV
	(%)	m/s	kPa
Doulton grade P6A	23	2.1×10^{-9}	152
Aerox "celloton" grade VI	46	2.9×10^{-8}	214
Kaolin-consolidated from a slurry and fired	45	6.2×10^{-10}	317
Kaolin-dust pressed and fired	39	4.5×10^{-10}	524

Table 8.2 Properties of the air-entry disks manufactured by soil moisture equipment corporation

Type of disks	Approximate pore diameter	Coefficient of permeability with respect to water k_d	Air-entry value AEV
	10^{-3} mm	m/s	kPa
0.5 bar high flow	6	3.11×10^{-7}	48–62
1 bar	2.1	3.46×10^{-9}	138–207
1 bar high flow	2.5	8.6×10^{-8}	131–193
2 bar	1.2	1.73×10^{-9}	241–310
3 bar	0.8	1.73×10^{-9}	317–483
5 bar	0.5	1.21×10^{-9}	>550
15 bar	0.16	2.59×10^{-11}	>1520

Table 8.3 Permeability and air-entry value measurements on HAE disk from soil moisture equipment corporation (Fredlund 1973; Rahardjo 1990)

Type of disks	Diameter of the disk	Thickness of the disk	Coefficient of permeability with respect to water k_d	Air-entry value AEV
	mm	mm	m/s	kPa
1 bar high flow	19	6.4	5.12×10^{-8}	115
	19	6.4	3.92×10^{-8}	130
	19	6.4	3.98×10^{-8}	110
	19	6.4	5.09×10^{-8}	130
	19	6.4	5.60×10^{-8}	150
	101.6	10	4.20×10^{-8}	>200
5 bar	56.8	6.2	1.30×10^{-9}	
15 bar	56.8	3.1	8.41×10^{-9}	
	57	3.1	6.82×10^{-10}	

8.2 JET-FILLED OR SMALL-TIP TENSIOMETER

A jet-filled or small-tip tensiometer can directly measure negative pore-water pressure in a soil specimen. The jet-filled tensiometer is more commonly used in the field, while the small-tip tensiometer is more commonly used in laboratory model tests. Both the jet-filled and small-tip tensiometers use a 1-bar HAE ceramic cup. Due to cavitation, it is only possible to measure negative pore-water pressures, as low as –90 kPa (Stannard 1990). According to Cassel and Klute (1986), an increase in the pressure gauge sensitivity may cause less water movement between soil and the tensiometer and provides higher accuracy. However, the low permeability of the ceramic cup may cause a longer response time.

Saturating the tensiometer before use is very important for reliable readings. The ceramic cup must first not be clogged. In order to check whether the ceramic cup is clogged, the ceramic cup is attached to the empty tensiometer and submerged under water overnight. If the ceramic cup works properly, water should be able to pass through the ceramic cup and fill up the tensiometer.

To reduce the high possibility of cavitation, the tensiometer must be filled up with deaerated water. After that, vacuum pressure of 80 kPa must be applied for 30–60 s from the top of the tensiometer to ensure that there is no air bubble inside the ceramic cup and the water.

The response time of the tensiometer can be checked by conducting an evaporation test where the ceramic cup is left to dry by evaporation. As water evaporates from the ceramic cup, the pore-water pressure becomes negative. When the pore-water pressure is around –80 kPa, it is then re-saturated by submerging the ceramic cup in water. The pore-water pressure should immediately revert to 0 kPa when the tensiometer is working properly. A longer response time may indicate clogging, entrapped air or calibration issue.

8.3 HIGH-CAPACITY TENSIOMETER

8.3.1 Introduction

High-capacity tensiometer or HCT (Ridley and Burland 1993; Guan and Fredlund 1997; Ridley and Burland 1999; Tarantino and Mongiovi 2002; He et al. 2006; Wijaya and Leong 2016) is perhaps the only device that can directly measure the pore-water pressure of the soil specimen less than –100 kPa (depending on the AEV of the ceramic disk). It is typically used to measure pore-water pressure as low as –1,500 kPa. Besides measuring pore-water pressure lower than –100 kPa, the HCT has two additional advantages:

1. Its size is only slightly bigger than the normal pressure transducer due to the requirement to attach the HAE disk, which makes it convenient and flexible to be attached to any apparatus.
2. It has a fast response time.

The schematic of a typical HCT is given in Figure 8.1. The HCT is able to measure matric suction as high as the AEV of the ceramic disk due to the following:

1. Small water reservoir about 2 mm³ (He et al. 2006) between the HAE ceramic disk and the diaphragm of the pressure transducer which reduces the possibility of cavitation occurring within the water reservoir. It has also been shown that cavitation is also possible in the ceramic disk (Mendes and Buzzi 2013).

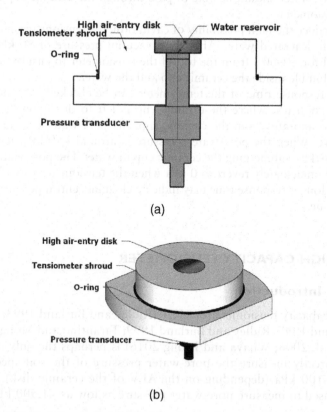

(a)

(b)

Figure 8.1 Schematic drawing of high-capacity tensiometer (from Wijaya and Leong 2016). (a) Section view of high-capacity tensiometer (Wijaya and Leong 2016). (b) Perspective view of high suction tensiometer.

2. High-water pressure (higher than the AEV of the HAE disk) is needed to dissolve any air inside the water reservoir and to saturate the ceramic disk.

8.3.2 Calibration

Most pressure gauges are not made to measure negative pore-water pressure. Thus, calibration of the HCT is done by applying positive water pressure. Under positive water pressure, the sensing diaphragm will flex in the inward direction. By assuming that the diaphragm follows Hooke's law under compression and tension, the calibration line can be extrapolated for negative pressure (tension).

Figure 8.2a shows the schematic diagram of calibrating HCT using a digital pressure/volume controller, while Figure 8.2b shows the results of the calibration. By applying different pressures to the HCT, a regression line of the voltage and the pressure readings can be made, as shown in Figure 8.2b. The regression line is assumed to be applicable to the negative range as well.

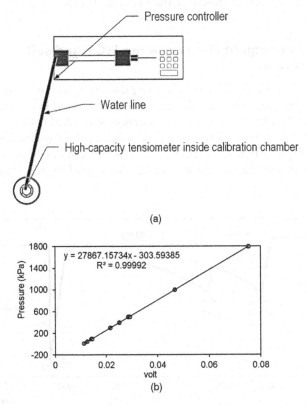

(a)

(b)

Figure 8.2 HCT calibration. (a) Schematic of calibrating HCT. (b) HCT calibration line.

8.3.3 Evaporation test

The capacity and the response time of the HCT can be tested by conducting an evaporation test. In the evaporation test, the HCT is exposed to the ambient environment (drying), and the water pressure reduces until it reaches the capacity of the HCT. The rate at which the water pressure reduces indicates the quality of the saturation of the HCT where a faster rate corresponds to a higher quality of saturation of the HCT. Figure 8.3 shows an example of HCT response during an evaporation test. At the beginning, the surface of the ceramic disk is slightly moist, and the reading remains constant at 0 kPa. Once the excess water on the surface of the ceramic disk is removed, the water pressure starts to decrease. If the ceramic disk is wetted before the negative water pressure reaches the capacity of the HCT, the water pressure will reduce to 0 kPa almost instantaneously. A sluggish reduction to 0 kPa indicates that the HCT was not saturated properly. Figure 8.3 shows the HCT subjected to several cycles of drying and wetting and in the final drying cycle, the water pressure was allowed to decrease until the HCT cavitates. Once the cavitation occurs, the water pressure will directly go back to around –100 kPa before slowly drifting back to 0 kPa.

8.3.4 Types of equilibrium between HCT and soil specimens

In order to read the negative pore-water pressure of soil, it is important to ensure that the water phase of the specimen is connected hydraulically with the water in the HAE ceramic disk. Lourenço et al. (2009) show that there are at least three types of HCT response during the equilibrium and are referred as Equilibrium A, B and C. This observation has been revisited in Wijaya and Leong (2016) and Wijaya (2017). Equilibrium A (Figure 8.4a)

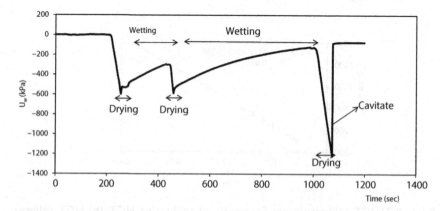

Figure 8.3 Evaporation test (from Wijaya and Leong 2016).

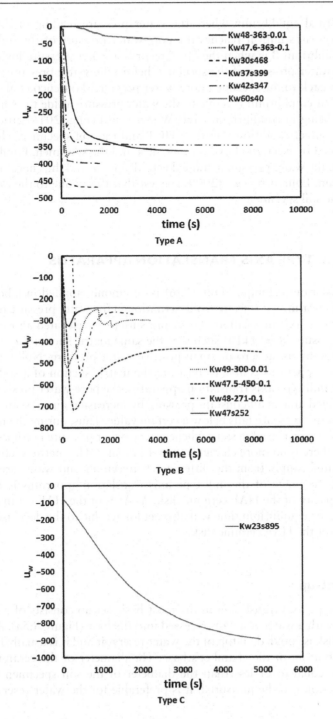

Figure 8.4 Typical equilibrium of kaolin specimen (from Wijaya and Leong 2016).

is the normal equilibrium where the water pressure reading of the HCT slowly decreases until it reaches the pore-water pressure of the soil specimen. Equilibrium B is when the HCT records a water pressure lower than the pore-water pressure of the specimen before the water pressure reading rebounds back to the specimen pore-water pressure. Lourenço et al. (2009) suggest that equilibrium B is due to the water pressure inside the soil specimen not being in equilibrium, while Wijaya and Leong (2016) attribute it to poor contact conditions between HCT and the soil specimen where air gets trapped between the HAE ceramic disk and the specimen. Equilibrium C is when the water pressure reading keeps dropping and thus never reaches equilibrium. Lourenço et al. (2009) suggest that this is due to the evaporation of the soil specimen.

8.4 NULL-TYPE AXIS TRANSLATION APPARATUS

Axis-translation technique (Hilf 1956) is commonly applied in laboratory testing of unsaturated soils to avoid cavitation in the water pressure measuring system. In axis translation, the air pressure (u_a) is elevated above atmospheric pressure $(u_a = 0 \text{ kPa})$ such that the same matric suction (u_a-u_w) can be achieved by using zero or small positive water pressure. Null-type axis translation apparatus is able to measure the matric suction of a soil specimen. In a null-type axis translation apparatus, the pore-water pressure (u_w) is maintained at a small constant pressure by increasing air pressure whenever the water pressure falls below a certain value. Thus, during the test, the air pressure is increased in steps whenever the water pressure becomes negative until there is no more change in water pressure. The matric suction can be calculated simply from the difference between air and water pressures. However, the range of the null-type axis translation apparatus is limited by the capacity of the HAE ceramic disk. As AEV of the HAE ceramic disk increases, the equilibrium time will also get longer due to the decreased permeability of the HAE ceramic disk.

8.4.1 Set-up

The null-type axis translation in the most basic set-up consists of a closed chamber with a water reservoir recessed into the base (Figure 8.5a). A HAE ceramic disk is epoxied on top of the water reservoir and it is usually flushed with the bottom of the chamber's base. The diameter of the ceramic disk should be equal to or less than the diameter of the soil specimen whose matric suction is to be measured. It is preferable for the water reservoir to consist of a spiral grooved channel so that water flow is unidirectional along the groove, making it easier to flush out any entrapped air. An air pressure supply inlet is provided either on the cover of the chamber or on the top side of the chamber. The air pressure supply into the chamber is controlled via a pressure valve that is connected to an air pressure transducer or gauge. The water reservoir at the base has an outlet at the bottom or the side to allow water to drain during saturation of the HAE ceramic disk and is connected to a water pressure transducer or gauge. Another water outlet at the base of the chamber can be provided to allow flushing water to drain from the chamber.

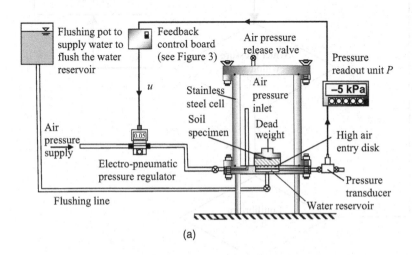

(a)

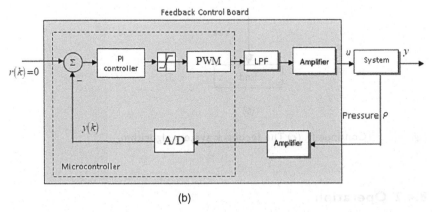

(b)

Figure 8.5 Active control null-type axis translation apparatus (from Leong et al. 2009). (a) Modified pressure plate apparatus with active control system. (b) Schematic drawing of the feedback control.

(Continued)

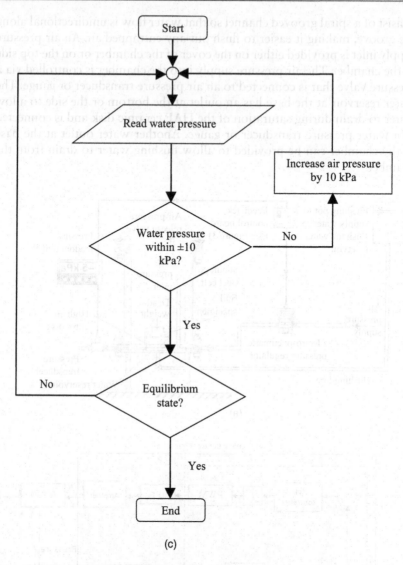

(c)

Figure 8.5 (Continued). (c) The feedback system algorithm.

8.4.2 Operation

The operation of the null-type axis translation consists of the following stages:

1. preparing the apparatus, and
2. measuring matric suction.

The preparation of the apparatus entails the following steps:

1. Remove the cover of the chamber and put in some de-aired water that is several cm above the ceramic disk. Replace the cover and secure it.
2. At the water reservoir outlet, connect a short tube with its free end submerged in a beaker of water and apply air pressure not exceeding the AEV of the ceramic disk. The water inside the chamber should flow through the HAE ceramic disk into the water reservoir, exit through the outlet tube into the beaker quickly, and eventually stop.
3. Monitor the outlet of the tube in the beaker for air bubbles for at least 10 minutes. It takes 10 minutes or more after the water has exited from the water reservoir for air bubbles to start appearing. The presence of minute air bubbles suggests that the HAE ceramic disk is not fully saturated. Repeat steps 1 to 3 to saturate the ceramic disk.
4. If steps 1 to 3 are not able to saturate the disk, connect the bottom inlet tube to an external de-aired water supply reservoir. Apply a low vacuum pressure of 10–30 kPa at the air pressure inlet valve and repeat steps 1–3. If step 4 also fails, it is highly likely that there is a fine crack on the ceramic disk or the epoxy seal around the perimeter of the ceramic disk has broken in parts. When this happens, the ceramic disk must be left to dry and then removed. The ceramic disk is either replaced or removed, dried and re-sealed into the water reservoir again. Steps 1–3 are again repeated.

When the ceramic disk is fully saturated, the null-type axis translation apparatus is considered prepared (ready) for matric suction measurement. For matric suction measurement, the following steps are followed:

1. A representative soil specimen of diameter larger than the diameter of the ceramic disk is placed onto the ceramic disk, and a 1 kg dead-weight is placed on the soil specimen to ensure good contact between soil specimen and ceramic disk.
2. Quickly replace the cover of the chamber and increase the air pressure to 10 kPa.
3. Note the reading of the water pressure transducer at the bottom of the water reservoir.
4. If the water pressure becomes increasing negative instead of positive, increase the air pressure by a further 10 kPa and observe.
5. Repeat Step 4 until the water pressure stabilises at zero or small positive air pressure.
6. When the water pressure reading stabilises near zero or small positive pressure, the matric suction of the soil specimen is given by the difference between air pressure and water pressure.

The operation of the null-type axis translation apparatus is labour intensive, as the air pressure needs to be adjusted frequently to avoid the water pressure going too highly negative (> –60 kPa) and causing cavitation in the water pressure measuring system. It is possible to automate the null-type axis translation apparatus to adjust the air pressure automatically when the water pressure falls below a certain value using a feedback control loop, as shown in Figure 8.5b. The algorithm for controlling the air pressure is shown in Figure 8.5c. More details on automating the null-type axis translation apparatus can be found in Leong et al. (2009). Typical responses of the air and water pressure readings of the null-type axis translation during matric suction measurement are shown in Figure 8.6.

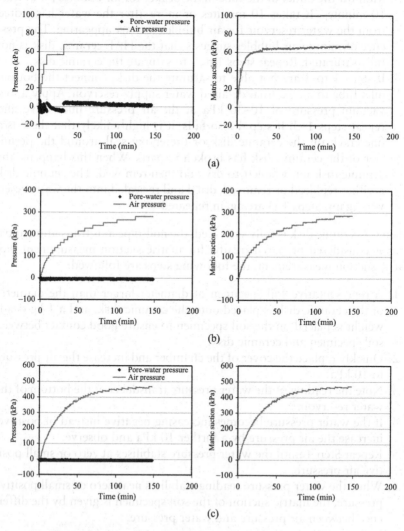

Figure 8.6 Active control null-type axis translation apparatus test results (from Leong et al. 2009). (a) K80S20 – w = 21.7%. (b) K80S20 – w = 14.5%. (c) K80S20 – w =11.1%.

8.5 HYGROMETER

The direct methods to measure matric suction are always limited in range due to the AEV of the ceramic disk which at present is limited to about 1,500 kPa. For matric suction beyond the residual suction of the soil, the water phase is no longer continuous, and moisture transfer is only possible through vapour. Provided there is negligible salt content in the pore water, matric suction beyond the residual suction range is taken to be the same as total suction. Experimental evidence shows that the water characteristic curves of soils and filter papers based on matric suction and total suction converge after suction of 1,000–3,000 kPa (Croney and Coleman 1948; Houston et al. 1994; Leong et al. 2002).

The hygrometer is an instrument that measures relative humidity (see Chapter 7). Instruments used in agronomy to measure relative humidity include the psychrometer, dewcell and dewprobe sensors and infrared gas analyser (Rosenberg et al. 1983). Previously, thermocouple psychrometers have often been used for the measurement of total suction in soils (Brown 1970; Krahn and Fredlund 1972; Richards 1974; van der Radt et al. 1987; Hamilton et al. 1979; Lee and Wray 1992; Ridley 1993). However, the psychrometer is known to show fluctuations in relative humidity readings due to non-constant air temperature. Currently, the chilled mirror hygrometer has become more common in the laboratory to measure total suction (Gee et al. 1992; Bulut et al. 2002; Leong et al. 2003; Schanz et al. 2004).

The chilled mirror hygrometer is included as Method D in ASTM D6836-16 for determining the soil-water characteristic curve from 1,000 kPa to 100,000 kPa. The chilled mirror hygrometer is also available commercially for laboratory use. Hence, the chilled mirror hygrometer is the only instrument considered here for the measurement of total suction.

8.5.1 Calibration

It is important to verify the calibration of the chilled mirror hygrometer. This is usually performed with standard solutions. Saturated solutions of magnesium chloride ($MgCl_2.6H_2O$), magnesium nitrate ($Mg(NO_3)^2.6H_2O$), sodium chloride (NaCl), potassium chloride (KCl), potassium nitrate (KNO_3) and potassium sulphate (K_2SO_4) are typically used as the standard solutions. The relative humidity of the saturated solutions as given in ASTM E104-20a and the relative humidity of the standard saturated solutions measured using a chilled mirror hygrometer, WP4, manufactured by Decagon Devices USA, which has since merged with Meter Group Inc. USA, are shown in Table 8.4.

If the range of suctions to be measured is known, it is better to use solutions within the range to check the calibration of the chilled mirror hygrometer. Various molal solutions of sodium chloride (NaCl) and potassium chloride (KCl) can be prepared to generate the required relative humidity. Procedures for preparing such salt solutions are given in ASTM E104-20a.

Table 8.4 Relative humidity from ASTM E104-20a of various solutions at 25°C

Solutions	Relative humidity from ASTM E104-20a (%)	Measured relative humidity by WP4 (%)
Magnesium Chloride (MgCl$_2$. 6H$_2$O)	32.8 ± 0.2	32.7
Magnesium Nitrate [Mg(NO$_3$)$_2$. 6H$_2$O]	52.9 ± 0.2	53.0
Sodium Chloride (NaCl)	75.3 ± 0.1	75.3
Potassium Chloride (KCl)	84.3 ± 0.3	84.3
Potassium Nitrate (KNO$_3$)	93.6 ± 0.6	93.9
Potassium Sulphate (K$_2$SO$_4$)	97.3 ± 0.5	97.7
Pure Distilled Water	100	99.9

Usually, the manufacturer would specify the calibration check needed before using their instrument. For the WP4, a one-point calibration check is recommended, and it would be better for the one-point to be within the suction range of the soil specimen to be measured. The WP4 will be described here, as it is the most common chilled mirror hygrometer used in geotechnical engineering. Similar chilled mirror hygrometers are available from other manufacturers but not specifically for use with soil samples. The decision to describe the WP4 here does not represent any endorsement of the product but is more of a case where there is more cumulative experience gained from using the WP4. The WP4 has gone through several upgrades. The current model is the WP4C, which has a measurement range of 33 kPa to 30,000 kPa. However, experiences of the measurement accuracy were mixed for suctions below 1,000 kPa and sometimes below 3,000 kPa, as the data points may not merge into the SWCC from the pressure plate test. The accuracy of the suction measurement is 50 kPa for suctions less than 5,000 kPa, and 1% for suctions from 5,000 to 30,000 kPa. The WP4C does not have an option to provide the relative humidity reading. Hence, if relative humidity reading is needed, Equation 7.5 can be used to convert the total suction measured to the relative humidity value. The operating temperature range is from 5°C to 40°C. Temperature during measurement can be set within 5°C–40°C. Temperature resolution is 0.1°C and accuracy is ±0.2°C. The manufacturer claims that the WP4C can give a reading within 5 minutes in the quick mode and between 10 and 15 minutes in the precise mode. It has been found that some soil specimens can take longer than 15 minutes to reach equilibrium. Readings are updated every 5 minutes until equilibrium.

8.5.2 Test procedures

The WP4 is shown schematically in Figure 8.7. The WP4 is a standalone unit with a sample drawer, an LCD display panel and buttons. The sample drawer is slid out to receive the soil specimen container. The soil specimen

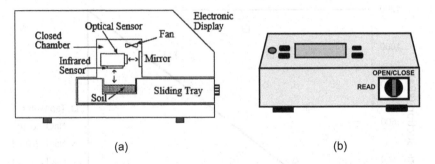

Figure 8.7 Schematic drawing of WP4. (a) Schematic drawing. (b) Front view.

container is either plastic or steel. The steel container is supposed to give better accuracy, as it allows the soil specimen to reach temperature equilibrium faster. The container is filled with the soil till half full and then placed into the slot of the drawer for the soil specimen container. The drawer is then pushed back into the unit, and the knob is turned to lift the soil specimen container to contact a seal at the top of the chamber. The sealed chamber is supposed to be airtight, and the temperature of the chamber can be controlled within 5°C and 40°C. A small internal fan circulates the air in the air space above the soil specimen to reduce equilibrium time. The WP4 will continuously record the relative humidity of the air space above the soil specimen and the surface temperature of the soil specimen displaying the reading every 5 minutes until equilibrium. The chamber contains a mirror whose temperature is precisely controlled by a thermoelectric (Peltier) cooler and detection of the instance at which condensation first appears on the mirror is by a beam of light directed onto the mirror and reflected into a photodetector cell. The purpose of the photodetector cell is to sense the change in reflectance when condensation first occurs on the mirror so that the temperature at which it occurs can be recorded. The temperature is given by a thermocouple attached to the mirror. Contamination of the mirror by either spillage of soil specimen or condensation in the sample chamber can lead to inaccurate readings.

8.5.3 Precaution

The chilled mirror hygrometer when measuring the suction of an aqueous solution gives the osmotic suction of the aqueous solution. When it is measuring the suction of a soil specimen, it is measuring both matric and osmotic suctions. It is a common practice to infer the third suction component when two other suction components are measured (e.g., Tsai and Petry 1995; Tang et al. 2002). The chilled mirror hygrometer should not be used to infer either matric or osmotic suction when either osmotic or matric suction is measured using other methods, as the error can be large (Leong et al.

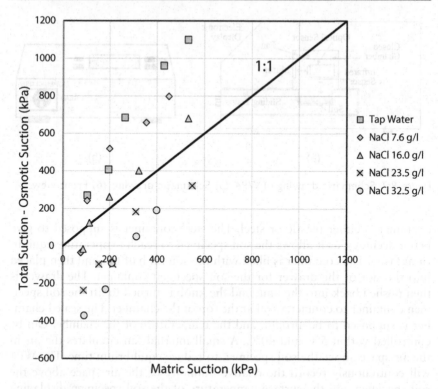

Figure 8.8 Comparison of inferred matric suction (total suction – osmotic suction) and measured matric suction for compacted soil samples using different solutions [Data from Leong et al. (2009)].

2007). A plot comparing the inferred matric suction given by (total suction – osmotic suction) and measured matric suction for compacted residual soil samples using different solutions is shown in Figure 8.8. In Figure 8.8, the total suction was measured using the chilled mirror hygrometer, the matric suction was measured using a null-type axis translation apparatus and the osmotic suction was determined from the electrical conductivity of the solution used to compact the soil samples. Figure 8.8 shows that the errors are large even for soil samples compacted using tap water.

REFERENCES

Blight, G. E. (1966). Strength characteristics of desiccated clays. *Journal of the Soil Mechanics and Foundations Division*, 92(6): 18–38.

Bulut, R., Hineidi, S. M., & Bailey, B. (2002). Suction measurements—filter paper and chilled mirror psychrometer. In *Proceedings of the Texas Section American Society of Civil Engineers, Fall Meeting*, Waco, 2–5 October.

Cassel, D. K. & Klute, A. (1986). *Water potential: Tensiometry. Methods of soil analysis, Part 1, Physical and mineralogical methods*, 2nd ed., A. Klute, Ed. Madison, WI: American Society of Agronomy, Soil Science Society of America.

Croney, D. & Coleman, J. D. (1948). Soil thermodynamics applied to the movement of moisture in road foundation. In *Proceedings of the 7th International Conference on Applied Mechanics*, Vol. 3, pp. 163–167.

Fredlund, D. G. (1973). *Volume change behaviour of unsaturated soils*. University of Alberta, Edmonton, Alta, Canada, vol. Ph. D., p. 490.

Fredlund, D. G. & Rahardjo, H. (1993). *Soil mechanics for unsaturated soils*. New York: Wiley.

Gee, G., Campbell, M., Campbell, G., & Campbell, J. (1992). Rapid measurement of low soil potentials using a water activity meter. *Soil Science Society of America Journal*, 56: 1068–1070.

Guan, Y. & Fredlund, D. G. (1997). Use of the tensile strength of water for the direct measurement of high soil suction. *Canadian Geotechnical Journal*, 34(4): 604–614.

He, L., Leong, E. C., & Elgamal, A. (2006). A miniature tensiometer for measurement of high matric suction. In G. A. Miller (Ed.), 147 edn. *Unsaturated soils 2006*. Reston Va, [Great Britain]: American Society of Civil Engineers, vol. 2, pp. 1897–1907.

Houston, S. L., Houston, W. N., & Wagner, A.-M. (1994). Laboratory filter paper suction measurements. *Geotechnical Testing Journal*, 17(2): 185–194.

Leong, E. C., He, L., & Rahardjo, H. (2002). Factors affecting the filter paper method for total and matric suction measurements. *Geotechnical Testing Journal*, 25(3): 321–332.

Leong, E. C., Lee, C. C., & Low, K. S. (2009). An active control system for matric suction measurement. *Soils and Foundation*, 49(5): 807–811.

Leong, E. C., Widiastuti, S., Lee, C. C., & Rahardjo, H. (2007). Accuracy of suction measurement. *Geotechnique*, 57(6): 547–556.

Lourenço, S. D. N., Gallipoli, D., Toll, D. G., & Evans, F. D. (2009). On the measurement of water pressure in soils with high suction tensiometers. *Geotechnical Testing Journal*, 32(6): 1–7.

Rahardjo, H. (1990). The study of Undrained and Drained Behaviour of Unsaturated Soil. In Department of Civil Engineering. University of Saskatchewan, Saskatoon, vol. Ph.D.

Ridley, A. M. & Burland, J. B. (1993). A new instrument for the measurement of soil moisture suction. *Géotechnique*, 43(2): 321–324.

Ridley, A. M. & Burland, J. B. (1999). Use of the tensile strength of water for the direct measurement of high soil suction: Discussion. *Canadian Geotechnical Journal*, 36(1): 178–180.

Schanz, T., Agus, S. S., & Tscheschlok, G. (2004). Determination of hydromechanical properties of Trisoplast. Research Report Bo-015/03. Laboratory of Soil Mechanics. Weimar, Germany: Bauhaus-University Weimar.

Tang, G. X., Graham, J., Blatz, J., Gray, M., & Rajapakse, R. K. N. D. (2002). Suctions, stresses and strengths in unsaturated sand–bentonite. *Engineering Geology*, 64(2–3): 147–156.

Tarantino, A. & Mongiovi, L. (2002). Design and construction of tensiometer for direct measurement of matric suction. In *Proceedings of 3rd International Conference on Unsaturated Soils, Recife*, vol. 1, pp. 319–324.

Tsai, C.-H. & Petry, T. M. (1995). Suction study on compacted clay using three measurement methods. *Transportation Research Record No, 1481*: 28–34.

Wijaya, M. (2017). Compression, shrinkage and wetting-induced volume change of unsaturated soils. In *Civil and Environmental Engineering*. Singapore: Nanyang Technological University, vol. Ph.D.

Wijaya, M. & Leong, E. C. (2016). Performance of high-capacity tensiometer in constant water content oedometer test. *International Journal of Geo-Engineering*, 7(1): 13.

Chapter 9

Indirect suction measurement methods

9.1 INTRODUCTION

In this chapter, several indirect suction measurement methods are described. All indirect suction measurement methods use one common measurement principle, that is, the use of a proxy to provide the suction value. The indirect methods described are the more common and accessible methods that can be readily adopted by a soil testing laboratory. These methods include the filter paper method, thermal conductivity sensor, electrical resistance sensor, capacitive sensor and electrical conductivity measurements. The suctions measured are total suction (filter paper), matric suction (filter paper, thermal conductivity sensor, electrical resistance sensor, capacitive sensor) and osmotic suction (electrical conductivity).

9.2 RELATED STANDARDS

1. ASTM D1125 Test Methods for Electrical Conductivity and Resistivity of Water.
2. ASTM D4542-15 Test Methods for Pore Water Extraction and Determination of the Soluble Salt Content of Soils by Refractometer.
3. ASTM D5298–16. Standard Test Method for the Measurement of Soil Potential (Suction) Using Filter Paper.
4. ASTM D6836–16. Standard Test Methods for Determination of the Soil Water Characteristic Curve for Desorption Using Hanging Column, Pressure Extractor, Chilled Mirror Hygrometer, or Centrifuge.
5. ASTM E230/E230M Specification and Temperature–Electromotive Force (emf) Tables for Standardized Thermocouples.
6. ASTM E337 Test Method for Measuring Humidity with a Psychrometer (the Measurement of Wet- and Dry-Bulb Temperatures).
7. ASTM E832-81 (2019), Standard Specification for Laboratory Filter Papers.

9.3 FILTER PAPER METHOD

The use of filter paper as a proxy to determine matric and total suction was developed in soil science and has been standardised for measurement of soil potential (suction) in ASTM D5298-16 (2016). The test procedure is simple, straightforward and does not require any special equipment. It involves the measurement of the water content of the filter paper that has equilibrated with a soil specimen in a closed container. It is generally accepted that when the filter paper is placed in direct contact with the soil, moisture transfer is by capillary flow and matric suction is measured. When the filter paper is not in contact with the soil, moisture transfer is by vapour transfer. The filter paper comes to equilibrium with the partial pressure of water vapour in the air of the container enclosing the filter paper and the soil, measuring total suction.

Gardner (1937) was probably the first to use filter paper to measure soil suction. The filter paper used was Schleicher & Schuell No. 589 (S&S 589) white ribbon filter paper. Calibration of the filter paper was first done by equilibrating both wet and dry filter papers above sulphuric acid solutions in closed containers for low water contents and then transferring the equilibrated filter papers to a centrifuge to obtain the desired suctions for high water contents. Since then, the filter paper method for suction measurement has become more widely used due to the accessibility of filter papers. The two most common filter papers used for suction measurement are Schleicher & Schuell No. 589 filter paper (McQueen and Miller 1968b; Al-Khafaf and Hanks 1974; McKeen 1980; Harrison and Blight 1998; Leong et al. 2002) and Whatman No. 42 filter paper (Fawcett and Collis-George 1967; Hamblin 1981; Chandler and Gutierrez 1986; Chandler et al. 1992; Harrison and Blight 1998; Deka et al. 1995; Leong et al. 2002; Power et al. 2008). Other filter papers that have been used include Whatman No. 1 filter paper (El-Ehwany and Houston 1990), Thomas Scientific No. 4705-F10 filter paper (Miller and Nelson 1992) and Fisher quantitative coarse (9.54 Å) filter paper (Houston et al. 1994). In fact, any adsorbent materials besides filter paper can be used. Sibley and Williams (1990) evaluated five different absorbent materials (Whatman No. 42, unwashed dialysis tubing, washed dialysis tubing, Millipore MF 0.025 pm and Millipore 0.05 pm) for suction measurements and concluded that the Whatman No. 42 filter paper is the most appropriate for suction measurement.

ASTM D5298-16 (2016) specifies that the filter paper used must be ash-free quantitative Type II filter papers (ASTM E832-81 2019). Examples of ash-free quantitative Type II filter papers are Whatman No. 42, Fisherbrand 9-790A and Schleicher & Schuell No. 589 White Ribbon.

9.3.1 Calibration

It is recommended that the calibration curve of filter paper (for a batch) be obtained before use (Likos and Lu 2002; Leong et al. 2002) and the filter paper calibrated according to the method (contact or non-contact) that it is used (Ridley and Wray 1996; Power et al. 2008). ASTM D5298-16 (2016) suggests that the calibration curve can be obtained by suspending a filter paper over a salt solution of various concentrations (minimum 50 mL) using an improvised platform of inert material such as plastic tubing or stainless-steel screen. This method of calibration corresponds to the non-contact filter paper method that measures total suction. Houston et al. (1994), Bulut et al. (2002), Leong et al. (2003) and Power et al. (2008) found that the calibration curves for the contact method which measures matric suction and the non-contact method which measures total suction are different. Discrepancies between suctions measured using contact and non-contact methods are more severe in the low suction range, <100 kPa. Bulut and Wray (2005), and Marinho and Oliveira suggest that the difference is due to insufficient equilibrium time for the non-contact filter paper. As the equilibrium time is increased, the suction measured using the non-contact method approaches slowly towards the suction measured using the contact method but never reaches it even after an excessively long equilibrium time. Such excessive equilibrium time is not suitable for practical purposes. Hence, it is not possible to calibrate filter paper for the low suction range (<100 kPa) using the non-contact method. In a footnote, ASTM D5298-16 (2016) mentioned that the filter paper may be calibrated by using the pressure membrane for the range 100–1,500 kPa and the ceramic plate for the range 10–100 kPa. Leong et al. (2020) calibrated the Whatman No. 42 filter paper in the low suction range (< 16 kPa) using the Tempe cell and capillary rise method.

Fawcett and Collis-George (1967) calibrated Whatman No. 42 filter papers, treated with 0.005% mercury chloride ($HgCl_2$), with soil samples from a soil column at equilibrium with a water table at suctions of 1 and 10 kPa. Hamblin (1981) calibrated Whatman No. 42 filter papers using a suction plate apparatus for suction up to 7 kPa. Deka et al. (1995) calibrated the Whatman No. 42 filter paper using a tension table for suctions from 1 to 65 kPa. The principles of a tension table are similar to those of a suction plate or the hanging column (Method A) described in ASTM D6836-16. It is possible to apply a maximum suction of up to 80 kPa using a tension table or hanging column with a vacuum pump (Parcevaux 1980). Leong et al. (2002), Power et al. (2008) and Elgabul (2013) calibrated Whatman No. 42 filter papers in a pressure plate for matric suctions less than 1,000 kPa.

Most of the calibration curves were developed from initially dry filter paper. Hence, the calibration equation usually corresponds to the wetting curve of the filter paper. As the filter paper is thin, the amount of hysteresis

Table 9.1 Calibration curves for Whatman No. 42 and Schleicher & Schuell No. 589 filter
        papers

(a) Whatman No. 42 filter paper

References	Calibration curves*
Hamblin (1981)	$\log \psi = 8.022 - 3.683 \log w_f$
Chandler and Gutierrez (1986)	$\log \psi = 4.84 - 0.0622 \, w_f \, w_f < 47$
Chandler et al. (1992)	$\log \psi = 6.05 - 2.48 \log w_f \, w_f \geq 47$
Greacen et al. (1987)	$\log \psi = 5.327 - 0.0779 \, w_f \, w_f < 45.3$
ASTM D5298-16	$\log \psi = 2.412 - 0.0135 \, w_f \, w_f \geq 45.3$
Leong et al. (2002)	$\log \psi = 4.945 - 0.0673 \, w_f \, w_f < 47$
	$\log \psi = 2.909 - 0.0229 \, w_f \, w_f \geq 47$

(b) Schleicher & Schuell No. 589 filter paper

References	Calibration curves*
McQueen and Miller (1968)	$\log \psi = 5.238 - 0.0723 \, w_f \, w_f < 54$
	$\log \psi = 1.8966 - 0.01025 \, w_f \, w_f \geq 54$
Al-Khafaf and Hanks (1974)	$\log \psi = 4.136 - 0.0337 \, w_f \, w_f < 85$
	$\log \psi = 2.0021 - 0.009 \, w_f \, w_f \geq 85$
McKeen (1980)	$\log \psi = 4.9 - 0.0624 \, w_f \, w_f < 66$
	$\log \psi = 1.25 - 0.0069 \, w_f \, w_f \geq 66$
Greacen et al. (1987)	$\log \psi = 5.056 - 0.0688 \, w_f \, w_f < 54$
ASTM D5298-16	$\log \psi = 1.882 - 0.0102 \, w_f \, w_f \geq 54$
Leong et al. (2002)	$\log \psi = 5.438 - 0.069 \, w_f \, w_f < 54$
	$\log \psi = 2.659 - 0.018 \, w_f \, w_f \geq 54$

* Note: ψ = suction in kPa, w_f = filter paper water content in %

between the drying and wetting filter paper is small in the range of about 2% of the water content (Leong et al. 2002).

Many calibration equations have been proposed in the literature for the filter paper. A summary of the calibration equations is listed in Table 9.1. If the filter paper used is not calibrated, it is recommended to use that given in ASTM D5298-16 (Leong et al. 2007).

9.3.2 Test procedures

As the water content of the filter paper is used to determine suction, any accidental transfer of fluid to the filter paper during water content determination will cause inaccurate measurement and, hence, wrong suction value. The filter paper should not be handled with bare hands; vinyl surgical non-powdered or similar gloves are recommended to be used when handling tools or containers during the filter paper water content determination (Bulut and Leong 2008). A pair of tweezers should be used for picking up, placing or transferring the filter paper. Hence, a strict protocol is needed when using the filter paper method to obtain a reliable suction measurement.

Although filter paper size is not rigidly fixed, the most used filter paper size is 55 mm in diameter. Such a filter paper weighs approximately 0.2 g. ASTM D5298-16 (2016) requires the weight of the filter paper to be weighed to a resolution of 0.0001 g. The filter paper from the box usually has some initial moisture content of about 3%–6% (Leong et al. 2020). ASTM D5298-16 (2016) recommended that the filter papers be dried for at least 16 hours or overnight in a drying oven and placed in a desiccator over a desiccant for storage before use. This is not necessary if the filter paper will be wetted to water content greater than 6% during moisture equilibration with the soil specimen. If organism growth or biological decomposition of the filter paper is expected, the filter paper can be pre-treated by dipping the filter paper into a 2% concentration of formaldehyde prior to use (ASTM D5298-16, 2016). Biological decomposition is possible if the filter paper is in a warm and moist environment for more than 14 days. Mercury chloride ($HgCl_2$) (Fawcett and Collis-George 1967; Hamblin 1981) and pentachlorophenol solutions (McQueen and Miller 1968; Al-Khafaf and Hanks 1974; Bulut 1986) have been used to pre-treat filter papers before use. However, such treatment may alter the characteristics of the filter paper and hence its calibration curve. Pre-treated filter papers should be used with caution.

The soil specimen used should be as large as possible. ASTM D5298-16 (2016) suggests that the mass of the soil specimen be between 200g and 400g. For total suction measurement, the soil specimen can be fragmented, and the filter paper placed above the soil specimen with a spacer which can be a wire screen, O-ring or other inert material with minimum surface area between the filter paper and the soil (Figure 9.1a). For matric suction measurement, the whole soil specimen should be used, and the filter paper can be sandwiched between two halves of the specimen (Figure 9.1b). In both cases, the soil specimen shall be placed in an airtight container made of rust-free metal or glass with minimal air space to reduce equilibration time and to minimise suction changes in the specimen. ASTM D5298-16 (2016) specifies two filter papers to be used in the total suction measurement and three filter papers to be used in the matric suction measurement. For the matric suction measurement, a filter paper is sandwiched between two slightly larger diameter filter papers (about 3–4 mm larger in diameter) to prevent soil contamination. The use of more than one filter paper will increase equilibration time and may have the unintended consequence of the filter paper not reaching suction equilibrium with the soil specimen (Leong et al. 2002).

The lid of the airtight container can be additionally sealed using tape and placed in an insulated chamber such as a cooler box or Styrofoam box, where temperature can be maintained within ±1°C when the variation of the external temperature is ±3°C (ASTM D5298-16). The equilibration period is suggested to be seven days and is expected to be less in ASTM D5298-16 (2016). The length of equilibration time depends on the degree of saturation of the soil specimen. Drier soil specimens may take a longer time.

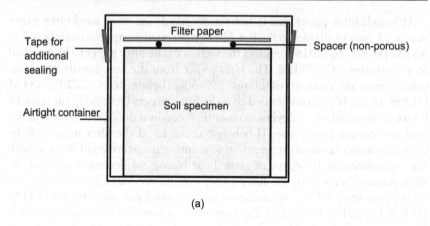

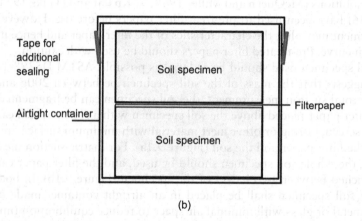

Figure 9.1 Filter paper method for measuring soil suction. (a) Non-contact filter paper. (b) Contact filter paper.

Experiments have shown that 14–15 days may be more appropriate (Leong et al. 2002; Bulut and Wray 2005).

At the end of the equilibration period, the water content of the filter paper is determined. The equipment needed for the water content determination should be prepared before removing the filter paper from the container. The filter paper is transferred to a metal (aluminium stainless steel) moisture container with lid using a pair of tweezers and gloved hands quickly (within 3s–5s) for weight determination to the nearest 0.0001 g. Plastic bags with ziplock may also be used in place of the metal moisture container, but greater care needs to be exercised. First, weigh the ziplock plastic bag to 0.0001 g. Place the filter paper quickly into the ziplock plastic bag and press the plastic bag flat on a clean and dry surface to remove the air before zipping it for weighing to 0.0001 g. After weighing, transfer the filter paper to an evaporation dish or moisture can for drying in the oven. After drying, repeat the previous steps to obtain the dry weight of the filter paper using a new ziplock plastic bag. The temperature of the oven should be the same as for soil moisture container determination – i.e., 105°C–110°C. If metal moisture container is used, the moisture container can be placed on a metal block with a flat surface to increase the cooling rate (ASTM D5298-16). Metallic moisture container when hot is known to weigh less than when cold due to adsorbed moisture and should be used with caution, as it will introduce errors in the water content determination of the filter paper.

9.3.3 Calculations

The final water content of the filter paper w_f is determined using Equation 9.1.

$$w_f = \frac{m_{ff} - m_{f0}}{m_{ff}} \tag{9.1}$$

where
m_{ff} = mass of filter paper after oven drying
m_{f0} = mass of filter paper before oven drying

Using w_f, the suction ψ can be obtained using a calibration equation of the form shown in Table 9.1 if a calibration of the filter paper was performed or using the ASTM D5298-16 equation (Leong et al. 2007).

9.3.4 Summary

It is recommended to use only the contact filter paper method to measure matric suction. When matric suction exceeds 1,000 kPa, contact and non-contact filter paper methods will yield similar suction values (Leong et al. 2016). If the non-contact filter paper method is used, it should only be used for suction above 1,000 kPa, and the matric suction calibration equation can be used to determine the suction.

9.4 THERMAL CONDUCTIVITY SENSOR

A thermal conductivity sensor (TCS) relies on the rate of heat dissipation of a porous material as it equilibrates with the surrounding soil to determine the matric suction. A typical TCS consists of a heater, a temperature-sensing unit and a ceramic block. An example of a TCS is the 229 Water Matric Potential Sensor by Campbell Scientific shown in Figure 9.2. Comparison

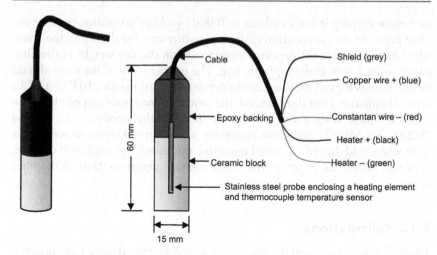

Figure 9.2 Schematic representation of the 229 Water Potential Sensor from Campbell Scientific.

Table 9.2 Comparison of 229 and FTC thermal conductivity sensors

Sensor	Company	Suction range (kPa)	Resolution (°C)	Accuracy	Dimensions
229	Campbell Scientific	10–2,500	±0.1°C	–	15 mm (D) × 60 mm (L)
FTC	GCTS Testing Systems	1–1,000	–	5% of reading (up to 500 kPa in wetting)	28 mm (D) × 60 mm (L)

of the 229 Water Matric Potential Sensor and the Fredlund thermal conductivity (FTC) the pressure plate at equilibrium sensor from GCTS Testing Systems is given in Table 9.2. The heater and temperature sensor are placed inside a needle probe and embedded inside the ceramic block. The cables extend out from the back of the block through a backing. The principle of the TCS is that the thermal conductivities of air (= 0.026 Wm⁻¹K⁻¹), water (≈0.60 Wm⁻¹K⁻¹) and the ceramic are different. The thermal conductivity of the 229 Water Potential Sensor when dry is about 0.35 Wm⁻¹K⁻¹ (Reece 1996). To do a measurement, the heater is turned on for a fixed period of time. The heat causes the temperature to rise in the ceramic block, and the temperature rise ΔT is recorded by the temperature sensor. The amount of heat generated by the heater is fixed by the duration of heating, but ΔT is dependent on the amount of water present in the ceramic block as negligible heat is dissipated by the air in the pores. An example of temperature rise ΔT with a reduction of water content (i.e., increase in matric suction) is shown in Figure 9.3. Therefore, thermal conductivity of the ceramic block

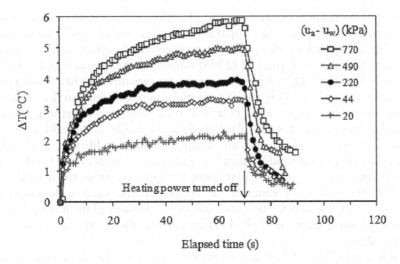

Figure 9.3 Temperature rises in a TCS at different matric suctions. (from Leong et al. 2011).

will increase with its water content, leading to a lower temperature rise ΔT. When the TCS is embedded in the soil, the water content of the ceramic block will equilibrate with the surrounding soil to reach matric suction equilibrium. Hence, knowing the calibration curve of the ceramic block enables the matric suction of the soil to be determined.

9.4.1 Calibration

The thermal transfer properties of a TCS depend on the interfaces between the different components that make up the TCS: the interface between the stainless-steel probe and the ceramic, the arrangement of the heating elements and temperature sensor in the probe, the homogeneity of the ceramic block. Differences in manufacturing quality of the TCS make it necessary to calibrate each sensor individually for a more accurate reading (Fredlund and Wong 1989; Reece 1996; Flint et al. 2002; Hu et al. 2007).

The TCS can be calibrated using a pressure plate apparatus. A soil mixture is prepared and placed in the pressure plate with the TCS embedded (e.g., Fredlund and Wong 1989; Reece 1996; Feng and Fredlund 1999; Fredlund et al. 2000; Flint et al. 2002). Hu et al. (2007) calibrated without a soil mixture using only kaolin paste between the tip of the FTC ceramic block and a pressure plate. The matric suction is increased, and when the reading of the TCS becomes stable over time, the reading of the TCS is deemed to correspond to that of the applied matric suction. The air pressure is then increased for the next matric suction calibration. This technique allows several TCSs to be calibrated at the same time. Variation of how the

TCS reading is taken exists. For example, Flint et al. (2002) released the air pressure of the pressure plate apparatus when the soil mixture came to equilibrium before making measurements to preclude the effects of air pressure on the TCS reading. Campbell (2006) thought that the heat transfer in the ceramic block is highly dependent on the air pressure and hence releasing the air pressure will allow the TCS to make measurements in the environment (atmospheric pressure) that it is used. Using this method, special attachments to the pressure plate need not be made as the cables can be left in the pressure plate apparatus during suction equilibrium. The cables are connected to the readout unit for reading after air pressure is released. However, using the pressure plate does not guarantee that the matric suction of the soil is the same as the applied matric suction. It has been observed that the matric suction of soil in the pressure plate at equilibrium is always lower than the applied matric suction (Gee et al. 2002; Creswall et al. 2008).

When using the TCS, it is important to know the response time of the TCS. Using the calibration procedures mentioned earlier will not allow the response time to be evaluated as the TCS is following a drying curve when air pressure is increased. An alternative procedure is to compact soil samples at various water content to calibrate the TCS (Leong et al. 2009). The TCS is embedded in the soil samples, and the response of the TCS with time is recorded. The matric suctions of the compacted samples are then determined by independent measurement, e.g., using a null-type axis translation apparatus or tensiometer (see Chapter 8). However, this method of calibration is highly labour intensive and may not be suitable in practice.

The following procedures are typically used when calibrating using a pressure plate apparatus:

1. Saturate the ceramic block of the TCS by immersing the ceramic sensor under water and apply a vacuum, the same way as saturating porous stones for saturated soil tests.
2. Prepare a saturated soil mixture. Ideally, the soil which the TCS is to be used with or kaolin may be used.
3. Prepare the pressure plate apparatus. Ensure that the ceramic plate is fully saturated.
4. Place the soil mixture into a steel or plastic ring (diameter of at least two times the diameter of the TCS and height of at least two times the length of the TCS) and place it into the pressure plate.
5. Embed the TCS to be calibrated, ensuring that the ceramic block is fully embedded into the soil.
6. Connect the wires to the readout unit. Special attachments to the pressure plate are needed to allow the wires to be brought out of the pressure plate.
7. Apply the lowest suction of the calibration range.

8. Take measurements of the TCS periodically to determine if the equilibrium condition has been reached. Close off the drainage valve of the pressure plate and release the suction before making a TCS measurement. The reading of the TCS is usually performed at a fixed heating time, e.g., for the TCS 229, this is at the 30s.
9. When equilibrium at the applied suction is attained, apply the next suction and repeat steps 7 and 8 until the upper limit of the calibration range.

Suitable functions to describe the calibration curves are exponential or power functions as given in Equations 9.2 and 9.3, respectively.

$$\psi = \exp\left(a.\Delta T + b\right) \tag{9.2}$$

$$\psi = a.\Delta T^b \tag{9.3}$$

where a and b are curve-fitting parameters.

The fitting parameters in Equations 9.2 and 9.3 can also be determined using normalised temperature change to be between 0 and 1 where the normalised temperature change ΔT_{norm} is given by Equation 9.4.

$$\Delta T_{norm} = \frac{\Delta T_{dry} - \Delta T}{\Delta T_{dry} - \Delta T_{wet}} \tag{9.4}$$

where ΔT_{dry} is the temperature change when the TCS ceramic block is dry, and ΔT_{wet} is the temperature change when the TCS ceramic block is saturated.

9.4.2 Test procedures

The TCS is normally used in the field but it is useful in a model laboratory test or as an alternative to measure matric suction of soil samples. The TCS is first saturated and left with its ceramic block in water. To make a suction measurement, a pre-bored hole about the size of the TCS is first made into the soil sample. The TCS is then inserted and the wires are connected to a readout unit. The TCS and soil sample should be covered or placed in an airtight container to prevent the soil sample from drying during measurement. Take readings periodically until there is no change in the reading. As the TCS starts from the fully saturated condition, the soil sample should be large enough such that its suction is not greatly affected by moisture exchange between the soil sample and TCS during measurement. Equilibration time depends on the suction of the soil sample, and the contact condition between the TCS and the soil sample (Leong et al. 2011). The matric suction is then obtained from the calibration equation.

9.5 ELECTRICAL RESISTANCE SENSORS

Electrical resistance sensors (ERSs) rely on measuring the electrical resistance of a porous material as it equilibrates with the surrounding soil to determine the matric suction. The development of the ERS dates back to 1940 (Bouyoucos and Mick 1940). Materials that have been used as porous materials include gypsum, fibreglass or nylon (Scalon et al. 2002). Gypsum is commonly used in early ERSs, and such sensors are still being manufactured commercially e.g., 227-L Soil Matrix Potential Block manufactured by Delmhorst. The ERS suffers from problems ranging from sensitivity to soil salinity and temperature, hysteresis, stability of calibration curve and durability of the block.

Another ERS is the Watermark Model 200 sensor manufactured by the Irrometer Company, Inc. A schematic drawing of the Watermark Model 200 sensor is shown in Figure 9.4a. It consists of two concentric stainless-steel electrodes embedded in a uniform granular matrix containing a gypsum wafer encased in a synthetic membrane and covered by a perforated stainless-steel shell with plastic end caps (Figure 9.4b). The perforated stainless-steel shell helps to ensure that the granular matrix is packed more uniformly automatically under uniform pressure. Both the synthetic membrane and the perforated stainless-steel shell make the sensor more robust for handling. The gypsum wafer buffers against salinity variations less than 0.2 S/m (electrical conductivity of saturated gypsum). The upper matric suction limit of the sensor is controlled by the air-entry value (AEV) of the granular matrix as the matrix remains saturated above the AEV. For Watermark sensors, the lowest suction limit is about 10 kPa (Armstrong and Thomson 1985; Thomson and Armstrong 1987; Spaans and Baker 1992). The highest matric suction limit of the sensor depends on the smaller pore

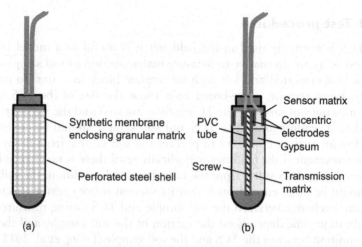

Figure 9.4 Schematic drawing of Watermark Model 200 sensor. (a) Schematic drawing. (b) Section view.

Table 9.3 Comparison of 227-L and Watermark 200 ERSs

Sensor	Company	Suction range (kPa)	Physical quantities (mm, g)	Accuracy	Operating temperature range
Gypsum Block	Delmhorst	10–1,000	22.5 (D) × 28.6 (L), 180	±(10% of reading + 2 kPa) from −100 to −5 kPa	> 0°C
Watermark Model 200	Irrometer Sensors	0–200			0 – 80°C

sizes of the granular matrix which has been specified by the manufacturer as 239 kPa. Comparison of the 227-L and Watermark 200 sensors is given in Table 9.3.

9.5.1 Calibration

Scalon et al. (2002) recommend that ERS should be soaked for about 24 hours and then oven-dried two to three times before calibration and use. The calibration equation given by the manufacturer should be verified even though some ERSs are marketed as "identical" with only a single calibration equation independent of soil types. For Watermark Model 200 sensors, McCann et al. (1992) and Spaans and Baker (1992) found their calibration equation to be different from that given by the manufacturer. In addition, Spaans and Baker (1992) found poor repeatability of the calibration curve. Hence, for better accuracy, each sensor should be calibrated individually before use.

The ERSs can be calibrated using a pressure plate apparatus (Thomson and Armstrong 1987; McCann et al. 1992). Sensors are embedded in a saturated soil in the pressure plate, and the drying curve is measured by increasing the air pressure incrementally similar to the procedures described for TCS. Alternatively, the sensors can be calibrated using a direct matric suction measurement device like the tensiometer (Spaans and Baker 1992) by embedding the ERSs and a tensiometer in the same soil.

Temperature effects are known to affect the accuracy of ERS. Generally, the correction for temperature effects is about 3% (Campbell and Gee 1986; Thomson and Armstrong 1987; Wang 1988). Equation 9.5 by Campbell and Gee (1986) may be used to correct for temperature effects.

$$R_c = R_m \left[1 + 0.03 \left(T_c - T_m \right) \right]$$ (9.5)

where

R_c = corrected resistance for calibration temperature T_c

R_m = measured resistance at temperature T_m

The Model 200SS Watermark Model 200SS sensor relies on the calibration equation (Equation 9.6) of the form given by Thomson and Armstrong (1987).

$$(u_a - u_w) = \begin{cases} 20R - 11 & R < 1\,k\Omega \\ -0.00279R^3 + 0.190109R^2 + 3.71485R + 6.73956 & R \geq 1\,k\Omega \end{cases}$$

$$(9.6)$$

9.5.2 Test procedures

The test procedures using the ERSs are similar to that for the TCS. ERSs should be saturated first before use following the saturation procedures mentioned earlier. For example, the Watermark Model 200 sensors' granular matrix is initially filled with air. The air in the granular matrix must be replaced by water to obtain correct readings. An alternative saturation procedure for the Watermark Model 2000 sensors is provided by Barani (2021):

1. Submerge the sensor to less than halfway for 30 minutes to allow capillary action to draw the water into the inner pores.
2. Let it dry for about 6–8 hours.
3. Repeat Steps 1 and 2 two more times.
4. Finally, submerge the sensor fully in water for about 6–8 hours.
5. The sensor should be fully saturated and is now ready for use.

To make a suction measurement on a soil sample, a pre-bored hole about the size of the ERS is first made into the soil sample. The ERS is then inserted, and the connections to a readout unit are made. The ERS and soil sample should be covered or placed in an airtight container to prevent the soil sample from drying during measurement. Take readings periodically until there is no change in the reading. As the ERS starts from the fully saturated condition, the soil sample should be large enough such that its suction is not greatly affected by any moisture exchange between the soil sample and the ERS during measurement. Equilibration time depends on the suction of the soil sample and the contact condition between the ERS and the soil sample (Leong et al. 2011). The matric suction is then obtained from the calibration equation.

9.6 CAPACITANCE SENSOR

Capacitance sensors (CSs) are similar to the TCS and ERS in that they rely on a porous matrix to provide a repeatable matric suction reading. The difference is that CS relies on measuring the dielectric permittivity of the porous material as it equilibrates with the surrounding soil to determine the matric suction. The CSs make use of the difference in dielectric permittivity ε of air (1) and water (80) in the pore space. The ε of dry soil is about 3 to 4 and that of saturated soil is about 40–50. The ε of ice is 5. Dielectric permittivity refers to the ability to hold a charge and does not measure water content directly. There are several different types of sensors that make use of dielectric permittivity to measure water content in soil. Examples are time-domain reflectometry (TDR), standing-wave technique and CSs for soil moisture. Early suggestions of using dielectric permittivity for matric suction measurement use TDR (Noborio et al. 1999; Or and Wraith 1999).

An example of a CS is the Teros-21 sensor manufactured by the Meter Group Inc. that is shown schematically in Figure 9.5. The width, length and thickness of the sensor are 35 mm, 96 mm and 15 mm, respectively. The suction measurement range is 5–100,000 kPa with a resolution of 0.1 kPa, and the operating temperature range is –40°C–60°C. The accuracy of the sensor is ±(10% of reading +2 kPa) for a suction range of 9–100 kPa. It consists of a solid-state circuit board sandwiched between two 8-mm thick and 32-mm in diameter ceramic wafers (porous matrix). Each ceramic wafer is protected by 0.3-mm-thick perforated stainless-steel plates on the outer surface that are grounded. The perforated stainless steel serves two purposes: to protect the ceramic wafer and as an electromagnetic shield, as the sensor uses an alternating current of 70 MHz to charge. The charge-release time is measured by the circuit. In principle, the electromagnetic field is confined to within the ceramic wafers and, hence, the sensor is independent of soil type. Effects of salinity of the soil do not affect the sensor reading, as long as it is below 10 dS/m. The sensor is factory calibrated from 20 to 80 kPa and, hence, is most sensitive in this suction range. It is less sensitive beyond 500 kPa due to the water retention characteristic of the ceramic wafer. According to the manufacturer, the amount of hysteresis between 20 and 100 kPa suction is less than 10 kPa.

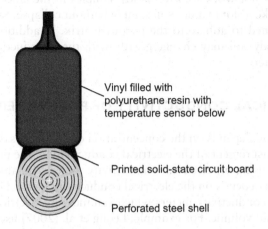

Vinyl filled with polyurethane resin with temperature sensor below

Printed solid-state circuit board

Perforated steel shell

Figure 9.5 Schematic drawing of Teros-21 capacitance sensor.

9.6.1 Calibration

The capacitance sensor has seen several improvements since it was first made commercially. For example, the current Teros-21 Gen 2 has shown improvements in design from its predecessors, MPS-1, MPS-2, MPS-6 and Teros-21 Gen 1. The suction measurement range of MPS-1 and MPS-2 is from 10 to 500 kPa, MPS-6 is from 9 to 100,000 kPa and Teros-21 is from 5 to 100,000 kPa. From MPS-2 onwards, measurement is done using 70 MHz instead of 5 MHz, as it was found that more stable readings were obtained at the higher frequency. The MPS-1 has a one-point factory calibration, the MPS-2 has a two-point factory calibration and the MPS-6 has six-point factory calibration (Decagon 2015). Since MPS-6, all later sensors have six-point factory calibrations done between 20 and 80 kPa suction. Hence, theoretically, no calibration of the sensors needs to be performed, but the sensors still suffer from non-uniformity, and at least a one-point verification for the sensors is needed to ensure accuracy.

9.6.2 Test procedures

The use of CS is the same as for TCS and ERS, the sensor should be saturated first before use. The response time and accuracy of the CS are dependent on the contact condition of the porous matrix with the soil. To install the CS into a soil sample, a slot slightly larger than the sensor needs to be first made into the soil sample. Moisten the removed soil and use it to pack around the sensor such that it covers the entire porous matrix before installing. Once installed, take an initial reading and then at half an hour, doubling the time interval for subsequent readings. The readings should show an increase in matric suction with time and reach a stable reading (equilibrium condition). The equilibrium time of the CS varies from a few hours to several days and does not depend on the matric suction of the soil sample to be measured (Tripathy et al. 2016).

The CS may not work well for sandy samples in the laboratory, as it is difficult to make a slot in a dry soil sample without collapse. Sandy soil cannot be moistened to adhere to the porous matrix. In addition, the matric suction of sandy soil may change greatly with the introduction of a small amount of water.

9.7 ELECTRICAL CONDUCTIVITY OF PORE WATER

Osmotic suction depends on the concentration of dissolved salts in the pore water. The measurement of the electrical conductivity of the pore water can be performed using an electrical conductivity meter. The amount of pore water required depends on the electrical conductivity meter. There are compact electrical conductivity meters which can measure electrical conductivity of small fluid volume. For example, Leong et al. (2007) used the Horiba B-173 which can measure the electrical conductivity of a drop of fluid

(~1 ml) but for a limited range of electrical conductivity. The measurement range of Horiba B-173 is from 0 to 19.9 mS/cm with an accuracy of ±2% full scale. Benchtop type of electrical conductivity meters has a wider range. For example, the Horiba ES-12 electrical conductivity meter has a measurement range of 0–199.9 mS/cm and an accuracy of 0.5% full scale.

The electrical conductivity can be used to indirectly obtain the osmotic suction of soil. There are many methods to extract the pore water from a soil sample (USDA 1954; Iyer 1990; Fares 2009). Examples of such methods are gas extraction, centrifuging, saturation extraction or dilution, immiscible liquid displacement, leaching and pressurised squeezing. Among the methods, extracting the pore water by dilute suspension of soil or saturated soil paste (saturated extract) and mechanical squeezing are more popular. Hence, these two methods will be described here.

9.7.1 Saturated extract

The saturated extract method assumes that all the salts present in the soil will dissolve or remain in the solution as distilled water is added. In the saturation extract method, the soil sample can be air-dried or used as is. Drying in the oven at 105°C is not recommended, as it will convert parts of gypsum to plaster of Paris, which is more soluble (Rhoades 1982). In either case, the initial dry density and water content of the soil sample need to be determined first. This can be done with a sub-sample. The preparation of a saturated soil paste is given in USDA (1954). Distilled water is added to the soil sample until water can be extracted with or without application of a small suction. During the addition of distilled water, stir the soil-water mixture using a spatula. Note the volume or weight of the distilled water added. Soil-water mixtures at saturation of 1:1, 1:2 and 1:5 by weight have been used to obtain the saturation extract. Some studies have shown that the increase in water content leads to a reduction in the electrical conductivity estimated for the soil pore water (e.g., Hogg and Henry 1984; He et al. 2013). After the addition of distilled water, the soil-water mixture is left to stand for 4–6 hours before extraction.

For extraction of the pore water, prepare a clean conical flask, a piece of filter paper, funnel and connector to a vacuum source as shown in Figure 9.6. A small hand-held vacuum pump may suffice as the vacuum source. Pour all the soil mixture into the funnel. If required, apply a small suction using the vacuum source. Collect as much of the extract as possible. Measure the electrical conductivity of the extract using an electrical conductivity meter. Electrical conductivity is affected by temperature, increasing approximately by 1.9% per °C. The measured electrical conductivity, $EC_{measured}$, can first be corrected to the reference temperature 25°C, $EC_{25°C}$, using Equation 9.7 (USDA 1954).

$$EC_{25°C} = f_t \cdot EC_{measured}, \tag{9.7}$$

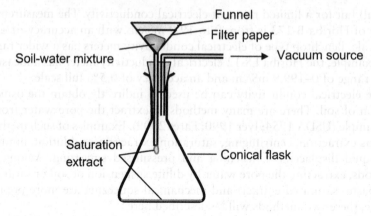

Figure 9.6 Set-up for filtering of soil-water mixture.

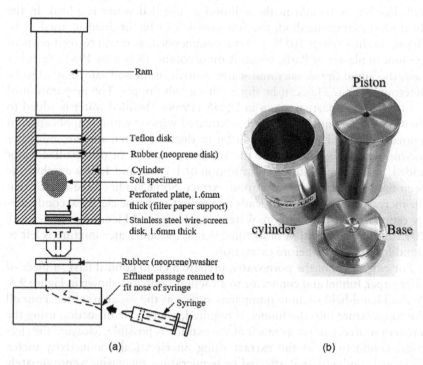

Figure 9.7 Pore fluid squeezer: (a) Schematic drawing (modified from Manheim 1966 and ASTM D4542 – 15). (b) Main components.

where the temperature correction factor f_t is given by Equation 9.8 (Sheets and Hendrickx 1995).

$$f_t = 0.4470 + 1.4034 \exp\left(-\frac{t}{26.815}\right) \qquad (9.8)$$

The electrical conductivity of the pore water in the original soil EC is then obtained by scaling $EC_{25°C}$ with the volumetric water content to give the EC of the original soil using Equation 9.9.

$$EC = \frac{\theta}{\theta_{measured}} EC_{25°C} \qquad (9.9)$$

where
$\theta_{measured}$ = volumetric water content of the soil-water mixture from which the saturation extract was obtained.

The osmotic suction can be obtained from the electrical conductivity using Figure 7.1 or Equation 7.4.

9.7.2 Mechanical squeezing

The saturation extract method is widely used because of its simplicity. However, there are criticisms of the method. Some studies have found that there is a difference between the electrical conductivity calculated using Equation 9.9 at different levels of dilutions (Khan 1967), while others show that the effect is negligible (Argo 1997; Smethurst et al. 1997).

Mechanical squeezing of a soil sample enables the free pore water from a sample with dissolved salts to be obtained from the soil (Fredlund et al. 2012). This method does not require any treatment of the soil prior to the pore-water extraction. The squeezing can be performed using a pore fluid squeezer. A pore fluid squeezer is shown in Figure 9.6. The pore fluid squeezer comprises three main parts: a base, a thick-walled cylinder and a piston (ram). The soil sample is placed into the cylinder and pressure is applied on the piston using a compression machine to extract the pore water. The pore water is collected from the effluent outlet using a syringe. If the base is made to fit a sampling tube, a sample within the sampling tube can be directly compressed to extract the pore water.

The procedures for extracting pore water by mechanical squeezing are given in ASTM D4542 – 15:

1. Clean the components of the mechanical squeezer thoroughly and rinse with distilled water and dry. Check that no rust or dirt should remain in any part or the mechanical squeezer.

2. Place a clean piece of filter paper (5–10-μm pore size) on top of the mechanical squeezer base.
3. Place the soil sample into the heavy cylinder. Drier soil samples will require more soil to be used.
4. Place the ram and apply pressure slowly until the first drop of water is seen coming out of the effluent outlet.
5. Insert a clean, disposable syringe (25 mL) in the effluent outlet to collect the subsequent pore water coming out of the effluent outlet.
6. Continue applying pressure until no more water is expelled or until the syringe is full. ASTM D4542 – 15 specifies a maximum squeezing pressure of 80 MPa based on the limits of pressure used by Kriukov and Komarova (1956) of 59 MPa and Manheim (1966) of 101 MPa. Krahn and Fredlund (1972) used a maximum pressure of 10 MPa.
7. Remove the syringe with the maximum pressure still being applied.
8. Transfer the extracted pore water to an electrical conductivity meter for measurement.

The osmotic suction can be obtained from the electrical conductivity using Figure 7.1 or Equation 7.4.

9.8 SUMMARY

A summary of the indirect suction measurement methods for total, matric and suction components is given in Table 9.4. The thermal conductivity, electrical resistance and conductance sensors for matric suction measurement are low-cost sensors that could be considered when a large number of

Table 9.4 Summary of indirect measurement methods

Suction component	Technique	Suction range (kPa)	Approx. Equilibration time
Total	Non-contact filter paper method	All	≥ 14 days
Matric	Contact filter paper	All	≥ 14 days
	Thermal conductivity sensor	10–2,500	Hours to days
	Electrical resistance sensor	10–200 (granular matrix) 10–1,000	Hours to days
	Capacitance sensors	5–100,000	Hours to days
Osmotic	Saturation extract technique	0–1,500	Minutes to hours
	Squeezing technique	0–1,500	Minutes to hours

measurements are needed at the same time. Although they are of lower accuracy than the direct measurement methods, they are better than the filter paper method, as they are less dependent on the operator, do not require a high-resolution weighing balance (0.0001g) and have a shorter equilibrium time, as little as 2 hours.

REFERENCES

Aitchison, G. D. (1964). Engineering concepts of moisture equilibria and moisture changes in soils. Statement of review panel, Ed., *Moisture equilibria and moisture changes in soils beneath covered areas*. Australia: Butterworths, pp. 7–21.

Argo, W. R., Weesies, N. J., Bergman, E. M., Marshal, M., & Biernbaum, J. A. (1997). Evaluating Rhizon soil solution samplers as a method for extracting nutrient solution and analysing media for container-grown crops. *Horticulture Technology*, 7: 404–408.

Armstrong, C. F., Ligon, J. T., & Thomson, S. J. (1985). Calibration of Watermark Model 200 soil moisture sensor. ASAE Paper No. 85-2077. ASAE, St, Joseph, MI 49085.

Barani, J. (2021 February 01). How to prepare and condition Watermark 200SS soil water tension sensors. https://www.baranidesign.com/faq-articles/2021/2/1/how-to-condition-install-watermark-200ss-soil-water-tension-sensors

Bouyoucos, G. J. & Mick, A. H. (1940). An electrical resistance method for the continuous measurement of soil moisture under field conditions. *Michigan Agricultural Experiment Station Tech. Bull.*, 172.

Brown, A. J. (1907). On the existence of semipermeable membrane enclosing the seeds of some of the Gramineae. *Annals of Botany*, 21: 79–87.

Bulut, R. & Leong, E. C. (2008). Indirect measurement of suction. *Geotechnical and Geological Engineering Journal*, 26(6): 633–644.

Bulut, R. & Wray, W. K. (2005). Free energy of water – suction – in filter papers. *Geotechnical Testing Journal*, 28(4): 355–364. https://doi.org/10.1520/GTJ12307

Bulut, R., Hineidi, S. M., & Bailey, B. (2002) Suction measurements-filter paper and chilled mirror psychrometer. In *Proceedings of the Texas Section, American Society of Civil Engineers, Fall Meeting*, Waco, Texas, USA (2002), pp. 2–5.

Campbell, G. S. & Gee, G. W. (1986). Water potential: Miscellaneous methods. In A. Klute (Ed.), *Methods of soil analysis, Part 1*, 2nd ed. Madison, WI: American Society of Agronomy, pp. 619–633.

Chandler, R. J. & Gutierrez, C. I. (1986). The filter-paper method of suction measurement. *Geotechnique*, 36(2): 265–268.

Chandler, R. J., Crilly, M. S., & Montgomery-Smith, G. (1992). A low cost method of assessing clay desiccation for low-rise buildings. *Proceedings of the Institution of Civil Engineers*, 92(2): 82–89.

Corwin, D. L. & Yemoto, K. (2019). Measurement of soil salinity: Electrical conductivity and total dissolved solids. *Soil Science Society of America Journal*, 83: 1–2. https://doi.org/10.2136/sssaj2018.06.0221

Al-Khafaf, S. & Hanks, R. J. (1974). Evaluation of the filter paper method for estimating soil water potential. *Soil Science*, 117(4): 194–199.

Crilly, M. S., Schreiner, H. D., & Gourley, C. S. (1991). A simple field suction measurement probe. In *Proceedings of the 10th African Regional Conference on Soil Mechanics and Foundation Engineering*, Lesotho, pp. 291–298.

Croney, D. & Coleman, J. D. (1961). Pore pressure and suction in soil. In *Conference on Pore Pressure and Suction in Soils*, London: Butterworths, pp. 31–37.

Decagon (2015). *MPS-2 & MPS-6—Dielectric water potential sensors. Operator's manual*. Pullman, WA: Decagon Devices, Inc.

Duran, A. J. G. (1986). Study of effect of contact on the filter paper technique in the measurement of soil suction. M.Sc. dissertation, Imperial College, London.

El-Ehwany, M. & Houston, S. L. (1990). Settlement and moisture movement in collapsible soils. *Journal of Geotechnical Engineering*, 116(10): 1521–1535.

Fares, A., Deb, S. K., & Fares, S. (2009). Review of vadose zone soil solution sampling techniques. *Environmental Reviews*, 17: 215–234. https://doi.org/10.1139/A09-010

Fawcett, R. G. & Collis-George, N. (1967). A filter-paper method for determining the moisture characteristics of soil. *Australian Journal of Experimental Agriculture and Animal Husbandry*, 7: 162–167.

Feng, M. & Fredlund, D. G. (2003). Calibration of thermal conductivity sensors with consideration of hysteresis. *Canadian Geotechnical Journal*, 40: 1048–1055.

Fredlund, D. G. (1992). How negative can pore-water pressures get? *Geotechnical News*, 9(3): 44–46.

Gardner, R. (1937). A method of measuring the capillary tension of soil moisture over a wide moisture range. *Soil Science*, 43: 277–283.

Greacen, E. L., Walker, G. R., & Cook, P. G. (1987). Evaluation of the filter paper method for measuring soil water suction. *International Conference on Measurement of Soil and Plant Water Status*, pp. 137–143.

Greenspan, L. (1977). Humidity fixed points of binary saturated aqueous solutions. *Journal of Research of the National Bureau of Standards Section A: Physics and Chemistry*, 81A(1): 89–96.

Gutierrez, C. I. G. (1985). *Assessment of filter paper technique for the measurement of soil water suction*, M.Sc. dissertation, Imperial College, London.

Hamblin, A. P. (1981). Filter-paper method for routine measurement of field water potential. *Journal of Hydrology*, 53: 355–360.

Hansen, H. C. (1926). The water-retaining power of the soil. *Journal of Ecology*, 14: 92–111.

Harrison, B. A. & Blight, G. E. (1998). The effect of filter paper and psychrometer calibration techniques on soil suction measurements. In *Proceedings of the Second International Conference on Unsaturated Soils*, International Academic Publishers, Beijing, China, Vol. 1, pp. 362–367.

He, Y., DeSutter, T., Hopkins, D., Jia, X., & Wysocki, D. A. (2013). Predicting ECe of the saturated paste extract from value of EC1:5. *Canadian Journal of Soil Science*, 93(5): 585–594. https://doi.org/10.4141/cjss2012-080

Hillel, D. (1980). *Fundamentals of soil physics*. San Diego, CA: Academic Press, p. 403.

Hogg, T. J. 7 Henry, J. L. (1984). Comparison of 1:1 and 1:2 Suspensions and extracts with the saturation extract in estimating salinity in Saskatchewan soils. *Canadian Journal of Soil Science*, 64(4): 699–704. https://doi.org/10.4141/cjss84-069

Houston, S. L., Houston, W. N., & Wagner, A. N. 1994. Laboratory filter paper suction measurements. *Geotechnical Testing Journal*, GTJODJ, 17(2): 185–194.

Hu, Y., Vu, H., & Fredlund, D. G. (2007). Calibration of thermal conductivity sensors for suction measurement. In *60th Canadian Geotechnical Conference & 8th Joint CGS/IAH: CNC Groundwater Conference* (Ottawa Geo 2007) [Proceedings] (21 October 2007): 1–17.

Iyer, B. (1990). Pore water extraction – Comparison of saturation extract and high-pressure squeezing. In K. B. Hoddinott & R. O. Lambs (Eds.), *Physico-chemical aspects of soils and related materials*, ASTM STP 1095. Philadelphia: American Society for Testing and Materials, pp. 159–170.

Khan, S. U. (1967). *Some aspects of equilibrium solutions for a solonetzic and chernozenie soil*, Ph.D. thesis, Department of Soil Science, University of Alberta, Edmonton.

Krahn, J. & Fredlund, D. G. (1972). On total, matric and somotic suction. *Journal of Soil Science, 114*(5): 339–348.

Kriukov, P. A. & Manheim, F. T. (1982). Extraction and investigative techniques for study of interstitial waters of unconsolidated sediments: A review. In K. A. Fanning & F. T. Manheim (Eds.), *The dynamic environment of the ocean floor*. Toronto: LexingtonBooks, pp. 3–26.

Kryukov, P. A. & Komarova, N. A. (1956). Issledovanie rastvorov pochv, ilov i gornykh porod (Studies on the interstitial waters of soils, muds and rocks). Mezhdunarodnomu kongressu pochvovedov, 2nd Komissiya, Doklady. 6: 151–184.

Lee, H. C. & Wray, W. K. 1992, Evaluation of soil suction instruments. In *Proceedings of the 7th International Conference on Expansive Soils*, Dallas, USA, pp 307–312.

Leong, E. C., Kizza, R., & Rahardjo, H. (2016). Measurement of soil suction using moist filter paper. In *3rd European Conference on Unsaturated Soils - E-UNSAT 2016*, Delage, P., Cui, Y. J., Ghabezloo, S., Pereira, J. M., and Tang, A. M. (Eds.), E3S Web of Conferences. https://doi.org/10.1051/e3sconf/20160910012

Leong, E. C., Tripathy, S., & Rahardjo, H. (2003). Total suction measurement of unsaturated soils with a device using the chilled-mirror dew-point technique. *Geotechnique, 53*(2): 173–182.

Leong, E. C., Widiastuti, S., Lee, C. C., & Rahardjo, H. (2007). Accuracy of suction-measurement. *Geotechnique, 57*(6): 547–556.

Leong, E. C., Wijaya, M., Tong, W. Y., & Lu, Y. (2020). Examining the contact filter paper method in the low suction range. *Geotechnical Testing Journal, 43*(6): 1567–1573. https://doi.org/10.1520/GTJ20190237

Leong, E. C., Zhang, X. -H. and Rahardjo, H. (2011). Calibration of a thermal conductivity sensor for field measurement of matric suction. *Geotechnique, 61*(1): 81–85. https://doi.org/ 10.1680/geot.9.P.008.

Likos, W. J. & Lu, N. (2002). Filter paper technique for measuring total soil suction. *Transportation Research Record*, 1786(1): 120–128. https://doi.org/10.3141/1786-14

Malazian, A., Hartsough, P., Kamai, T., Campbell, G. S., Cobos, D. R., & Hopmans, J. W. (2011). Evaluation of MPS-1 soil water potential sensor. *Journal of Hydrology*, 402(1–2): 126–134, https://doi.org/10.1016/j.jhydrol.2011.03.006

Manheim, F. T. (1966a). A hydraulic squeezer for obtaining interstitial water from Consolidated and unconsolidated sediments. U. S. Geol. Surv. Profess. Paper 550-C, C256.

Manheim, F. T. (1966b). A Hydraulic Squeezer for Obtaining Interstitial Water from Consolidated and Unconsolidated Sediment. U.S. Geological Survey Professional Paper 550-C, pp. 256–261.

Marinho, F. & Oliveira, O. (2006)., The filter paper method revisited. *Geotechnical Testing Journal*, 29(3): 1–9.

Marinho, F. A. M. (1994). *Shrinkage behaviour of some plastic clays*, Ph.D. thesis, University of London.

McCann, I. R., Kincaid, D. C., & Wang, D. (1992). Operational characteristics of the Watermark Model 200 soil water potential sensor for irrigation management. *Applied Engineering in Agriculture* 8(5): 605–609. ASAE, 2950 Niles Rd., St. Joseph, MI 49085

McKeen, R. G. (1980). Field studies of airport pavements on expansive soils. In *4th International Conference on Expansive Soils*, pp. 242–261.

McKeen, R. G. (1985). Validation of procedures for pavement design on expansive soil behaviour. Report No. DOT/FAA/PM-85/15. Washington, DC, USA.

McKeen, R. G. (1988). Soil characterization using suction measurements. In *25th Proceedings of Paving and Transp. Conference*, University of Mexico, Albuquerque, N.M.

McQueen, I. S. & Miller, R. F. (1968a). Determination of a soil moisture potential. In Water in the unsaturated zone: Proceedings of the Wageningen Symposium, P. E. Rijtema & H. Wassink (Eds.), International *Association of Science and Hydrology Publication* No. 82, pp. 147–155.

McQueen, I. S. & Miller, R. F. (1968b). Calibration of a wide-range gravimetric method for measuring moisture stress. *Soil Science*, 106(3): 225–231.

Miller, D. J. & Nelson, D. (1992). Osmotic suction as a valid stress state variable in unsaturated soils. In *Proceedings of the 7th International Conference on Expansive Soils*, Dallas, USA, pp. 179–184.

Moncur, M. C., Blowes, D. W., & Ptacek, C. J. (2013). Pore-water extraction from the unsaturated and saturated zones. *Canadian Journal of Earth Sciences*, 50(10): 1051–1058. https://doi.org/10.1139/cjes-2012-0165

Noborio, K., Horton, R., & Tan, C. S. (1999). Time domain reflectometry probe for simultaneous measurement of soil matric potential and water content. *Soil Science Society of America Journal*, 63: 1500–1505.

Or, D. & Wraith, J. M. (1999). A new soil matric potential sensor based on time domain reflectometry. *Water Resources Research*, 35: 3399–3407.

Power, K. C., Vanapalli, S. K., & Garga, V. K. (2008). A revised contact filter paper method. *Geotechnical Testing Journal*, 31(6): 461–469. https://doi.org/10.1520/GTJ101099

Rhoades, J. D. (1982). Soluble salts. In Methods of soil analysis, Part 2, Chemical and Microbiological Properties, 2nd ed., *ASA-SSSA, Agronomy Monograph*, 9: 167–179.

Ridley, A. M. & Wray, W. K. (1996). Suction measurement: A review of current theory and practices. In *First International Conference on Unsaturated Soils*, Rotterdam, the Netherlands: A. A. Balkema, pp. 1293–1322.

Ridley, A. M. (1993). *The measurement of soil moisture suction*, Ph.D. thesis, University of London.

Scanlon, B. R., Aandraski, B. J., & Bilskie, J. (2002). Miscellaneous methods for measuring matric or water potential. In J. H. Dane & G. C. Topp (Eds.), *Methods of soil analysis, part 4, physical methods*. Madison, WI: American Society of Agronomy, pp. 643–670.

Schröder, H. (1911). Über die selektiv permeable hülle des weizenkornes. *Flora*, 102: 186–208.

Sheets, K. R., & Hendrickx, J. M. H. (1995). Noninvasive Soil Water Content Measurement Using Electromagnetic Induction. *Water Resources Research*, *31*(10): 2401– 2409, https:/doi.org/10.1029/95WR01949.

Sibley, J. W. & Williams, D. J. (1990). A new filter material for measuring soil suction. *Geotechnical Testing Journal*, *13*(4): 381–384.

Sibley, J. W., Smyth, G. K., & Williams, D. J. (1990). Suction-moisture content calibration of filter papers from different boxes. *Geotechnical Testing Journal*, *13*(3): 257–262.

Spaans, E. J. A. & Baker, J. M. (1992). Calibration of Watermark soil moisture sensors for soil matric potential and temperature. *Plant Soil*, 143: 213–217.

Stocker, O. (1930). Über die Messung von Bodensaughkräften und ihren verhältnis Suden Wurzelsaughräften. *Zeitschrift für Botanik.*, *23*: 27–56.

Swarbrick, G. E. (1995). Measurement of soil suction using the filter paper method. In *Unsaturated Soils: Proceedings of the 1st International Conference on Unsaturated Soils*, A. A. Balkema, Rotterdam, Netherlands, pp. 653–658.

Thomson, S. J. & Armstrong, C. F. (1987). Calibration of the Watermark 200 soil moisture sensor. *Applied Engineering Agriculture*, *3*: 186–189.

Tripathy, S., Al-Khyat, S., Cleall, P. J. Baille, W., & Schanz, T. (2016). Soil suction measurement of unsaturated soils with a sensor using fixed-matrix porous ceramic discs. *Indian Geotechnical Journal*, *46*: 252–260. https://doi.org/10.1007/s40098-016-0200-z

van der Raadt, P., Fredlund, D. G., Clifton, A. W., Klassen, M. J., & Jubien, W. E. (1987). Soil suction measurement at several sites in Western Canada, Transportation Res. Rec. 1137, Soil Mech. Considerations in Arid and Semi-Arid Areas, Transportation Res. Board, Washington, DC, pp. 24–35.

Wagenet, R. J. & Jurinak, J. J. (1978). Spatial variability of soluble salt content in a mancos shale watershed. *Soil Science*, *126*: 342–349.

William, O. B. & Sedgley, R. H. (1965). A simplified filter paper method for determining the 15-atmosphere percentage in soils. *Australian Journal of Experimental Agriculture and Animal Husbandry*, *5*: 201–205.

FURTHER READING

Walshire, L., Berry, W., & Taylor, O.-D. (2020). Redesigned filter paper method: Protocol and assessment using reconstituted samples. *Geotechnical Testing Journal*. https://doi.org/10.1520/GTJ20190420

Chapter 10

Soil-water characteristic curve

10.1 BACKGROUND

The soil-water characteristic curve (SWCC) relates the water content of a soil with its suction(s). The first SWCC was plotted in soil physics by Edgar Buckingham in 1907 for six soils ranging in texture from sand to clay. The SWCC has since become an important relationship in many other disciplines and as such there are many alternative names for the SWCC (Ng and Menzies 2007; Fredlund et al. 2001):

1. Soil-water retention curve (SWRC)
2. Soil-moisture retention curve
3. Moisture retention curve
4. Water retention curve
5. Suction-water content relationship
6. Retention curve

In this book, the term "SWCC" is preferred, as it is more commonly used in geotechnical engineering (Fredlund et al. 2001). The SWCC is considered one of the most important properties of unsaturated soils, as it can be used to estimate the effect of negative pore-water pressure on shear strength, permeability and compressibility of unsaturated soils. In order to guide the engineer in designing the test programme for SWCC, there are several questions which need to be considered:

1. Does the analysis require only drying SWCC, wetting SWCC or both drying and wetting SWCC, along with its scanning curves?
2. What is the soil state (i.e., void ratio, density, water content, etc) which is needed in the analysis?
3. Does the soil have bimodal pore-size distribution?
4. Does the soil have significant change in volume during the test?
5. What is the suction range of interest?

DOI: 10.1201/b22304-10

6. What is the available equipment to measure matric suction and volume change?
7. What is the time constraint in obtaining the SWCC?

By answering all of these questions, suitable tests can be performed to obtain the appropriate SWCC.

10.2 RELATED STANDARDS

1. ASTM-D6836-02. 2008. Standard Test Method for Determination of Soil Water Characteristic Curve for Desorption Using Hanging Column, Pressure Extractor, Chilled Mirror Hygrometer, or Centrifuge. ASTM Book of Annual Standards.
2. BS EN ISO 11274. 2014. Soil Quality. Determination of the Water-Retention Characteristic. Laboratory Methods.

10.3 SWCC CONVENTION

To date, many SWCCs have been determined. However, the SWCC presented takes many forms. In determining SWCC, it is not sufficient to ask, "What is the SWCC of soil?" without giving further information on "which" SWCC is required. Generally, SWCC is plotted with volumetric water content on the y-axis and suction on the x-axis. The following SWCC descriptions should be adopted for clarity:

1) Water content – gravimetric water content based SWCC (w-SWCC), volumetric water content based SWCC (θ-SWCC) and degree of saturation based SWCC (S-SWCC)
2) Volume measurement – SWCC without volume measurement (w-SWCC) and SWCC with volume measurement (θ-SWCC and S-SWCC)
3) Path – initial drying, main drying, main wetting and scanning SWCC
4) Shape – unimodal and bimodal (or multimodal) SWCC

Aside from the general description, it is also important to specify the initial condition of the soil, such as initial dry density, initial water content, initial soil suction and the soil structure. Incorrect initial soil conditions may give a misleading SWCC.

Standard parameters that can be obtained from the SWCC are as follows (Fredlund et al. 2001; Ng and Menzies 2007):

1. Initial/saturated gravimetric water content (w_s), volumetric water content (θ_s), and degree of saturation ($S_{r,s}$)
2. Air-entry value (AEV) which is only obtainable directly from S-SWCC

3. Residual matric suction (s_r) and residual water content (w_r)
4. Suction at the intersection points
5. Slope of each linear segment of the curve
6. Inflection point suction and water content (s_{inf} and w_{inf}, respectively)
7. Curve-fitting equation and parameters

The x-axis of SWCC represents the soil suction which can be either matric suction or total suction. At low suction, the x-axis of the SWCC is matric suction and at high suctions where the air phase becomes continuous, the operative suction is total suction. The transition suction where this occurs depends on the soil type. However, it is quite common to not differentiate the suction into matric or total suction. In soil physics- and agronomy-related disciplines, the unit used for soil suction is pF which is equal to the logarithm to the base of ten of the absolute value of the column height of water in centimetres that equilibrates with the suction of the soil (Fredlund et al. 2001). However, in geotechnical engineering practice, stress unit (i.e., kPa) is more common and relevant. The suction range of interest can be as low as 0.1 kPa up to around 1 GPa. Thus, it is quite common to use log scale for the suction x-axis.

10.4 THEORY

The typical shape of the SWCC is sigmoid, as shown in Figure 10.1. Such an SWCC is now more commonly known as unimodal SWCC as opposed to bimodal or multimodal SWCC which consists of more than one sigmoid.

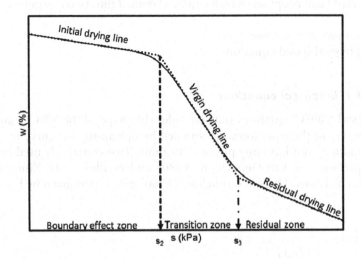

Figure 10.1 Typical shape of SWCC.

The unimodal drying SWCC (desorption SWCC) can be discretised into three linear segments and zones (Wijaya and Leong 2016; Kohgo 2003) which are the initial drying line (boundary effect zone), virgin drying line (transition zone) and residual drying line (residual zone). The suction at the intersection of initial drying and virgin drying lines (s_2) is commonly referred to as AEV for a soil that does not change in volume on drying as s_2 is the same for w-SWCC, θ-SWCC and S-SWCC. However, for a soil that experiences a change in volume on drying (e.g., most fine-grained soils), s_2 is AEV only for S-SWCC (Fredlund et al. 2011; Wijaya et al. 2015). The suction at the intersection of virgin and residual drying lines (s_3) is commonly referred to as residual matric suction (s_r) and the corresponding water content as the residual water content (w_r). Similarly, the unimodal wetting SWCC (adsorption SWCC) can be discretised into three linear segments which are starting from the high suction, initial wetting line, virgin wetting line and residual wetting line. The suction at the intersection of the initial and virgin wetting lines is commonly referred to as water-entry value (WEV).

To obtain S-SWCC and θ-SWCC, the volume of the soil specimen at each suction equilibrium is required. When volume measurement is difficult to carry out during the SWCC test, it is possible to use a shrinkage curve to replace the volume measurement.

10.4.1 SWCC equations

The SWCC test provides discrete data points where each data point represents a pair of suction and the corresponding equilibrium water content. However, a continuous relationship between suction and water content is more desirable for application. Hence, a number of unimodal SWCC equations have been proposed which can be classified into two categories:

1. Empirical equations
2. Physical-based equations

10.4.1.1 Empirical equations

Empirical SWCC equations only consider the shape of the SWCC, and the parameters of the equations are obtained by optimising the curve fit to the data and may not have any physical meaning. Two commonly used empirical equations are van Genuchten (1980) and Fredlund and Xing (1994) SWCC equations. The van Genuchten (1980) equation is given by Equation 10.1.

$$w(s) = \frac{w_s - w_r}{\left[1 + (a.s)^n\right]^m} + w_r \tag{10.1}$$

The Fredlund and Xing (1994) SWCC equation is given by Equation 10.2.

$$w = w_{sat} \frac{C(s)}{\left\{ \ln \left[\exp(1) + \left(\dfrac{s}{a_f} \right)^{n_f} \right] \right\}^{m_f}} \qquad (10.2a)$$

$$C(s) = 1 - \frac{\ln\left(1 + \dfrac{s}{\Psi_r} \right)}{\ln\left(1 + \dfrac{10^6}{\Psi_r} \right)}, \qquad (10.2b)$$

where w_{sat} is the saturated gravimetric water content, s is the matric suction, C(s) is a correction function to force the SWCC to have zero water content at $\Psi = 1$ GPa, Ψ_r is an empirical parameter which represents the residual matric suction s_r, a_f, n_f and m_f are the curve-fitting parameters. More empirical equations are shown in Table 10.1.

Table 10.1 Unimodal SWCC empirical equation

Reference	Equation	Input parameter
van Genuchten (1980)	$w(s) = \dfrac{(w_s - w_r)}{\left[1 + (a_v s)^{n_v} \right]^{m_v}} + w_r$	General notation: w_s = saturated water content w_r = residual water content s = soil suction
Brooks and Corey (1964)	For $s < s_2$: $w(s) = w_s$ For $s \geq s_2$ $w(s) = \left(\dfrac{s_2}{s} \right)^{a_b}$	a_v, n_v, m_v = van Genuchten (1980) empirical parameters s_2 = suction at the intersection between initial drying line and virgin drying line
Fredlund and Xing (1994)	$w(s) = \dfrac{w_s C(s)}{\left\{ \ln \left[\exp(1) + \left(\dfrac{s}{a_f} \right)^{n_f} \right] \right\}^{m_f}}$ $C(s) = 1 - \dfrac{\ln\left(1 + \dfrac{s}{\psi_r} \right)}{\ln\left(1 + \dfrac{10^6}{\psi_r} \right)}$	a_b = Brooks and Corey (1964) empirical parameter a_f, n_f, m_f, ψ_r = Fredlund and Xing (1994) empirical parameters b_d, c, d_d = Feng and Fredlund (1999) empirical parameters
Feng and Fredlund (1999)	$w(s) = \dfrac{w_s b_d + c \cdot s^{d_d}}{b_d + s^{d_d}}$	

Although empirical equations are convenient for use, there are several disadvantages (Wijaya and Leong 2016):

1. The parameters are determined by optimising the curve fit to the data. Sometimes, optimisation does not work without constraining the range of the parameters.
2. Different parameters in the equation may give the same SWCC which causes difficulties when relating the non-unique parameters to other soil properties.

10.4.1.2 Physical-based equations

Physical-based equations use parameters obtained directly from the curve. Thus, the parameters of physical-based equations are unique. Compared to empirical equations, physical-based equations are more complicated in form and cumbersome to use.

Examples of physical-based equations proposed for unimodal and bimodal SWCCs are given in Table 10.2 and Table 10.3, respectively. More details of bimodal SWCCs are given later in the chapter.

Wijaya and Leong (2016) proposed a physical-based equation that can be applied to both unimodal and bimodal SWCCs. The proposed equation is shown in Equation 10.3.

$$w(s) = w_{sat} + m_1 \log\left(\frac{s}{s_1}\right) + \sum_{i=2}^{n} T_i\left(\log s, \log s_i, m_i, m_{i-1}, c_i\right) \qquad (10.3a)$$

$$T_i\left(x, x_i, m_i, m_{i-1}, k_i\right) = \left(m_i - m_{i-1}\right)\frac{1}{2}\left\langle\left(x - x_1\right) + \frac{1}{k_i}\ln\left\{\frac{\cosh\left[k_i\left(x - x_i\right)\right]}{\cosh\left[k_i\left(x_i - x_0\right)\right]}\right\}\right\rangle \qquad (10.3b)$$

$$c_i = \frac{2}{\log\left(\frac{s_{i+}}{s_{i-}}\right)} \qquad (10.3c)$$

where m_i is the slope of segment i, s_i is the intersection between segment i and segment i-1, c_i is the curvature parameter that controls the curve joining segment i and segment i-1, s_{i+} is the point of convergence between the curve and segment i, s_{i-} is the point of convergence between the curve and segment i-1 and s_1 represents the minimum x-axis value and can be simply taken as 0.1 or 1 kPa. All m_i and s_i values can be determined directly from straight lines drawn as linear approximations of the curve.

Table 10.2 Unimodal SWCC equation that uses the unique parameters approach (Wijaya 2017)

Reference	Equation	Input parameter
Pham and Fredlund (2008)	$w = S_2 f_2 + S_3 + m_3 \log\left(\dfrac{10^6}{s}\right)$ $f_i\left(s, s_i, c_{pi}\right) = f_i = \dfrac{s_i^{c_{pi}}}{s_i^{c_{pi}} + s^{c_{pi}}}; g_i\left(s, s_i, c_{pi}\right) = g_i = 1 - f_i\left(s, s, c_{pi}\right)$ $f'_i\left(s, s_i, c_{pi}, m_{i-1}, m_i\right) = \dfrac{df_i\left(s, s_i, c_{pi}\right)}{\log s} \alpha = f'_i = f_i \cdot g_i \left(m_{i-1} - m_i\right) \dfrac{\ln(10)}{2 c_{pi}}$ $S_i = f_i \left(m_i - m_{i-1}\right) \log\left(\dfrac{s}{s_2}\right) + f'_i$ When it is desired force $w = 0$ at $s = 10^6$ kPa $m_3 = \dfrac{w_s + \left(m_2 - m_1\right) \log s_2 - m_2 \log s_3}{\log\dfrac{10^6}{s_3}}$	General notation: s = matric suction s_1 = matric suction at w_{sat} For Unimodal SWCC: s_2 = matric suction at the intersection point of segments 1 and 2 s_3 = matric suction at the intersection point of segments 2 and 3 s_{min} = matric suction at zero water content w_s = saturated water content m_1 = slope of segment 1 m_2 = slope of segment 2 m_3 = slope of segment 3
Gould et al. (2012)	$w = w_{sat} + S_2 + S_3 + \dfrac{m_1 + m_3}{2} \log\dfrac{s}{s_{min}}$ $f\left(x, y, c_{Gi}\right) = c_{Gi} \log\dfrac{x}{y} \tan^{-1}\left(c_{Gi} \log\dfrac{x}{y}\right) - \dfrac{1}{2} \ln\left[1 + \left(c_{Gi} \log\dfrac{x}{y}\right)^2\right]$ $S_i = -\dfrac{m_{i-1} - m_i}{c_{Gi}\pi}\left[f\left(s, s_i, c_{Gi}\right) - f\left(s_{min}, s_i, c_{Gi}\right)\right]$	Additional parameter for Pham and Fredlund (2008): c_{p2} = parameter that controls the curvature at s_2 c_{p3} = parameter that controls the curvature at s_3 Additional parameters for Gould et al. (2012): c_{G2} = parameter that control the curvature at s_2 c_{G3} = parameter that control the curvature at s_3

Table 10.3 Bimodal SWCC equations that used unique parameters approach (Wijaya 2017)

Reference	Equation	Input parameter
Gitirana and Fredlund (2004)	$$w = f_{n+1} + \sum_{i=2}^{n} \frac{f_i - f_{i+1}}{1 + \left(\dfrac{s}{\sqrt{s_i \cdot s_{i+1}}}\right)^{d_i}}$$ $$f_i = \frac{\tan\theta_i\left(1 + r_i^2\right)\ln(s/s_i)}{\left(1 - r_i^2 \tan^2\theta_i\right)} +$$ $$(-1)^{i-1} \times \frac{\left(1+\tan^2\theta_i\right)}{\left(1 - r_i^2\tan^2\theta_i\right)}\sqrt{r_i^2 \ln^2(s/s_i) + \frac{a^2\left(1 - r_i^2\tan^2\theta_i\right)}{1+\tan^2\theta_i}} + w_i$$ $$d_i = 2\exp\left[\frac{1}{\ln(s_{i+1}/s_i)}\right]; \quad r_i = \tan\left(\frac{\lambda_{i-1} - \lambda_i}{2}\right);$$ $$\theta_i = -\frac{\lambda_{i-1} + \lambda_i}{2} \quad \lambda_i = \arctan\left[\frac{\left(w_i - w_{i+1}\right)}{\ln(s_{i+1}/s_i)}\right]$$ Unimodal equation one bending point; i = 2 For two bending point (n = 2)and bimodal SWCC (n = 4):	General notation: AEV = air-entry value Ψ_r = curve-fitting parameter which represents residual matric suction sm_{-i} = matric suction at inflection point of segment i s = matric suction s_1 = Matric suction at w_{sat} For Bimodal SWCC: s_2 = matric suction at the intersection point of segments 1 and 2 s_3 = matric suction at the intersection point of segments 2 and 3 s_4 = matric suction at the intersection point of segments 3 and 4 s_5 = matric suction at the intersection point of segments 4 and 5 $s_6 = s_{min}$ = s at zero water content $w_1 = w_{sat}$ w_2 = w at s_2 w_3 = w at s_3 w_4 = w at s_4 w_5 = w at s_5 $w_6 = 0$

Satyanaga et al. (2013)	$$w = \left\{ w_r + \sum_{i=2}^{n} (w_{i-1} - w_{i+1}) \left\langle erf \left[\frac{\ln\left(\dfrac{s_i - s}{s_i - s_{m-i}}\right)}{s_{si}} \right] \right\rangle \right\} \left[1 - \frac{\ln\left(1 + \dfrac{s}{\Psi_r}\right)}{\ln\left(1 + \dfrac{10^6}{\Psi_r}\right)} \right]$$ $$erf = 1 - \int_{-\infty}^{x} \frac{1}{\sqrt{2\pi}} \exp\left(-\frac{x^2}{2}\right) dx$$ For unimodal SWCC n = 2 while for bimodal SWCC n = 3

Li (2009) or Li et al. (2014)	$$w(s) = (w_{sat} - w_3)\frac{\sqrt{s_2 s_3}}{s}^{n_1/\log\left(s_3/s_2\right)}\frac{\sqrt{s_2 s_3}}{s^{n_1/\log\left(s_3/s_2\right)} + \sqrt{s_2 s_3}}$$ $$+ (w_{sat} - w_3)/\lambda_1\frac{(l_1 s_3)^{mL_1}}{s^{mL_1} + (l_1 s_3)^{mL_1}}$$ $$+ (w_4 - w_5)\frac{\sqrt{s_4 s_5}}{s}^{n_2/\log\left(s_5/s_4\right)}\frac{\sqrt{s_4 s_5}}{s^{n_2/\log\left(s_5/s_4\right)} + \sqrt{s_4 s_5}}$$ $$+ (w_4 - w_5)/\lambda_2\frac{(l_2 s_5)^{mL_2}}{s^{mL_2} + (l_2 s_5)^{mL_2}}$$ Additional constrain: $$(w_{sat} - w_3)(1 + \lambda_1) + (w_4 - w_5)(1 + \lambda_2) = w_{sat}$$

m_1 = slope of segment 1
m_2 = slope of segment 2
m_3 = slope of segment 3
m_4 = slope of segment 4
m_5 = slope of segment 5
a = additional curve-fitting parameter of Gitirana and Fredlund (2004)
S_{si} = additional curve-fitting parameter of Satyanaga et al. (2013)
$n_i, l_i, mL_i, \lambda_i$ = curve-fitting parameters of Li (2009) equation

10.4.2 Hysteresis

Soil on drying and wetting experiences hysteresis where the water content of the soil on the wetting path is less than the water content on the drying path at the same suction (Pham et al. 2005; Haines 1930; Hillel 2003; Ng and Menzies 2007; Fredlund et al. 2000; Klausner 1991). Figure 10.2b shows a schematic of the hysteresis phenomenon. Initial drying SWCC is obtained for a soil drying from the fully saturated condition to the dry condition beyond the residual suction. Main wetting SWCC is obtained for a soil wetting from the dry condition beyond the residual suction to zero suction condition. At zero suction, if the soil is dried again, it will give the main drying SWCC. The gap on the y-axis between the initial drying SWCC and the main drying SWCC is the residual air content and represents the voids in the soil that are occluded. In nature, it may not be possible for the occluded voids to become saturated. When the soil is wetted (or dried) at any point on the main drying (wetting) SWCC, it will give the scanning curve. The main drying and main wetting SWCCs form the bounds enclosing the zone where scanning curves can occur. Thus, a soil undergoing multiple drying and wetting cycles may be on the main drying SWCC, main wetting SWCC or scanning curve. There are a number of hypotheses about the occurrence of hysteresis in SWCC (Hillel 2003; Klausner 1991):

1. Non-uniform pore-size distribution which caused irregularities of the void passages leading to the ink bottle effect.
2. The contact angle at an advancing interface during the wetting process is different from that at a receding interface during the drying process.

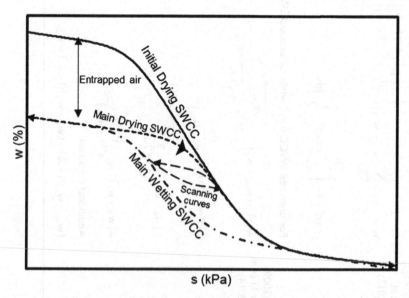

Figure 10.2 Drying and wetting SWCC shows hysteresis.

3. Entrapped air in the soil.
4. Non-reversible changes of volume during drying and wetting or aging.

Some hysteresis models for SWCC are presented in Table 10.4. Mualem's (1977) hysteresis model can be used to estimate the main wetting SWCC from the main drying SWCC and can be easily incorporated into any SWCC equation. However, the shape of the main wetting curve is predetermined regardless of the availability of the main wetting SWCC.

Feng and Fredlund's (1999) hysteresis model can be used when there are at least two data points to calibrate the parameters of the main wetting SWCC equation, while the extended Feng and Fredlund (1999) hysteresis model (Pham et al. 2005) can be used to estimate the main drying and the main wetting SWCC from the initial drying SWCC. Pham et al. (2005) conducted statistical studies on the hysteresis of SWCC and conclude that:

1. Occluded air bubble accounts for about 5%–15% of the degree of saturation and an average of 10% can be assumed. Thus, the degree of saturation where main drying and main wetting SWCC converge is about 90%.
2. The ratio of slopes of the main drying and main wetting SWCCs on a semilogarithmic plot R_{SL} is between 1 and 2, depending on soil type, as shown in Table 10.4.
3. The horizontal distance between the main virgin drying line and main virgin wetting line (D_{SL}) is between 0.2 and 0.5 log suction depending on soil type as shown in Table 10.4.

The preceding conclusions can be applied to any hysteresis model. However, most of the soils used by Pham et al. (2005) do not undergo volume change during drying and wetting. Hence, the conclusions may not be applicable to clay and highly expansive soils.

10.4.3 The effect of dry density on SWCC

As the water can only fill up the void space in a soil, the SWCC depends on the dry density of the soil. Wijaya and Leong (2017) show that the effect of dry density on the SWCC will only shift the initial portion of the w-SWCC downwards (Figure 10.3a) but will shift the entire S-SWCC to the right (Figure 10.3b).

Thus, the effect of density is easier to account in w-SWCC than in S-SWCC. For w-SWCC, the dry density only affects w_s of the initial drying line while the virgin drying line and residual drying line are unaffected. For soils which do not experience volume change due to a change in matric

Table 10.4 SWCC hysteresis model

Reference	Equation	Input parameter
Inverted Mualem (1977) equation (Pham et al. 2005)	$w^e = \left(\dfrac{w - w_{min}}{w_u - w_{min}} \right)$ $w_w^e = 1 - \left[1 - w_d^e(s) \right]^{1/2}$ Main drying curve: $w_d(s) = \dfrac{w_u b_d + c \cdot s^{d_d}}{b_d + s^{d_d}}$ Main wetting curve: $w_w(s) = \dfrac{w_u b_w + c \cdot s^{d_w}}{b_w + s^{d_w}}$ c and w_s are obtained from drying SWCC	General notation: w = water content (can be either gravimetric, volumetric or degree of saturation) S_i = degree of saturation of the initial drying curve S_d = degree of saturation of the main drying curve S_w = degree of saturation of the main wetting curve S_s = degree of saturation of the initial drying curve at the lowest suction (usually is 100%) S_u = degree of saturation that becomes the meeting point between main drying and main wetting curve w_d = water content at drying SWCC w_w = water content at wetting SWCC s = soil suction w^e = normalised water content w_{min} = meeting point between drying and wetting SWCC at higher soil suction w_u = meeting point between drying and wetting SWCC at lower soil suction
Feng and Fredlund (1999)	2 data points (s_{w1}, w_{w1}) and (s_{w2}, w_{w2}) are required to obtain b_w and d_w which can be defined as (Pham et al. 2003): $d_w = \dfrac{\log \left[\dfrac{(w_{w1} - c)(w_u - w_{w2})}{(w_u - w_{w1})(w_{w2} - c)} \right]}{\log \left(s_{w2} / s_{w1} \right)}$ $b_w = \dfrac{(w_{w1} - c) s_{w1}^{d_w}}{w_u - w_{w1}}$	w_d^e = normalised water content of the drying SWCC w_w^e = normalised water content of the wetting SWCC c = empirical parameter for Feng and Fredlund (1999) for both drying and wetting SWCC

Extended Feng and Fredlund (1999) equation (Pham et al. 2005)	**Initial drying curve:** $$S_i(s) = \frac{S_s b_d + c \cdot s^{d_d}}{b_d + s^{d_d}}$$ **Main drying curve:** $$S_d(s) = \frac{S_u b_d + c \cdot s^{d_d}}{b_d + s^{d_d}}$$ **Main wetting curve:** $$S_w(s) = \frac{S_u b_w + c \cdot s^{d_w}}{b_w + s^{d_w}}$$ **Additional relationship:** $$b_w = \left[\frac{b_d}{\left(10^{D_{SL}}\right)^{d_d}} \right]^{\frac{1}{R_{SL}}} \; ; R_{SL} = \frac{d_d}{d_w}$$ $$D_{SL} = \log\left(\frac{b_d^{1/2 d_d}}{b_w^{1/d_w}} \right)$$	b_d, d_d = empirical parameters for Feng and Fredlund (1999) for drying SWCC b_w, d_w = empirical parameters for Feng and Fredlund (1999) for wetting SWCC w_{w1} = first water content located at the wetting SWCC s_{w1} = suction of the w_{w1} w_{w2} = second water content located at the wetting SWCC s_{w2} = suction of the w_{w2} Recommended value for R_{SL} and D_{SL} (Pham et al. 2005) | Soil type	R_{SL}	D_{SL}
---	---	---
Sand	2	0.2
Sandy loam	2.5	0.25
Silt loam and clay loam	1.5	0.5
Compacted silt and compacted sand	1	0.35	 S_u is approximately $0.95 S_s$ to $0.85 S_s$, and $0.9 S_s$ can be taken as reasonable value (Pham et al. 2005)

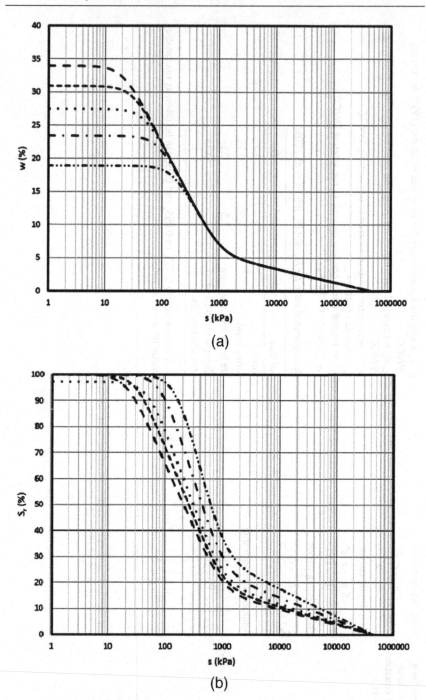

Figure 10.3 Effect of density on SWCC. (a) w-SWCC under different density. (b) S-SWCC under different density.

suction (coarse-grained soil), the S-SWCC can be obtained by using w-SWCC and initial void ratio e_0, as shown in Equation 10.4.

$$S = \frac{w \cdot G_s}{e_0} \tag{10.4}$$

However, for soils which experience volume change due to a change in matric suction, the shrinkage curve is needed to provide the equilibrium void ratio (e) at each suction to construct S-SWCC. Several equations have been proposed to account for density in the SWCC, as shown in Table 10.5.

Equation (10.3) can be extended to account for the effect of density on w-SWCC (Wijaya and Leong 2017) as shown in Equation 10.5a.

$$\begin{aligned}
w(s) &= w_{s,f} - m_1 \log \frac{s}{s_1} \\
&- T_2 \left(\log s, \log s_1, \log s_{2,f}, m_2, m_1, k_2 \right) \\
&- T_3 \left(\log s, \log s_1, \log s_3, m_3, m_2, k_3 \right)
\end{aligned} \tag{10.5a}$$

where $w_{s,f}$ is the saturated gravimetric water content post-density change, $s_{2,f}$ is the matric suction at the intersection between segment 1 and segment 2 post-density change and $s_{2,f}$ and $w_{s,f}$ can be obtained, respectively, using Equations 10.5b and 10.5c.

$$s_{2,f} = s_{2,0} \cdot 10^{\frac{w_{s,0} - w_{s,f}}{m_2 - m_1}} \tag{10.5b}$$

$$w_{s,f} = \left(\frac{S_{rf} e}{G_s} \right) \tag{10.5c}$$

where $w_{s,0}$ is the initial saturated gravimetric water content pre-density change, $s_{2,0}$ is the matric suction at the intersection between segment 1 and segment 2 pre-density change, S_{rf} is usually assumed to be 100% (which may not be due to the presence of occluded air voids), G_s is the specific gravity and e is the void ratio of the soil.

For coarse-grained soils such as sand, it is more convenient to express e in Equation (10.5c) in terms of relative density D_r, maximum void ratio e_{max} and minimum void ratio e_{min} as shown in Equation (10.5d). Both e_{max} and e_{min} can be obtained from laboratory tests following ASTM-D4254-14 (2014).

$$w_{s,f} = \left(\frac{S_r}{G_s} \right) \left[e_{max} \left(1 - D_r \right) + e_{min} D_r \right] \tag{10.5d}$$

Table 10.5 SWCC equations which account for density

Reference	Equation	Parameters
Gallipoli et al. (2003) – modified van Genuchten (1980) Equation	$$S_r = \left\{ \cfrac{1}{1 + \left[s\phi(e)^{\psi} \right]^{n_g}} \right\}^{m_g}$$	w = gravimetric water content w_{sat} = saturated gravimetric water content S_r = degree of saturation n_0 = initial porosity e = void ratio e_0 = initial void ratio s = matric suction s_r = residual matric suction s_p = matric suction of virgin drying line at 0 water content m_2 = slope of virgin drying line a_f, n_f, m_f = Fredlund and Xing (1994) curve-fitting parameters n_g, m_g = van Genuchten (1980) curve-fitting parameters A_s, B_s = Salager et al. (2010) curve-fitting parameters ϕ, ψ = Gallipoli et al. (2003) curve-fitting parameters A_t, B_t = Tarantino (2009) curve-fitting parameters ζ = Zhou et al. (2012) curve-fitting parameters
Tarantino (2009) – van Genuchten (1980) Equation	$$S_r = \left[1 + \left(\cfrac{e}{A_t} \right)^{1/B_t} s \right]^{-B_t/n_g}$$ ϕ and Ψ require additional SWCCs to be determined	
Salager et al. (2010) – modified Fredlund and Xing (1994) equation	$$w = \cfrac{w_{sat}}{\left\{ \ln \left[\exp(1) + (s/a_f)^{n_f} \right] \right\}^{m_f}} \left\{ 1 - \cfrac{\ln\left[1 + (s/s_r) \right]}{\ln\left[1 + (10^6/s_r) \right]} \right\}$$ $$m_f = 3.67\ln\left[\cfrac{w_{sat}}{m_2 \ln\left(\cfrac{s_p}{a_f} \right)} \right] \left\{ 1 - \cfrac{\ln\left[1 + \cfrac{a_f}{s_r} \right]}{\ln\left[1 + \cfrac{10^6}{s_r} \right]} \right\}$$	

$$n_f = 3.72 \frac{1.3^{m_f+1}}{\left\{1 - \dfrac{\ln\left[1+\dfrac{a_f}{s_r}\right]}{\ln\left[1+\dfrac{10^6}{s_r}\right]}\right\}^{m_f}} - \frac{m_2}{w_{sat}} - \frac{a_f}{1.31^{m_f}(a_f+s_r)\ln\left[1+\dfrac{10^6}{s_r}\right]}$$

$$a_f = A_s e_0^{B_s}$$

A_s and B_s require additional SWCCs to be determined

Zhou et al. (2012)

$$S_e = -\int \frac{S_e}{e_i}(1-S_e)^\zeta\, de_i$$

S_e is determined from other SWCC equations. Solution based on Simpson's rule is given as:

$$\ln\frac{e_f}{e_i} \approx \frac{S_{e,f}-S_{e,0}}{6}\left[f(S_{e,0})\right] + 4f\left(\frac{S_{e,0}+S_{e,f}}{2}\right) + f(S_{e,f})$$

$$f(x) = \frac{1}{x(1-x)^\zeta}$$

ζ requires additional SWCCs to be determined

Zhou et al. (2014) – modified Fredlund and Xing (1994) equation

$$S_r = 1 - \frac{\ln\left[1+(s/s_r)\right]}{\ln\left[1+(10^6/s_r)\right]}\left\{1 - \frac{\ln\left[1+(10^6/s_r)\right]}{\left\{\ln\left[\exp(1)+(s\cdot n_0/a_f)^{n_0 n_f}\right]\right\}^{n_0 n_f}}\right\}$$

$$w = G_s \frac{e}{\left\{\ln\left[\exp(1)+(s\cdot n_0/a_f)^{n_0 n_f}\right]\right\}^{n_0 n_f}}\left\{1 - \frac{\ln\left[1+(s/s_r)\right]}{\ln\left[1+(10^6/s_r)\right]}\right\}$$

10.4.4 Bimodal/multimodal SWCC

Some soils exhibit bimodal or multimodal SWCCs due to the presence of macropores and micropores in the soils. Such a phenomenon is commonly observed in soils with bimodal grain-size distribution (Zhang and Chen 2005; Satyanaga et al. 2013). In a soil with bimodal grain-size distribution, the arrangement of coarse grains and fine grains can lead to large pores (macropores) and small pores (micropores), respectively, creating a dual-porosity soil (Burger and Shackelford 2001; Zhang and Chen 2005). However, compaction or other features such as cracks in the soil may also give rise to dual porosity as well (Li 2009; Satyanaga et al. 2013; Li et al. 2014). Figure 10.4 shows a bimodal SWCC consisting of five segments where segments 1 and 2 are attributed to the macropores, while segments 3, 4 and 5 are attributed to the micropores. It is also possible for a SWCC to appear to be weakly bimodal due to small difference in sizes between micropores and macropores, inaccuracy of the SWCC test or use of different methods to obtain the SWCC. Thus, it is important to know a priori if the SWCC of a soil is unimodal or bimodal, as a bimodal SWCC requires smaller suction increments to capture the AEVs of the macropores and micropores accurately. Zou and Leong (2019) proposed a classification tree to determine if a soil with bimodal grain-size distribution has a unimodal SWCC or bimodal SWCC. Knowing a priori if the SWCC is unimodal or bimodal enables better design of the suction increments for the SWCC test. The classification tree is given in Figure 10.5a and the parameters used in the classification tree are defined in Figure 10.5b.

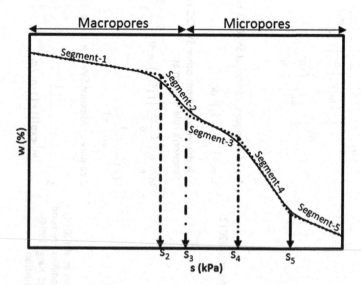

Figure 10.4 Bimodal SWCC.

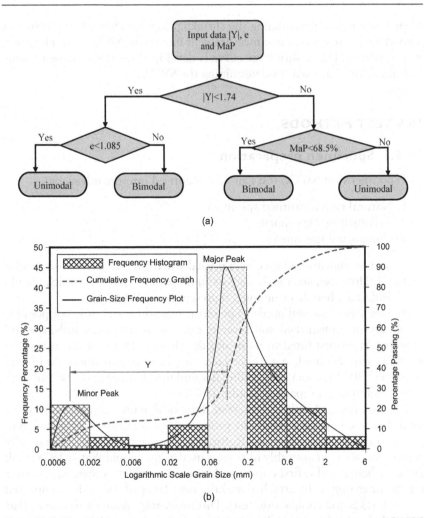

Figure 10.5 Classification tree and parameters for determining the type of SWCC
for a soil with bimodal grain-size distribution (from Zou and Leong, 2019).
(a) Classification tree. (b) Parameters Y and MaP (percentage for major peak)
for classification tree.

10.4.5 Using shrinkage curve as an alternative volume measurement

Measuring the volume of soil specimen during the SWCC test is sometimes
not possible due to several reasons:

1. Volume change of the specimen is non-uniform and the specimen
   becomes highly distorted making accurate volume measurement difficult.
2. Excessive handling of the specimen may cause disturbance to the specimen and may "damage" delicate specimen.

In such situations, determining the shrinkage curve independently provides a good alternative to volume measurement during the SWCC test. However, it is important that a shrinkage curve is developed for the specimen that is "identical" to that used for determining the SWCC.

10.5 TEST METHODS

10.5.1 Specimen preparation

Soil specimens for SWCC test can be categorised into the following:

1. Natural or undisturbed specimen
2. Reconstituted specimen
3. Compacted specimen

A natural or undisturbed specimen is used when SWCC of the in situ soil is required. However, such soils may exhibit spatial variability which should be considered when determining a representative SWCC.

In some cases, the soil needs to be reconstituted because it is not possible to obtain an undisturbed soil sample, e.g., sands. For such soils, the soil specimen is reconstituted to its in situ dry density. Different reconstitution methods can be used, e.g., air, water or vacuum pluviation (Vaid and Negussey 1988; Lagioia et al. 2006), dry and moist tamping (Frost and Park 2003) and under compaction (Ladd, 1973).

When compacted soil is of interest, static or dynamic compaction can be used. Static compaction tends to produce more uniform specimens compared to dynamic compaction. To further improve the quality of statically compacted soils, an assembly of rings is used as the mould (Satija 1978; Goh 2012), where a soil is first compacted into a ring which is then flipped over for the next ring to be attached and the next layer of soil to be compacted (see 6.4.3.3 Static compaction test). This multi-ring mould is to ensure that none of the soil layers are over-compacted. For dynamic compaction, it is difficult to obtain an undisturbed soil specimen by trimming a soil compacted in the standard mould. Hence, it is quite common to dynamically compact the soil directly into a specimen mould so that the soil specimen can be obtained without trimming. Different compaction methods, efforts and water contents will lead to different dry density and soil structure thus producing different soils and different SWCCs. Even when dry density is maintained constant, the soil structure/fabric is different and effectively, a different soil is obtained.

10.5.2 Test procedures

In general, the methods for determining the SWCC test can be broadly divided into two as follows:

Procedure A: Changing suction of the soil specimen and measuring its
    water content
Procedure B: Changing water content of the specimen and measuring its
    suction

In procedure A, a method to impose the suction is needed such as by using a
suction table, axis-translation apparatus (Figure 10.6), vapour equilibrium
technique or centrifuge.

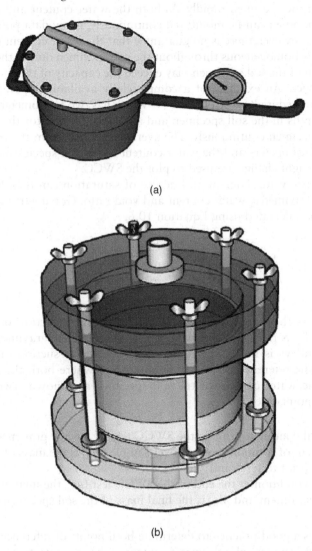

(a)

(b)

Figure 10.6 Axis-translation apparatuses. (a) Pressure plate apparatus. (b) Tempe cell.

Procedure A usually requires a longer time, as it is required to ensure that the soil specimen achieves equilibrium at the imposed suction (usually indicated by negligible changes in the weight of the specimen).

In procedure B, the water content of the soil specimen is changed for instance by evaporation (non-steady state or steady state via controlled relative humidity) and measuring its suction using devices such as tensiometer, null-axis-translation apparatus or chilled mirror hygrometer (Schindler 1980). It is faster than procedure A as evaporation can change the water content of the specimen rapidly. As both the water content and suction of the soil specimen can be monitored continuously, many data points can be obtained. However, there is no guarantee that the water content of the soil specimen is homogeneous throughout the soil specimen during the test, and the suction of the soil specimen may exceed the capacity of the suction measuring device. An example of a commercially available apparatus implementing procedure B is the HYPROP® which has two tensiometers installed at two depths in the soil specimen and a balance to monitor the weight of the soil specimen continuously. The averaged suction from the readings of the two tensiometers and the water content of the soil specimen calculated from its weight change are used to plot the SWCC.

Volumetric water content and degree of saturation can then be derived from the gravimetric water content and void ratio. Gravimetric water content can be calculated using Equation 10.6.

$$w_i(\%) = \left[\frac{M_i}{M_{ref}} \left(\frac{w_{ref}}{100} + 1 \right) - 1 \right] 100\% \tag{10.6}$$

where M_i is the weight of specimen at a particular suction or point of interest, M_{ref} is the reference mass, w_{ref} is the reference gravimetric water content and w_i is the water content at a particular suction or point of interest. The reference point can be any point where both the mass and gravimetric water content of the soil specimen are known. Two suitable reference points are

1. initial condition prior to the SWCC test (w_{ref} is the gravimetric water content of trimmings or left-over sample, M_{ref} is the mass of the specimen prior to test), and
2. final condition at the end of the SWCC test (w_{ref} the final gravimetric water content and M_{ref} is the final mass of the soil specimen).

It is always a good practice to determine both points of reference. The first point of reference enables the water content of the specimen to be calculated (and hence SWCC to be plotted) while the test is still ongoing, but it is less accurate as the water content was determined from trimmings or left-over sample compared to the second point of reference where water content of the actual soil specimen is determined. However, the second point of reference should be used to obtain the more accurate SWCC.

If volume measurement is performed during the test, the void ratio of the specimen can be calculated using Equation 10.7.

$$e_i = V_i \left(1 + \frac{W_{ref}}{100} \right) \frac{G_s}{M_{ref}} - 1 \tag{10.7}$$

where V_i is the volume of specimen at particular suction or point of interest, G_s is the specific gravity of the soil and e_i is the void ratio at particular suction or point of interest.

For most sandy and gravelly soils, the instantaneous total volume can be approximated by the initial total volume V_0, as most sandy and gravelly soils do not show volume change with change in matric suction. ASTM-D6836-02 (20) recommends five methods to determine the SWCC (A, B, C, D and E):

Method A: Reducing pore-water pressure (Hanging column)
Method B: Axis-translation principle with volumetric measurement (pressure plate)
Method C: Axis-translation principle with gravimetric measurement (pressure plate)
Method D: Adjusting water content and measuring the total suction (hygrometer)
Method E: Matric suction is applied by applying a centrifugal field (centrifuge)

BS EN ISO 11274:2014 recommends four methods to determine the SWCC (a, b, c, d):

1. Method a: Reducing pore-water pressure (suction table)
2. Method b: Reducing pore-water pressure (hanging column)
3. Method c: Axis-translation principle with gravimetric measurement (pressure plate)
4. Method d: Axis-translation principle with gravimetric measurement (pressure membrane)

In BS EN ISO 11274:2014, the results are expressed on a volume basis based on the initial volume of the soil specimen used in the test.

Preferably one method should be used to determine the complete SWCC of a soil. However, ASTM D6836-02 (2016) allows two or more methods (Methods A to E) to be used to obtain the SWCC of a soil. The range of suction applicable for each method is shown in Table 10.7.

10.5.3 Suction intervals

The recommended suction intervals in ASTM D6836-02 (2016) for Methods A to E are summarised in Table 10.8. The recommended suction intervals in ASTM D6836-02 (2016) may be excessive to determine a unimodal SWCC and insufficient to determine bimodal or multimodal SWCC. Zou and Leong (2019) recommend the suction intervals given in Table 10.9 to determine unimodal and bimodal SWCCs.

10.5.4 SWCC test on coarse-grained soils

SWCC test on coarse-grained soil is considerably fast, and the ceramic disk used to test a coarse-grained soil is usually only up to 100 kPa. The volume change of coarse-grained soils due to the change in suction is negligible and thus can be ignored. Since the volume change is negligible, there will be a negligible amount of water coming out from the specimen until the applied suction is higher than the AEV and the shapes of w-SWCC, θ-SWCC and S-SWCC are similar.

The test is commonly performed on a reconstituted specimen which makes it possible to repeat the test. The saturation of the specimen can be done by either sprinkling water from the top of the specimen or by using a burette to allow the water to flow up the specimen from the bottom of the specimen. The water level in the burette should be maintained equal to or slightly higher than the specimen height.

Coarse-grained soils have AEV as low as less than 1 kPa up to around 20 kPa, depending on the grain-size distribution and the density of the specimen. Thus, suction range of interest can be from 0.1 kPa up to 100 kPa. The suction interval can be very difficult to determine as the differences between AEV and s_r can be very small. An incorrect suction interval may give a misleading AEV. Figure 10.7 shows the same SWCC-w with different suction intervals. The AEV is located between 0.7 kPa and 1.3 kPa. Thus, accurate determination requires the suction interval to be around 0.1 kPa. Using a higher suction interval will lead to misleading information on the shape of the SWCC and the location of the AEV and s_r.

To accurately pinpoint the location of the AEV, a trial-and-error approach is recommended as follows:

1. Start by applying a higher suction interval (i.e., 1 kPa).
2. Once the AEV is exceeded, an obvious difference can be seen in the amount of water that flows out during the suction application.
3. The soil is then re-saturated, and 0.1 kPa suction is reapplied to remove excess water.
4. Increase suction up to the largest suction prior to AEV that has been applied from the previous test.

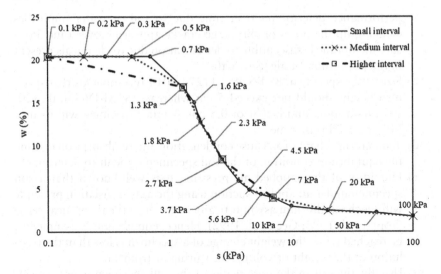

Figure 10.7 SWCC of sandy soils determined by using different suction intervals.

5. The third and following suction is by using a smaller interval (i.e., half of the previously used suction interval).
6. The test is repeated until a reasonably accurate AEV can be obtained.

10.5.5 SWCC test on fine-grained soils

SWCC test duration for fine-grained soil is much longer compared to that for coarse-grained soils. For fine-grained soils, AEV may range from around less than 100 kPa up to more than 1,000 kPa. Thus, suction range of interest can be from 1 kPa to 1,000,000 kPa.

However, it is also important to consider the fluctuation of suction at the site. In locations where the groundwater table is relatively shallow, the range of suction of interest can be quite low. On the other hand, when the ground-water level is quite deep, the range of suction of interest can be very high. Countries in temperate zones may have a very deep groundwater level in summer and a very shallow groundwater level in spring. Thus, a larger range of suction is expected.

10.5.6 Accuracy

The accuracy of the SWCC is dependent on a number of factors:

a) Condition of the soil.
b) Condition of the apparatus especially where ceramic plates or membranes are used. The ceramic plates may not be fully saturated or have minor defects not visible to the naked eye that may affect the test results. Generally, new ceramic plate performs well, but their

performance deteriorates over time due to soil or water impurities clogging up the pores or shrinkage cracks appearing over time. Pressure membrane is susceptible to deterioration due to chemicals present in the pore water or algae growth.

c) Size of the specimen. BS EN ISO 11274:2014 recommends that specimen height should not exceed 5 cm whereas ASTM-D6836-02 (20) specifies a minimum height of 0.5 cm. A taller specimen will incur a longer equilibrium time.

d) Temperature change can cause condensation and re-absorption making the equilibrium condition of the soil specimen difficult to determine.

e) Humidity of the supplied air pressure. It is well-known that a soil specimen under an applied suction using the axis-translation principle will reach equilibrium asymptotically or never in the theoretical sense (Topp et al. 1993; Gee et al. 1992). Hence, equilibrium is assumed to be reached when the weight change of a specimen is less than the resolution of the weight or volume measuring instrument.

f) Usually, the salt in the pore water of the soil specimen is not considered. Hence, the osmotic pressure is not normally considered when determining the SWCC.

g) Good capillary contact must always be maintained between the soil specimen and the suction table, ceramic plate or membrane. Poor contact results in water contents higher than expected for a given suction.

h) Air may become entrapped when wetting a soil specimen, resulting in a lower water content at a given suction.

10.6 ESTIMATION OF SWCC

Due to the required testing duration, it is desired to estimate SWCC by using basic soil properties. Grain-size distribution is commonly used as the basic information in estimating SWCC (and is commonly referred to as pedo-transfer function).

However, SWCC is not a unique curve, as it is affected by the initial void ratio of the soil or whether it is for the drying or wetting path. Thus, it is important to verify whether the estimation method which is used complies with the behaviour of SWCC.

In the following section, estimation methods such as pedo-transfer function and the one-point estimation method will be described.

10.6.1 Pedo-transfer function

Pedo-transfer function is used to estimate SWCC by using soil properties such as grain-size distribution. Examples of pedo-transfer function are given in Table 10.6.

Table 10.6 Pedo-transfer function

Reference	Equation	Parameters
Aubertin et al. (2003) MK Model	$S_r = S_c + (1-S_c)S_a^*$ $S_a^* = \min\left\{1, -0.5\left[\left((1-S_a) + \left((1-S_a)\right)\right]\right]^{1/1}\right\}$ $S_c = 1 - \left[\left(\left(\dfrac{h_{c0}}{h_s}\right)^2 + 1\right)^{-m} \exp\left[-m\left(h_{c0}/h_s\right)^2\right]\right]$ $S_a = a_c\left[1 - \dfrac{\ln\left(1 + h_s/h_r\right)}{\ln\left(1 + h_0/h_r\right)}\right]\dfrac{\left(h_{c0}/h_n\right)^{2/3}}{e^{1/3}\left(h_s/h_n\right)^{1/6}}$ $h_n = 1$ cm $h_0 = 10^7$ cm Coarse-grained soils: 0.75 $h_{c0}(cm) = \dfrac{1}{\left[1 + 1.17\log(C_U)\right]eD_{10}}$ $m = \dfrac{1}{C_U}; a_c = 0.01; h_r(cm) = 0.86h_{c0}^{1.2}$ $C_U = \dfrac{D_{60}}{D_{10}}$ For Fine-grained soils: $h_{c0}(cm) = \dfrac{0.15\rho_s}{e} LL^{1.45}$ $m = 0.0003; a_c = 0.0007$	S_r = degree of saturation, unitless S_c = saturation associated with the capillary component, unitless S_a = saturation associated with the adhesive component, unitless h_{c0} = equivalent capillary height which is related to an equivalent pore diameter and the solid surface area (cm) h_s = soil suction represented as a head (cm) h_r = residual soil suction represented as a head (cm) h_n = normalisation parameter (1 cm) h_0 = suction head represented as dry soil condition (cm) a_c = adhesion coefficient, unitless m = pore-size coefficient, unitless e = void ratio D_{10} = diameter corresponding to 10% passing on grain-size distribution curve D_{60} = diameter corresponding to 60% passing on grain-size distribution curve C_U = uniformity coefficient LL = Liquid limit (%) ρs = density of the soil solids (kg/m³)

Table 10.7 Suction range for different methods

ASTM-D6836-02 (2016)		BS EN ISO 11274:2014	
Method	Suction range (kPa)	Method	Suction range (kPa)
A	0 – 80	a	0–50
B	0 – 1,500	b	0–20
C	0 – 1,500	c	5–1,500
D	1,000 – 100,000	d	33–1,500
E	0 – 120		

Table 10.8 Summary of ASTM D6836-02 (2008) methods for determining SWCC

	Method A[1]	Methods B & C	Method D	Method E[2]
Suction level applied (kPa)	0.05	10	Not specified	0.5
	0.2	50		2
	0.4	100		8.5
	1	300		34
	2	500		120
	4	1000		
	6	1500		
	10			
	15			
	20			
	40			

1 CONVERTED FROM **5, 20, 40, 100, 200, 400, 600, 1000, 1500, 2000** AND **4000** MM OF WATER.
2 ESTIMATED FROM **angular velocities of 100, 200, 400, 800 and 1500 rpm and may vary with centrifuge.**

Figure 10.8 shows the ability of the Modified Kovacs (MK) Model (Aubertin et al. 2003) in estimating w-SWCC under different density by using sand specimens data from Wijaya and Leong (2017). Figure 10.8 shows that the MK Model gives a reasonable w-SWCC regardless of density despite the estimated w-SWCCs do not perfectly merge into a single virgin drying curve. It is important to note that the MK Model estimates the degree of saturation instead of gravimetric water content. Thus, directly converting S-SWCC into w-SWCC is only possible for coarse-grained soil or soil which does not change in volume when suction is changed. When the soil changes in volume, estimation of the shrinkage curve is required.

Table 10.9 Recommended suction intervals from Zou and Leong (2019)

Method	A		B & C		D		E*	
SWCC Type	Unimodal	Bimodal	Unimodal	Bimodal	Unimodal	Bimodal	Unimodal	Bimodal
Suction Level Applied (kPa)	0.2	0.2	1	1	500	500	0.2	0.2
	1	0.5	10	4	1,000	1,000	1	0.5
	10	1	100	10	10,000	2,000	10	1
	30	2	500	40	50,000	5,000	40	2
	80	5	1,500	100	100,000	10,000	120	5
		10		200		20,000		10
		30		500		50,000		50
		80		1,500		100,000		120

* Suction levels are suggestions only and depend on centrifuge.

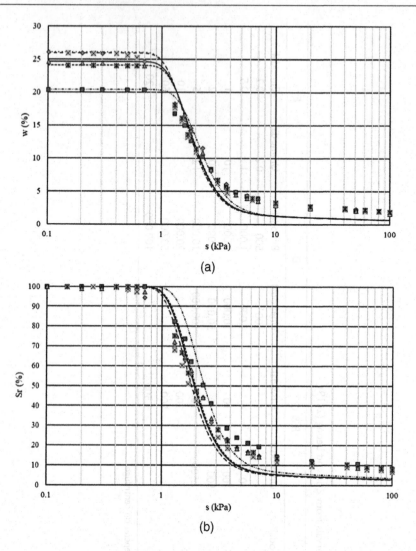

Figure 10.8 Comparison between estimated SWCC using MK Model (Aubertin et al., 2003) and SWCC data from Wijaya and Leong (2017). (a) w-SWCC of sand specimens under different density. (b) S-SWCC of sand specimens under different density.

10.6.2 One-point method

One-point method (e.g., Chin et al. 2010) is a "bridge" between a complete laboratory test and a pedo-transfer function. The problem with pedo-transfer function is its difficulty in estimating the correct SWCC, and it is possible for the entire estimated SWCC to be completely different from the actual SWCC. In the one-point method, only one experimental point of the

SWCC is needed (and thus, there is at least one point that is correct). Chin et al. (2010) recommend different equations to estimate Fredlund and Xing's (1994) equation parameters for fine-grained and coarse-grained soils based on the percentage of soil passing standard sieve No. 200 (P_{200}). Fine-grained soils are defined as those soils with $P_{200} \geq 30\%$.

For $P_{200} \geq 30\%$:

$$a_f = -2.4(x) + 722 \qquad (10.8a)$$

$$n_f = 0.07(x)^{0.4} \qquad (10.8b)$$

$$m_f = 0.015(x)^{0.7} \qquad (10.8c)$$

$$\Psi_r = 914\exp\left[-0.002(x)\right] \qquad (10.8d)$$

For $P_{200} < 30\%$

$$a_f = 0.53(D_{50})^{-0.96} \qquad (10.9a)$$

$$n_f = x \qquad (10.9b)$$

$$m_f = -0.23\ln(x) + 1.13 \qquad (10.9c)$$

$$\Psi_r = 100kPa, \qquad (10.9d)$$

where x is a curve-fitting parameter which ranges from 0 to 300.8 and is obtained by curve-fitting the experimental SWCC point. Chin et al. (2010) recommend the experimental SWCC point to be determined at suction of 100 kPa or 500 kPa (preferably 500 kPa) for fine-grained soils and 10 kPa for coarse-grained soils.

REFERENCES

ASTM-D4254-14 (2014). *Standard test methods for minimum index density and unit weight of soils and calculation of relative density.* West Conshohocken, PA: ASTM Book of Annual Standards.

ASTM-D6836-02 (2008). *Standard test method for determination of soil water characteristic curve for desorption using hanging column, pressure extractor, chilled mirror hygrometer, or centrifuge.* West Conshohocken, PA: ASTM Book of Annual Standards.

Aubertin, M., Mbonimpa, M., Bussière, B., & Chapuis, R. P. (2003). A model to predict the water retention curve from basic geotechnical properties. *Canadian Geotechnical Journal,* 40(6): 1104–1122.

Brooks, H. & Corey, A. T. (1964). *Hydraulic properties of porous media.* Fort Collins, CO: Colorado State University. Hydrology Paper No. 3.

BS EN ISO 11274 2014 *Soil quality. Determination of the water-retention characteristic. Laboratory methods.* UK: BSI.

Chin, K. B., Leong, E. C., & Rahardjo, H. (2010). A simplified method to estimate the soil-water characteristic curve. *Canadian Geotechnical Journal*, 47(12):1382–1400.

Feng, M. & Fredlund, D. G. (1999). Hysteretic influence associated with thermal conductivity sensor measurements. *Proceedings from Theory to the practice of Unsaturated Soils Mechanics in Association with the 52nd Canadian Geotechnical Conference and the Unsaturated Soil Group*, Regina, Sask 14: 14–20.

Fredlund, D. G., Rahardjo, H., Leong, E. C., & Ng, C. W. W. (2001). Suggestions and recommendations for the interpretation of soil-water characteristic curves. In *Proceedings of the Fourteen Southeast Asian Geotechnical Conference*, Hong Kong, vol. 1: pp. 503–508.

Fredlund, D. G., Shuai,F., & Feng, M. (2000). Use of a new thermal conductivity sensor for laboratory suction measurement. In *Proceedings of the Asian Conference in Unsaturated Soils, UNSAT ASIA 2000.* Singapore: A.A. Balkema, pp. 275–280.

Fredlund, D. G. & Xing, A. (1994). Equations for the soil-water characteristic curve. *Canadian Geotechnical Journal, 31*(4): 521–532.

Frost, J. & Park, J. (2003). A critical assessment of the moist tamping technique. *Geotechnical Testing Journal, 26*(1): 57–70. https://doi.org/10.1520/GTJ11108J

Gallipoli, D., Wheeler, S. J., & Karstumen, M. (2003). Modelling the variation of degree of saturation in a deformable unsaturated soil. *Géotechnique, 53*(1): 105–11.

Haines, W. B. (1930). Studies in the physical properties of soil – V:The hysteresis effect in capillary properties and the modes of water distribution associated therewith. *Journal of Agricultural Science, 20*(1): 97–116.

Hillel, D. (2003). *Introduction to environmental soil physics.* San Diego, CA: Academic Press, Elsevier Inc.

Klausner, Y. (1991). *Fundamentals of continuum mechanics of soils.* New York: Springer-Verlag.

Kohgo, Y. (2003). Review of constitutive models for unsaturated soils and initial-boundary value analysis. In *Proceedings of the 2nd Asian Conference on Unsaturated Soils.* Karube, D., Iizuka, A., Kato, S., Kawai, K., & Tateyama, K. (eds.) UNSAT-ASIA 2003, Osaka-Japan.

Ladd, R. (1978). Preparing test specimens using undercompaction. *Geotechnical Testing Journal, 1*(1): 16–23. https://doi.org/10.1520/GTJ10364J

Lagioia, R., Sanzeni, A., & Colleselli, F. (2006). Air, water and vacuum pluviation of sand specimens for the triaxial apparatus. *Soils Foundation, 46*(1): 61–67.

Li, X. (2009). Dual-porosity structure and Bimodal Hydraulic property functions for unsaturated coarse granular soils. In *Civil engineering.* Hong Kong University of Science and Technology, Hong Kong, vol. Ph.D.

Li, X., Li, J. H., & Zhang, L. M. (2014). Predicting bimodal soil-water characteristic curves and permeability functions using physically based parameters. *Computers and Geotechnics, 57*: 85–96.

Mualem, Y. (1977). Extension of the similarity hypothesis used for modeling the soil water characteristics. *Water Resources Research, 9*: 1324–1331.

Ng, C. W. W. & Menzies, B. (2007). *Advance unsaturated soil mechanics and engineering*. Abingdon, Oxon: Taylor and Francis.

Pham, H. Q., Fredlund, D. G., & Barbour, S. L. (2003). A practical hysteresis model for the soil-water characteristic curve for the soils with negligible volume change. *Géotechnique*, 53: 293–298.

Pham, H. Q., Fredlund, D. G., & Barbour, S. L. (2005). A study of hysteresis models for soil-water characteristic curves. *Canadian Geotechnical Journal*, 42(6): 1548–1568.

Salager, S., El Youssoufi, M. S., & Saix, C. (2010). Definition and experimental determination of a soil-water retention surface. *Canadian Geotechnical Journal*, 47(6): 609–622.

Satyanaga, A., Rahardjo, H., & Leong, E. C. (2013). Water characteristic curve of soil with bimodal grain-size distribution. *Computers and Geotechnics*, 48: 51–61. https://doi.org/10.1016/j.compgeo.2012.09.008.

Schindler, U. (1980). Ein Schnellverfahren zur Messung der Wasserleitfähigkeit im teilgesättigten Boden an Stechzylinderproben. *Archiv für Acker- und Pflanzenbauund Bodenkunde*, 24(1): 1–7 (In german).

Tarantino, A. (2009). A water retention model for deformable soils. *Géotechnique*, 59: 751–762.

Topp, G. C., Galganov, Y. T., Ball, B. C., & Carter, M. R. (1993). Soil water desorption curves. In M. R. Carter (Ed.), *Soil sampling and methods of analysis*. Ann Arbor, MI: Lewis Publishers, pp. 569–579.

Vaid, Y. & Negussey, D. (1988). Preparation of reconstituted sand specimens. In R. Donaghe, R. Chaney, & M. Silver (Eds.), *Advanced triaxial testing of soil and rock*. West Conshohocken, PA: ASTM International, pp. 405–417.

Van Genuchten, M. T. (1980). A closed-form equation for predicting the hydraulic conductivity of unsaturated soils. *Soil Science Society of America Journal*, 44(5): 892–898.

Wijaya, M. (2017). Compression, shrinkage and wetting-induced volume change of unsaturated soils. In *Civil and environmental engineering*. Singapore: Nanyang Technological University, vol. Ph.D.

Wijaya, M. & Leong, E. C. (2016). Equation for unimodal and bimodal soil-water characteristic curves. *Soils and Foundations*, 56(2): 291–300.

Wijaya, M. & Leong, E. C. (2017). Modelling the effect of density on the unimodal soil-water characteristic curve. *Géotechnique*, 67(7): 637–645.

Wijaya, M., Leong, E. C., & Rahardjo, H. (2015). Effect of shrinkage on air-entry value of soils. *Soils and Foundations*, 55(1): 166–180.

Zhou, A. N., Sheng, D., & Carter, J. P. (2012). Modelling the effect of initial density on soil-water characteristic curves. *Géotechnique*, 62: 669–680.

Zhou, W.-H., Yuen, K.-V., & Tan, F. (2014). Estimation of soil-water characteristic curve and relative permeability for granular soils with different initial dry densities. *Engineering Geology*, 179(0): 1–9.

Zou, L. & Leong, E. C. (2019). A classification tree guide to soil-water characteristic curve test for soils with bimodal grain -size distribution, *Geotechnical Engineering Journal of the SEAGS & AGSSEA*, 50(1): 28–36.

FURTHER READING

Wijaya, M. & Leong, E. C. (2017). Modelling the effect of density on the unimodal soil-water characteristic curve. *Géotechnique*, 67(7): 637–645.

Chapter 11

Permeability

Steady-state methods

11.1 BACKGROUND

Permeability is one of the important hydraulic properties of unsaturated soils. It governs seepage (transient and steady-state) and affects the magnitude and distribution of pore-water pressures in unsaturated soils. Water flow in unsaturated soils is only through the water phase, and, hence, the water phase must be continuous in order for flow in the liquid form to occur. The rate of water flow through saturated and unsaturated soils is governed by Darcy's law (Fredlund et al. 2012). In saturated soils, the coefficient of permeability K_s is a function of the void ratio e (Taylor 1948; Lame and Whitman 1979). However, the coefficient of permeability for unsaturated soil depends on any two of three possible volume-mass properties: void ratio e, gravimetric water content w and degree of saturation S (Lloret and Alonso 1980; Fredlund 1981).

When determining the permeability of unsaturated soils, it is important to review the various standards for determining permeability for saturated soils. A summary of the range of saturated coefficient of permeability in the standards is given in Table 11.1. Currently, ASTM D7664-10 is the only standard for measurement of unsaturated permeability of soils. The methods given in ASTM D7664-10 are suitable for determining permeability of unsaturated soils from the saturated coefficient of permeability to a permeability of 10^{-11} m/s. Hence, all the conditions specified in the other standards for determination of saturated permeability are equally applicable for unsaturated soils.

The methods described in ASTM D7664-10 are considered non-steady-state or transient methods which will be covered in the next chapter. In this chapter, the steady-state method of determining the permeability function of unsaturated soil is discussed. Such a test can be conducted using a rigid-wall permeameter or a flexible-wall permeameter. Because of the possibility of specimen shrinkage from the wall of the rigid-wall permeameter at high suctions, a flexible-wall permeameter is preferred. The triaxial cell offers a convenient way to modify it for permeability measurement of unsaturated soils. Such a system will be described in this chapter.

DOI: 10.1201/b22304-11

Table 11.1　Validity range for determination of saturated coefficient of permeability in standards

Standard	Range of saturated coefficient of permeability (m/s)
ASTM D2434 – 22	$> 10^{-7}$
ASTMD5084 – 16a	$10^{-11}-10^{-6}$
ASTM D5856-15	$< 10^{-5}$
BS 1377-5: 1990	$10^{-2}-10^{-5}$
BS 1377-6: 1990	Not explicitly mentioned
BS EN ISO 17892-11:2019	Full range

Table 11.2　Recommended maximum hydraulic gradient in standards

Hydraulic conductivity (m/s)	ASTM D5084 – 16a	ASTM D5856 – 15	ASTM D2434–22	BS 1377–5: 1990	BS 1377–6: 1990	BS EN ISO 17892-11:2019
> 1e-5	-	-	Low hydraulic gradient (0.05)	0.2 – 1*	Not explicitly mentioned except for clays soil, i > 20+	1
1e-5 – 1e-6	2	2				2
1e-6 – 1e-7	5	5				5
1e-7 – 1e-8	10	10	N.A.	N.A		10
1e-8 – 1e-9	20	20	N.A.			20
< 1e-9	30	30				$\geq$ 30

* for k_s range of $10^{-2} - 10^{-5}$ m/s
i = hydraulic gradient.

11.2 RELATED STANDARDS

1. ASTM D2434 – 22 Standard Test Methods for Measurement of Hydraulic Conductivity of Coarse-Grained Soils.
2. ASTM D5084 – 16a Standard Test Methods for Measurement of Hydraulic Conductivity of Saturated Porous Materials Using a Flexible Wall Permeameter.
3. ASTM D5856-15 Standard Test Method for Measurement of Hydraulic Conductivity of Porous Material Using a Rigid-Wall, Compaction-Mold Permeameter.
4. ASTM D6035/D6035M Standard Test Methods for Determining the Effect of Freeze-Thaw on Hydraulic Conductivity of Compacted or Intact Soil Specimens Using a Flexible Wall Permeameter.
5. ASTM D6836 Standard Test Methods for Determination of the Soil Water Characteristic Curve for Desorption Using Hanging Column, Pressure Extractor, Chilled Mirror Hygrometer, or Centrifuge.

6. ASTM D7664 Standard Test Methods for Measurement of Hydraulic Conductivity of Unsaturated Soils.
7. BS 1377-5: 1990 Methods of Test for Soils for Civil Engineering Purposes Part 5: Compressibility, Permeability and Durability Tests.
8. BS 1377-6: 1990 Methods of Test for Soils for Civil Engineering Purposes Part 6: Consolidation and Permeability Tests in Hydraulic Cells and with Pore Pressure Measurement.
9. BS EN ISO 17892-11:2019 Geotechnical Investigation and Testing. Laboratory Testing of Soil Permeability Tests.

11.3 THEORY

During measurement of permeability, it is important to ensure that the water content or degree of saturation of the soil does not change. This condition can be assured using the axis-translation technique (Hilf 1956) by controlling air and water pressures independently such that the matric suction of the soil specimen is kept constant during the permeability measurement. A summary of permeameters using such a technique to determine the air and water coefficients of permeability for unsaturated soils is given in Table 11.3. Air coefficient of permeability is measured sometimes as the air permeability function is an inverse of the water permeability function (Fredlund et al. 2012). It is possible to convert the air permeability function to the water permeability function, but this book concentrates only on the water phase; the flow of air is outside the scope of this book and will not be discussed further.

In applying the axis-translation technique, the porous elements in the triaxial cell will need to be replaced with high air-entry ceramic disk to separate the air and water phases. During permeability measurement, the matric suction (hence, water content or degree of saturation) needs to be maintained constant and thus the permeability measurement can only be done under the constant-head condition. Under variable head conditions, the air pressure must also vary in tandem to maintain constant matric suction which can be very complicated. The principle of the permeability test is to first equilibrate the soil specimen to the required matric suction at a specified net normal stress, as illustrated in Figure 11.1a. During measurement (Figure 11.1b), the water pressures across the soil specimen are adjusted such that the average pressure remains the same as before. Using Figure 11.1b for illustration, the water pressure was adjusted to be 190 kPa at the top of the specimen and 210 kPa at the bottom of the specimen such that the water pressure difference is 20 kPa, giving a hydraulic gradient i across the soil specimen as i = 20 kPa/(9.81 kN/m^3 x 0.1m) = 20.4 while the average pore-water pressure in the specimen remains at 200 kPa (= (190 kPa + 210 kPa)/2). Hence, water will flow from the bottom of the soil specimen to the top of the soil specimen under a hydraulic gradient of 20.4. When the

Table 11.3 Summary of permeameter systems used.

References	Range of k_w measured (m/s)	Range of k_a measured (m/s)	Suction range (kPa)	Measurement of volume change of specimen	Soil tested
Klute (1965)	Not mentioned	Not mentioned	0–90	Not measured	Not mentioned
Barden and Pavlakis (1971)	10^{-12}–10^{-10}	10^{-8}–10^{-3}	0–95	Not measured	Compacted soils
Fleureau and Taibi (1994)	up to 10^{-8}	up to 10^{-8}	0–80	Not measured	Undisturbed clay, loam and sand-kaolin mixture
Dane et al. (1998)	Not measured	up to 10^{-4}	0–4	Not measured	Sand
Huang et al. (1998)	10^{-11}–10^{-8}	Not measured	0–90	Using non-contacting transducers	Silty sand
Gan and Fredlund (2000)	10^{-10}–10^{-5}	Not measured	0–100	Using confining fluid volume	Residual and saprolitic soils
Samingan et al. (2003)	10^{-12}–10^{-6}	10^{-7}–10^{-12}	0–300	Not measured	Residual soils
Goh, et al. (2015).	10^{-9}–10^{-10}	Not measured	100–300	Not measured	Sand-kaolin

Specimen dimensions: Diameter = 100 mm, Length, L= 100 mm

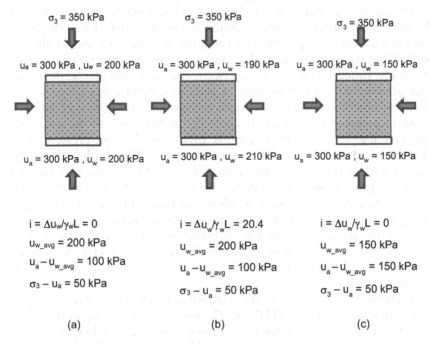

$i = \Delta u_w/\gamma_w L = 0$

$u_{w\_avg} = 200$ kPa

$u_a - u_{w\_avg} = 100$ kPa

$\sigma_3 - u_a = 50$ kPa

$i = \Delta u_w/\gamma_w L = 20.4$

$u_{w\_avg} = 200$ kPa

$u_a - u_{w\_avg} = 100$ kPa

$\sigma_3 - u_a = 50$ kPa

$i = \Delta u_w/\gamma_w L = 0$

$u_{w\_avg} = 150$ kPa

$u_a - u_{w\_avg} = 150$ kPa

$\sigma_3 - u_a = 50$ kPa

(a)   (b)   (c)

Figure 11.1 Principle of constant-head permeability test for unsaturated soil
using axis-translation technique. (a) Start. (b) During measurement.
(c) Next suction.

volume of water flowing into the soil specimen, Q_{in}, is equal to the volume
of water flowing out of the specimen, Q_{out}, the steady-state flow condition is
reached, and the water coefficient of permeability can be computed using
Darcy's law:

$$k = \frac{Q}{i \cdot A \cdot t} \tag{11.1}$$

where
Q = average of Q_{in} and Q_{out}
t = time to obtain Q
A = cross-sectional area of soil specimen

11.4 TEST METHOD

The soil specimen tested can be intact/undisturbed or reconstituted. For soil
specimen preparation, see Chapter 3. Suitable specimen size to be used for
permeability measurement is usually thin so that the steady-state condition

can be attained sooner. Most standards specify height to diameter ratio of 1:1 but BS 1377:1990: Part 6 suggests that the height/diameter ratio of 1/2.5 to 1/4 can be used but should not exceed 1. The specimen height is limited by the largest particle size which should not be more than 1/6 of the specimen height. For testing unsaturated soil, the specimen size should allow a sufficient quantity of water to flow out of or into the soil specimen so that the water volume change of the specimen is measurable. Hence, the specimen size should not be too small. A suitable specimen size is one with a diameter of at least 50 mm and height to diameter ratio of at least 1:2.

As air and water pressures are controlled at both ends of the soil specimen, the high-air entry (HAE) ceramic disk must allow air pressure to be distributed across the soil specimen cross-section. This can be affected by having part of the cross-sectional area of the HAE ceramic disk covered with a porous element that is permeable to air as well, such as a sintered bronze. Possible arrangements are shown in Figure 11.2. However, any of these arrangements will reduce the cross-sectional area for water flow and increase the duration of the test. A better alternative is to etch shallow grooves into the HAE ceramic disk, as shown in Figure 11.3. A modified triaxial cell to conduct a permeability test for unsaturated soils is shown in Figure 11.4.

A diffused air volume indicator is shown in Figure 11.4 for removing air that diffused through the HAE ceramic disk and reappearing in the water line. This is required when the test duration is very long and measuring the water volume of the specimen is needed. Details of the diffused air volume indicator can be found in Fredlund et al. (2012).

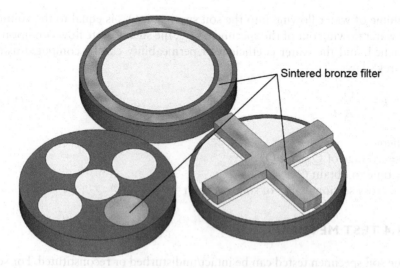

Sintered bronze filter

Figure 11.2 Possible arrangements of HAE ceramic disk with sintered bronze filter on the platens.

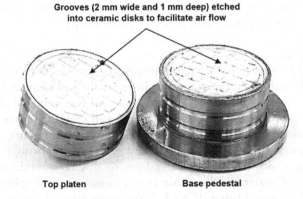

Grooves (2 mm wide and 1 mm deep) etched
into ceramic disks to facilitate air flow

Top platen   **Base pedestal**

Figure 11.3 Suggested HAE ceramic disk for permeability test of unsaturated soils.

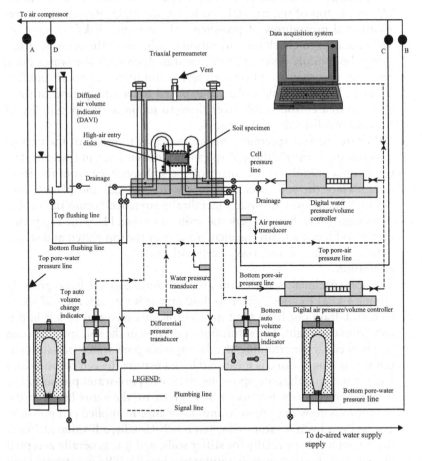

Figure 11.4 Schematic drawing of flexible-wall permeameter system for unsaturated soils (from Agus et al. 2003).

The test procedures can be divided into four stages: setting up, saturation, consolidation and measurement of water permeability.

1. Setting up
   The apparatus is first prepared to ensure that the HAE ceramic disks are fully saturated and the water lines are flushed of air bubbles. The air valves of the triaxial cell are shut and the triaxial cell without the soil specimen is filled with water. A cell pressure (at the air-entry pressure of the ceramic disk or the maximum air pressure that is applied during the test) is applied and water flows through the HAE ceramic disks. Allow the water lines to discharge underwater in a beaker of water to check for the presence of air bubbles. This is repeated several times. If air bubbles cannot be effectively removed in this way, half-fill the triaxial cell and ensure that the top platen is submerged in the water as well. Apply a vacuum of say 20 to 30 kPa at the top of the triaxial and leave overnight. Repeat the afore-mentioned procedures of pressuring the ceramic disks without the soil specimen to check the saturation condition of the ceramic disks. Once the ceramic disks are fully saturated, connect the water lines, ensuring that they are filled with water and there are no air bubbles in them. If the ceramic disks are exposed for a long period of time, place wetted cotton wools over them to minimise the possibility of the ceramic disks desaturating.

   Prepare the soil specimen; ensure that it is trimmed to the correct dimensions. Measure the soil specimen's dimensions and weight. Determine the initial water content of the soil specimen from the trimmings. Place the soil specimen into the triaxial cell and place a rubber membrane over it using a membrane stretcher. Secure the rubber membrane to the bottom pedestal with an O-ring. Place the top platen on the soil specimen, lift the membrane over the top platen and secure with an O-ring. Fill the triaxial cell with water.

2. Saturation
   The soil specimen is then saturated by using a small back pressure (about 5 kPa) to allow water to flow upwards through the soil specimen and drain out from the top. The soil specimen can be left in this condition overnight. The saturation condition of the soil specimen can be checked by determining the Skempton's pore-water pressure parameter B. The B value is measured by increasing the cell pressure and then monitoring the corresponding change in pore-water pressure. The pore-water pressure parameter B is given by the ratio between the change in pore-water pressure and the change in applied cell pressure. The soil specimen is saturated when B is 1. Usually, a B value of 1 cannot be obtained, especially for stiffer soils, and it is generally accepted that the soil specimen is fully saturated when B > 0.95 in saturated soil tests (Black and Lee 1973).

3. Consolidation

The soil specimen is then consolidated isotropically to the desired net normal stress and matric suction. If the change in net normal stress and matric suction is large, it can be done in several steps (minimum two steps) to reach the desired values. Double drainage should be adopted to reduce the consolidation time, as it might take several days at high matric suction values. The consolidation process can be monitored by recording the volume of water flowing out from the soil specimen with time. Consolidation is deemed to be completed when there is negligible water flow either into or out of the soil specimen. At this stage, the soil specimen is said to be in equilibrium at the applied net normal stress and matric suction.

4. Measurement of water permeability

The water coefficient of permeability is determined by applying a hydraulic gradient across the soil specimen by changing the water pressures applied to the top and bottom of the specimen such that the average water pressure remains the same as before the change (see Figure 11.1). The direction of water flow can be upwards or downwards but upward water flow is preferred as there is less likelihood for water to enter the air pressure lines with upward water flow. The maximum hydraulic gradient depends on the permeability of the soils (see Table 11.1). High hydraulic gradient is required to reduce the test duration as the rate of water flow is very slow. It is important to monitor the water inflow and outflow to establish the steady-state flow condition. An example is shown in Figure 11.5.

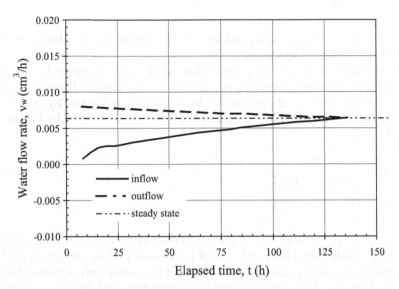

Figure 11.5 Example of water inflow and outflow during permeability test to establish steady-state flow condition.

At steady-state flow condition, the water coefficient of permeability can be obtained using Darcy's law (Equation 11.1). The impedance of the ceramic disks needs to be considered in the determination of the water coefficient of permeability of the soil when the water coefficient of permeability of the soil is close to that of the ceramic disks or lower. The impedance of the ceramic disk can be considered by treating the ceramic disk – soil – ceramic disk as a three-layered system. For one-dimensional flow, the flow velocity v through each layer must be the same as given by Equation 11.2.

$$v = v_b = v_s = v_t \tag{11.2}$$

where v_b, v_s and v_t are the flow velocities through the bottom ceramic disk, soil and top ceramic disk, respectively.

The loss in total head across the three layers H is equal to the sum of the total head loss in each layer, i.e.:

$$H = H_b + H_s + H_t \tag{11.3}$$

where H_b, H_s and H_t are the total head losses through the bottom ceramic plate, soil and top ceramic plate, respectively.

Using Equations 11.2 and 11.3 and applying Darcy's law, the coefficient of permeability of the three-layered system is given by Equation 11.4.

$$k = \frac{h_t + h_s + h_b}{\dfrac{h_t}{k_t} + \dfrac{h_s}{k_s} + \dfrac{h_b}{k_b}} \tag{11.4}$$

where h_b, h_s and h_t are the thicknesses of the bottom ceramic plate, soil and top ceramic plate, respectively.

The water coefficient of permeability of the soil specimen k_w can be obtained if the coefficients of permeability of the ceramic disks, k_t and k_b and the thicknesses of the ceramic disks, h_b and h_t, are known. By rearranging Equation 11.4, the water coefficient of permeability is given by Equation 11.5.

$$k_w = \frac{h_s}{\left[\dfrac{h}{k} - \left(\dfrac{h_t}{k_t} + \dfrac{h_b}{k_b} \right) \right]} \tag{11.5}$$

Equation 11.5 shows that if the coefficients of permeability of the disks are much higher than the coefficient of permeability of the soil specimen, k_w is equal to k, which is measured. However, if the coefficients of permeability of the disks are of the same order or lower than the coefficient of permeability of the soil specimen, the measured k is for the disks rather than for the soil specimen. Hence it is important to check when the water coefficient of permeability of the soil approaches that of the ceramic disks.

A temperature correction factor can be applied to the water coefficient of permeability of the soil as the viscosity of water changes with temperature. Usually, the water coefficient of permeability is determined at the standard temperature of 20°C.

REFERENCES

Barden, L. & Pavlakis, G. (1971). Air and water permeability of compacted unsaturated cohesive soils. *Journal of Soil Science, 22*: 302–318.

Benson, C. H. & Gribb, M. M. (1997). Measuring unsaturated hydraulic conductivity in the laboratory and field. In S. Houston & D. G. Fredlund (Eds.), *Unsaturated soil engineering practice, ASCE, geotechnical special publication,* No. 68: 113–168.

Black, D. K. & Lee, K. L. (1973). Saturating laboratory samples by back pressure. *Journal of the Soil Mechanics and Foundation Division, ASCE, 99*(SMII): 75–93.

Brooks, R. H. & Corey, A. T. (1964). Hydraulic properties of porous medium. Hydrology Papers. No. 3. Fort Collins, Co: Colorado State University.

Childs, E. C. & Collis-George, G. N. (1950). The permeability of porous materials. *Proceeding of the Royal Society, 210A*: 392–405.

Dane, J. H., Hofstee, C., & Corey, A. T. (1998). Simultaneous measurement of capillary pressure, saturation and effective permeability of immiscible liquids in porous media. *Water Resources Research, 34*(12): 3687–3692.

Fleureau, J. M. & Taibi, S. (1994). New apparatus for the measurement of water-air permeabilities. In *Proceeding of Geo-environmental Conference*, Edmonton, Canada.

Fredlund, D. G. (1981). Panel discussion: Ground water and seepage problems. In *Proceedings of the 10th International Conference on Soil Mechanics and Foundation Engineering*, Stockholm, Sweden, vol. 4, pp. 629–641.

Fredlund, D. G. & Rahardjo, H. (1993). *Soil mechanics for unsaturated soils.* New York: John Wiley & Sons, Inc.

Fredlund, D. G., Rahardjo, H., & Fredlund, M. D. (2012). *Unsaturated soil mechanics in engineering practice.* New York: John Wiley & Sons, Inc..

Gallage, C., Kodikara, J., & Uchimura, T. (2013). Laboratory measurement of hydraulic conductivity functions of two unsaturated sandy soils during drying and wetting processes. *Soils and Foundations, 53*(3): 417–430.

Gan, J. K. M. & Fredlund, D. G. (2000). A new laboratory method for the measurement of unsaturated coefficients of permeability of soils. In H. Rahardjo, D. Toll, & E. C. Leong (Eds.), *Unsaturated soils for Asia.* Rotterdam: Balkema, 381–386.

Goh, S. G., Rahardjo, H., & Leong, E. C. (2015). Modification of triaxial apparatus for permeability measurement of unsaturated soils. *Soils and Foundations, Japanese Geotechnical Society*, February, *55*(1): 63–73.

Huang, S., Fredlund, D. G., & Barbour, S. L. (1998). Measurement of the coefficient of permeability for a deformable unsaturated soil using a triaxial permeameter. *Canadian Geotechnical Journal, 35*: 411–425.

Iwata, S., Tabuchi, T., & Warkentin, B. P. (1995). *Soil-water interactions.* New York: Marcel Dekker., Inc.

Kaye, G. W. C. & Laby, T. H. (1973). *Tables of physical and chemical constant and some mathematical functions*. 14th Edition. London: Longman.

Klute, A. (1965). Laboratory measurement of hydraulic conductivity of unsaturated soils. In C. A. Black, D. D. Evans, J. L. White, L. E. Esminger, & F. E. Clark (Eds.), *Method of soil analysis. Monograph 9*. Madison, Wisconsin: American Society of Agronomy, 1: 253–261.

Lambe, T. W. & Whitman, R. V. (1968). *Soil mechanics*. New York: John Wiley & Sons, Inc.

Lloret, A. & Alonso, E. E. (1980). Consolidation of unsaturated soils including swelling and collapse behaviour. *Géotechnique*, 30(4): 449–477.

Rahimi, A., Rahardjo, H., & Leong, E. C. (2015). Effect of range of soil-water characteristic curve measurements on estimation of permeability function. *Engineering Geology*, February, 185: 96–104.

Richards, L. A. (1931). Capillary conduction of liquids through porous medium. *Journal of Physics*, 1: 318–333.

Samingan, A. S., Leong, E. C., & Rahardjo, H. (2003). A flexible wall permeameter for measurements of water and air coefficients of permeability of residual soils. *Canadian Geotechnical Journal*, June, 40(3): 559–574.

Taylor, D. W. (1948). *Fundamentals of soil mechanics*. New York: Wiley.

FURTHER READING

Leong, E. C. & Rahardjo, H. (1997). Permeability functions for unsaturated soils. *ASCE Journal of Geotechnical and Geoenvironmental Engineering*, 123(12): 1118–1126.

Chapter 12

Permeability

Transient-state methods

12.1 BACKGROUND

Measurement of the coefficient of permeability of unsaturated soils by the steady-state methods, though more accurate than the transient-state (unsteady-state) methods, is more difficult and time-consuming. The condition for steady state is achieved when the inflow is equal to outflow. In some cases, it is near impossible to obtain the steady-state condition. To partially alleviate the problem of long testing time, several transient-state methods have been developed to measure the coefficient of permeability of unsaturated soils (permeability function). In some of these methods, a mixed-phase formulation is used to obtain both the soil-water characteristic curve and the permeability function of the soil (e.g., Garnier et al. 1997; Fujimaki and Inoue 2003). As this book concentrates on only the water phase, such mixed-phase formulations or methods involving evaporation (Wind 1968; Schindler et al. 2010; Wendroth et al. 1993; Minasny and Field 2005) are not included.

A review of transient-state methods to measure unsaturated permeability can be found in Benson and Gribb (1997). Transient-state methods include Bruce-Klute adsorption, sorptivity, outflow, instantaneous profile and thermal methods. There are many variations of the outflow method. These include multistep, one-step and continuous outflow methods. Due to the advances in computer modelling, it is possible to combine experimental results and numerical modelling to better estimate unsaturated permeability (Kool et al. 1987). Examples are the HYPROP method (Schindler et al. 2010; Tian et al. 2019), XMSO-EVA method (Schelle 2011) and TWRI method (Wayllace and Lu 2012). Such methods are outside the scope of this book.

The transient-state methods involved two types of main apparatuses: a soil column which could be placed vertically or horizontally (Bruce-Klute adsorption, instantaneous profile, thermal) or a cell containing a soil specimen (sorptivity, outflow). The soil column and cell apparatus have been adopted in ASTM D7664-10 (reapproved 2018) for the measurement of permeability of unsaturated soils. In the standard, the soil column for the instantaneous profile method is described as Method A, and the cell for the multistep outflow method is described as Method B. In Method A, the additional instruments required are suction and water content measurement devices. In Method B, the cell adopted is similar to a pressure plate apparatus where matric suction is controlled using the axis-translation technique (see Chapter 11). Additional instruments required include flow volume measurement and, optionally, suction measurement.

A Method C is also described in ASTM D7664-10. Method C involves the use of a centrifuge to impose a steady-flow condition. The centrifuge used can be as big as a geotechnical centrifuge or as small as a benchtop centrifuge. In Method C, a special permeameter will have to be constructed, and the test requires a device to provide an infiltration flow and devices to measure matric suction and water content of the soil specimen in the centrifuge at steady-state flow condition (Zornberg and McCartney 2010; McCartney and Zornberg 2010). However, it was found that the steady-state condition was not fulfilled in the centrifuge permeameter at low G-levels, and a transient-state analysis should be performed instead (Parks et al. 2011). Transient analysis using a centrifuge analysis can provide "acceptable" characterisation of the soil-water characteristic curve and permeability function over wide suction and water content ranges. Other limitations of measuring permeability of unsaturated soils in a geotechnical centrifuge include limit of reading for time, total mass, inflow mass, outflow mass, rate of inflow, fluid held by surface tension in effluent reservoir and on reservoir base plate, evaporation effects (which increases with the increasing G-level) and sidewall leakage (Timms et al. 2014). In view of the requirements and limitations of the centrifuge permeameter, this method is omitted from this chapter.

12.2 RELATED STANDARDS

1. ASTM D2434 – 22 Standard Test Methods for Measurement of Hydraulic Conductivity of Coarse-Grained Soils.
2. ASTM D5084 – 16a Standard Test Methods for Measurement of Hydraulic Conductivity of Saturated Porous Materials Using a Flexible Wall Permeameter.
3. ASTM D5856-15 Standard Test Method for Measurement of Hydraulic Conductivity of Porous Material Using a Rigid-Wall, Compaction-Mold Permeameter.
4. ASTM D6035/D6035M Standard Test Methods for Determining the Effect of Freeze-Thaw on Hydraulic Conductivity of Compacted or Intact Soil Specimens Using a Flexible Wall Permeameter.
5. ASTM D6836 Standard Test Methods for Determination of the Soil Water Characteristic Curve for Desorption Using Hanging Column, Pressure Extractor, Chilled Mirror Hygrometer, or Centrifuge.
6. ASTM D7664 Standard Test Methods for Measurement of Hydraulic Conductivity of Unsaturated Soils.

12.3 INSTANTANEOUS PROFILE METHOD

The instantaneous profile method using a soil column is described in ASTM D7664-10 Method A. The instantaneous profile method makes use of mass continuity to evaluate the instantaneous water content and suction profiles of the soil column one-dimensional (1D) water flow condition. The soil column can start at different initial conditions, and different boundary conditions can be imposed during the test. Four conditions are described in ASTM D7664-10. These are infiltration from the top of a dry soil column (Method A1), imbibition from the bottom of a dry soil column (Method A2), drainage from the bottom of a saturated soil column (Method A3) and evaporation from the top of a saturated soil column (Method A4). During the test, the suctions and water contents at various levels in the soil column are monitored. Depending on the conditions imposed and the soil type, the test duration can range from one week to several weeks.

As a 1D water flow condition is assumed, the void ratio of the soil should not change during the test, and side wall leakage should not be present. Hence, the instantaneous profile method using a soil column precludes high plasticity soils which may experience change in void ratio during the test. Clays which have very low permeability are also not possible to test using the instantaneous profile method because of the prohibitively long testing time and the requirement to use a long soil column. As the soil column gets taller, it is increasingly more difficult to ensure that the soil column is homogeneous. If suction is directly measured during the test, the suction range is limited to being less than 80 kPa (limitation of tensiometer).

The aforementioned requirements/limitations are specifically listed in ASTM D7664-10 under each sub-method. Methods A1 and A2 are suitable for fine-grained sands and for low-plasticity silts, but the permeability measurement range is restricted within the matric suction range of tensiometers, i.e., 0 to 80 kPa. Methods A3 and A4 are suitable for fine- and coarse-grained sands. Methods A3 can measure permeability for the matric suction range from 0 to 200 kPa, while Method A4 can measure permeability for the matric suction range from 0 to 1000 kPa. For Method A4, desiccation shrinkage and cracking may occur as evaporative flux is applied on the top surface of the soil column.

12.3.1 Test set-up

The soil column container can be a cylinder made of non-reactive metals, acrylic or PVC (ASTM D7664-10). The use of acrylic is preferred as the water movement can be seen (Yang et al. 2004; Krisdani 2009). The soil column can be prepared inside the container by compaction, wet tamping or pluviation. Hence, the container should be strong enough to withstand the stresses imposed during soil column preparation. A typical soil column set-up is shown in Figure 12.1. There should be provisions for the installation

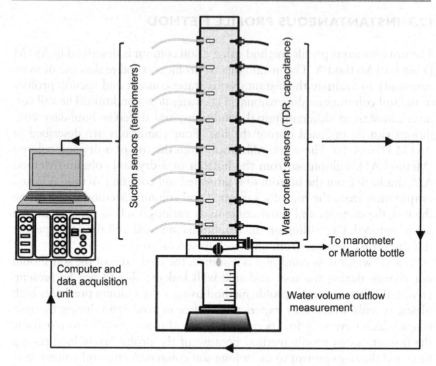

Figure 12.1 Typical soil column set-up for instantaneous profile method.

of suction and water content sensors along the soil column. Typical sensors for direct suction measurement are small-tip tensiometers which have quick response to suction changes (Yang et al. 2004; Krisdani et al. 2009), but these sensors are limited to suctions less than 80 kPa. Typical sensors for water content measurement are time-domain reflectometry (TDR) sensors (Yang et al. 2004; Krisdani et al. 2009) or capacitance sensors (CSs; Caicedo 2017). The holes are made in the cylinder so that the sensor cable or non-sensing part of the sensor can exit the cylinder. For tensiometer, the diameter of the hole should be bigger than the diameter of the sensing tip so that the tensiometer can pass through from the outside of the cylinder. A plug can be provided to make a watertight seal. The use of a cylinder with a larger diameter will reduce the need for the plug design to consider the curvature of the cylinder surface or else an additional block will need to be introduced at the cylinder surface to provide a flat contact surface for the plug to achieve a watertight seal (see Figure 12.2). For the TDR sensor, the hole should be large enough to accommodate the connector. The TDR probe is disconnected and placed on the inside of the cylinder, and then the connector is fixed from the outside of the cylinder. Usually, an additional O-ring at the connector is sufficient to achieve a watertight seal. For the CSs, a hole large enough for the cable to pass through is sufficient. A plug similar

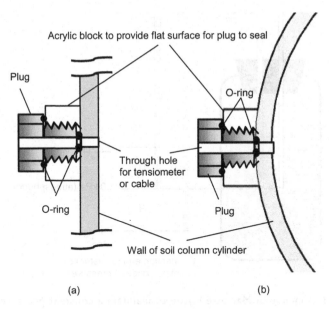

Acrylic block to provide flat surface for plug to seal

Plug

O-ring

Through hole
for tensiometer
or cable

O-ring

Plug

Wall of soil column cylinder

(a)   (b)

Figure 12.2 Plug for watertight seal of small-tip tensiometer or cable. (a) Side view.
(b) Top view.

to that for the tensiometer can be used (Figure 12.2). The minimum internal
diameter of the cylinder recommended by ASTM D7664-10 is 200 mm, as
it can serve to minimise side wall leakage, as well as detect preferential flow
of the water due to macro features in the soil column. ASTM D7664-10
recommends that the top sensors should be within 10 mm of the surface of
the soil column and the bottom sensors within 5 mm of the bottom of the
soil column, and at least three sensors (or sets of sensors) evenly spaced in
between. The height of the soil column should be at least 0.5 m for coarse-
grained soils and 1 m for fine-grained soils in order to capture representative distribution of matric suction and water content with the height of the
soil column. A taller soil column is required if the test needs to achieve the
unit hydraulic gradient condition in the upper part of the soil column. The
bottom of the soil column should be free draining. A perforated plate made
from metal or acrylic with a piece of filter paper or high permeability geo-
textile at the bottom of the soil column will usually suffice to prevent loss of
soil particles and the avoidance of creating a capillary break.

At the base of the soil column, provision for measuring water outflow
with time (Methods A1 and A3) and/or imposing a constant head condition
(Methods A2 and A4) are required. For measuring water outflow (methods
A1 and A3), a weighing balance, a graduated cylinder or other means can
be used to record the volume of water outflow from the base with time. For
Method A3, a Mariotte bottle may be used to maintain a constant head
condition and to measure the volume of water outflow at the same time.

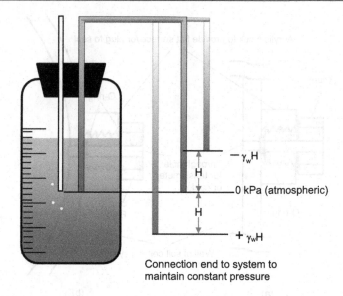

$-\gamma_w H$

H

0 kPa (atmospheric)

H

$+\gamma_w H$

Connection end to system to
maintain constant pressure

Figure 12.3 Principle of Mariotte bottle to maintain a constant pressure at inlet.

The principle of the Mariotte bottle is shown in Figure 12.3. For Methods A2 and A4, a manometer system or a Mariotte bottle can be used to maintain a constant head condition at the base of the soil column. The Mariotte bottle provides a convenient means of determining the volume of water inflow into the soil column, but if the manometer is used, an inflow volume measurement system such as the one mentioned in Chapter 11 needs to be added.

Infiltration at the top of the soil column can be applied by a peristaltic pump system where the amount of water flowing onto the top surface of the soil column can be measured using a weighing balance or graduated cylinder (Method A1). However, the flow rate into the soil column is usually very small (less than the saturated permeability of the soil). Provision needs to be made to distribute the infiltration uniformly across the top surface of the soil column while ensuring that the evaporation rate is low. A piece of filter paper can be placed on the top surface of the soil column on to which the inflow line drips (Yang et al. 2004) or a cup to receive the water from the inflow line and distribute it through a series of cotton fibre wicks draped across the soil surface (ASTM D7664-10). To impose an evaporative flux boundary on the top surface of the soil column, an infrared lamp to provide a constant temperature to the soil surface and an electric fan to circulate the air to the soil surface can be used (ASTM D7664-10). It is also possible to provide a constant relative humidity by circulating vapour from a salt solution through the top surface of the soil column (Figure 12.4). Verification of the temperature and relative humidity at the top of the soil column can be

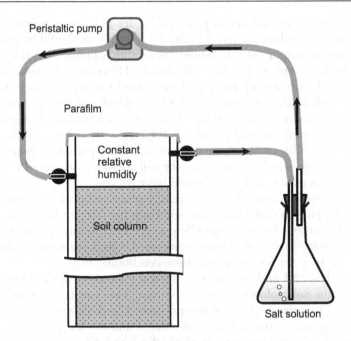

Figure 12.4 Maintaining constant relative humidity at the top surface of the soil column using salt solution.

performed by monitoring the temperature and relative humidity using a capacitive psychrometer with temperature sensor.

12.3.2 Test procedures

The test procedures can be divided into four stages: sample preparation, soil column preparation, column test and finishing.

12.3.2.1 Sample preparation

Conduct basic soil properties on the soil sample to obtain grain size distribution, maximum and minimum dry densities and specific gravity. Air dry the sample and prepare about twice the mass of the sample required to fill the soil column.

12.3.2.2 Soil column preparation

First check that the instruments, suction and moisture content sensors, are calibrated and working. Usually, small-tip tensiometers are used to measure matric suction. These tensiometers should be saturated and tested as explained in Chapter 8.

Initially, a dry soil column is prepared by dry pluviation. Select a target dry density to conduct the test. Ideally, the soil column should be prepared in lifts of equal height until the whole soil column is prepared. A convenient lift height is the vertical spacing of the sensors. However, if the vertical spacing exceeds 15 cm, a convenient lift height should be selected such that the maximum lift height does not exceed 15 cm. Weigh out the sand into bags or containers to meet the target dry density for each lift height. Apply suitable dry pluviation method to achieve the target dry density for each lift. The soil in each lift can be compacted by static compaction, a piston compactor or a drop weight. Ensure that the soil column is secured if piston or drop weight is used to compact the soil layer.

At each sensor placement level, carefully place and position the sensors. Tighten the plug to seal. Place some soil from the next lift height to cover/protect the sensors and continue filling to the next sensor placement level. Repeat until the entire soil column is prepared. Check the senor readings to ensure that they reached the equilibrium condition.

For Methods A3 and A4, saturate the soil column by allowing water to flow into the soil column from the bottom. This can be done using the manometer to impose a small water pressure. Check that the wetting front is moving up slowly through the soil column and the soil is not disturbed. Alternatively, the soil column for Method A3 and A4 can be prepared by wet pluviation. Regardless of whether a dry or wet pluviation method was used, it is good practice to allow the soil column to drain and to re-saturate again to ensure that there are no trapped or occluded air bubbles. This can be repeated several times. The other advantage of Methods A3 and A4 over Methods A1 and A2 is that the working condition of the matric suction and moisture sensors can be checked during this process.

12.3.2.3 Column test

12.3.2.3.1 Infiltration

In Method A1, the initial condition of the soil column is dry. A constant infiltration rate is applied to the top surface of the soil column and a free drainage boundary is applied to the bottom of the soil column. Permeability is determined at various matric suctions by imposing different infiltration rates. The number of infiltration rates chosen should be at least three, and all shall be less than the saturated permeability of the soil. It is convenient to choose the infiltration rate based on a factor of the saturated permeability of the soil. The saturated permeability of the soil can first be determined using test methods for saturated soils. The infiltration rate should start from the smallest infiltration rates. Convenient rates are 1,000, 100 and 10 times lower than the saturated permeability of the soil.

Prepare the top surface of the soil column as described in test set-up to ensure a uniform distribution of the water. Start the pump and the data recorder. Record inflow volume, matric suctions at the different depths of the soil column and water content at the different depths of the soil column. Continue the test until there are no further changes in the matric suctions and water contents (steady-state condition).

12.3.2.3.2 Imbibition

In Method A2, the initial condition of the soil column is dry. A zero-water pressure condition is applied to the bottom of the soil column while the top of the soil column is open to the atmosphere. Shut the drainage valve off at the bottom of the soil column and connect the manometer with inflow volume measurement or Mariotte bottle to the bottom of the soil column. Set the water pressure to zero (atmospheric pressure) at the bottom of the soil column. Start the data recorder and open the drainage valve. Record inflow volume, matric suctions at the different depths of the soil column and water content at the different depths of the soil column. Continue the test until inflow volume becomes zero, i.e., water stops entering the bottom of the soil column due to capillary rise.

12.3.2.3.3 Drainage

In Method A3, the initial condition of the soil column is saturated and under hydrostatic water pressure condition with the top of the soil column at zero pore-water pressure. Shut the drainage valve off at the bottom of the soil column and connect the manometer with outflow volume measurement or Mariotte bottle to the bottom of the soil column. Set the water pressure to zero at the bottom of the soil column. Start the data recorder and open the drainage valve. Record outflow volume, matric suctions at the different depths of the soil column and water content at the different depths of the soil column. Continue the test until outflow volume becomes zero, i.e., the change in 30 minutes is less than 1%.

12.3.2.3.4 Evaporation

In Method A4, the initial condition of the soil column is saturated and under hydrostatic water pressure conditions with the top of the soil column at zero pore-water pressure. Turn the drainage valve off at the bottom of the soil column and connect the manometer or Mariotte bottle to the bottom of the soil column. Set the water pressure to zero at the top of the soil column. Open the drainage valve to ensure that no water is flowing out from the soil column and the pore-water pressure condition in the soil column is hydrostatic. Turn the drainage valve off. Apply a constant relative humidity boundary condition at the top of the soil column and start the data recorder. Record matric suctions at the different depths of the soil column and water content at the different depths of the soil column. Continue the test until there is no change in matric suction and water content with time. The test-stopping criterion can be taken as the changes being less than 1% of the measured values within 24 hours (ASTM D7664-10).

12.3.2.4 Finishing

The water content distribution of the soil column at the end of the test can be determined by removing soil samples from different depths of the soil column and drying in the oven for water content determination. This can be used to check against the water contents measured by the water content sensors.

12.4 MULTISTEP OUTFLOW METHOD

Multistep outflow methods can be performed using a pressure plate apparatus. The change in matric suction is performed by applying the axis-translation technique. The pressure plate apparatus can be treated as a rigid-wall permeameter (Method B1) such as using a Tempe cell or volumetric pressure plate apparatus. Tempe cell and volumetric pressure plate are mentioned in Chapter 10. However, if a standard pressure plate apparatus is used, the water pressure that can be applied below the high air entry (HAE) ceramic disk is small. As for a flexible-wall permeameter (Method B2), the standard triaxial apparatus can be modified to become a flexible-wall permeameter (Goh et al. 2015) or the triaxial permeameter described in Chapter 11 can be used. In both test methods, the water outflow with time needs to be measured.

12.4.1 Test set-up

The addition of water outflow measurement is required to make the standard pressure plate apparatus suitable for use in Test Method B. The water outflow volume measurement can be done using a burette or a digital pressure/volume controller with a reading resolution of 1 mL or better. The connection tubing from the pressure plate apparatus to the water volume measurement device should be incompressible, as it affects the water volume measurement.

The HAE ceramic disk and the water reservoir below the ceramic disk should be fully saturated before the test. The saturation of the pressure plate apparatus is similar to that in Chapter 10.

As Test Method B has a long test duration, flushing of diffused air from the water reservoir below the HAE ceramic disk is needed, as it may interfere with the water outflow volume measurement. The flushing of diffused air can be affected using a diffused air volume indicator (Samingan et al. 2003; Fredlund et al. 2012).

12.4.2 Test procedures

The test procedures can be divided into four stages: sample preparation, preparation of apparatus, permeability test and finishing.

12.4.2.1 Sample preparation

Conduct basic soil properties on the sample to obtain grain size distribution, maximum and minimum dry densities, and specific gravity. Both intact/undisturbed and reconstituted soil specimens may be used for Test Method B. The preparation of the soil specimen is given in Chapters 3 and 10.

12.4.2.2 Preparation of apparatus

The permeability of the HAE ceramic disk should be tested periodically to ensure that it is not clogged or defective. For the rigid-wall permeameter, the measurement of permeability of the saturated HAE ceramic disk can be conducted by first saturating the HAE ceramic disk in a vacuum desiccator under distilled water at least overnight and then assembling it into the pressure plate apparatus. Half-fill the rigid-wall permeameter and apply high air pressure and measure the volume of water exiting from the permeameter with time. The saturated permeability of the HAE disk can be determined using the constant head permeability equation given in Chapter 11. The pressure gradient is given by the air pressure across the thickness of the ceramic plate. For the flexible-wall permeameter, the measurement of the saturated permeability of the HAE ceramic disks is given in Chapter 11. ASTM D7664-10 recommends that if there is a decrease in saturated permeability of the HAE ceramic disks by five times from the saturated permeability when new, the HAE ceramic disk should be discarded or cleansed.

12.4.2.3 Permeability test

In ASTM D7664-10, the HAE ceramic disk is dried in the oven at 110°C and placed dry in the set-up. The soil specimen is then placed on top of the HAE ceramic disk, the pressure chamber is closed and then the whole system of soil specimen and HAE ceramic disk is saturated. This is a very difficult and slow process in practice as the permeability of the HAE ceramic disk is very low and the air in the system will have to be removed through the small drainage tubing. It is very difficult to ensure that all the air in the system is fully removed by this process.

Instead, the permeameter system can first be ensured to be fully saturated, i.e., the HAE ceramic disk, water reservoir below the HAE ceramic disk and the water drainage lines are fully saturated. For the rigid-wall permeameter, the soil specimen in the ring can be placed on top of the HAE ceramic disk and the bottom drainage line connected to a water burette filled with water that is allowed to flow into the soil specimen by upward seepage and capillary action. When air bubbles ceased to appear from the surface of the soil specimen, the soil specimen can be further saturated by closing the rigid-wall permeameter and applying a small vacuum (~–20 kPa) at the air pressure inlet. For the flexible-wall permeameter, the saturation of the soil specimen can follow the procedures as mentioned in Chapter 11. The degree of saturation in the flexible-wall permeameter can be checked using Skempton pore-water pressure parameter B. A B parameter greater than 0.9 is accepted as full soil saturation in ASTM D7664-10 and 0.95 in BS 1377-6 (1990).

In ASTM D7664-10, the test starts with the measurement of the saturated permeability of the soil specimen. For the rigid-wall permeameter, this can only be done if there is water above the soil sample. The saturated permeability test is done as a constant head permeability test and the equations in Chapter 11 are applicable. For the permeability of the unsaturated soil, the test starts at equal water and air pressures. The maximum air pressure applied should not exceed the HAE of the ceramic disk. ASTM D7644-10 suggests 300 kPa at the start of the test. The air pressure is then maintained constant while the water pressure is reduced in steps to apply the matric suctions.

12.4.2.4 Rigid-wall permeameter

The test is started with the soil specimen in the fully saturated condition. The matric suctions for the whole test should be planned. The maximum matric suction is governed by the air-entry value (AEV) of the ceramic disk and the final water pressure below the ceramic disk. For example, if the AEV of the ceramic disk is 500 kPa and the final water pressure below the ceramic disk is 50 kPa, and the target test matric suctions are 100, 200, 300 and 400 kPa, then the initial air and water pressures can be set at 450 kPa. To achieve the matric suctions of 100, 200, 300 and 400 kPa, the air pressure is kept constant at 450 kPa while the water pressure is reduced to 350, 250, 150 and 50 kPa, respectively, in the test.

The test is started with both air and water pressures at their maximum value (i.e., 450 kPa). The matric suction is applied by keeping air pressure constant and decreasing the water pressure. The water starts to flow out from the soil specimen when the matric suction is applied. At each matric suction, measure the volume of the water outflow from the soil specimen with time. The measurements should be done at regular intervals to obtain the nonlinear shape of the water outflow volume with time. The test at each matric suction can stop when the water outflow becomes negligibly small. In ASTM D7644-10, negligibly small water outflow is defined when the water outflow volume in one hour is less than 1% of the water outflow volume in the previous hour. The next matric suction is then applied and the measurement of water outflow with time is then repeated. This process is carried out until the last matric suction.

At the highest applied matric suction, the matric suction can be decreased in the reverse manner and volume of water inflow can be measured with time to obtain the unsaturated permeability for the wetting path.

For the rigid-wall permeameter, it is possible to measure the change in height of the specimen and/or to apply a net vertical stress during the test by adding a piston and a load cap on top of the specimen. Examples of tests conducted in such a manner are given in Wayllace and Lu (2012), and Fredlund and Houston (2013).

12.4.2.5 Flexible-wall permeameter

The test procedures using the flexible-wall permeameter are largely similar to those for the rigid-wall permeameter except that there is an additional stress state variable to control which is the net confining stress. The test is usually conducted at a constant net confining stress. The test is started with the soil specimen in the fully saturated condition. The matric suctions for the whole test should be planned as explained for the rigid-wall permeameter. The maximum matric suction is governed by the AEV of the ceramic disk and the final water pressure below the ceramic disk. For the flexible-wall permeameter, the net confining stress, i.e., the difference between cell pressure and air pressure must be maintained throughout the test.

The test is started by applying equal air and water pressures that are both at the maximum values (e.g., 450 kPa). If the air pressure is adjusted, the cell pressure needs to be adjusted accordingly to maintain constant net confining pressure. At no time during the test should the confining pressure be smaller than the air pressure, as the soil specimen may be damaged when this happens. The matric suction is applied by keeping air pressure constant and decreasing the water pressure. Keeping the air pressure constant is sensible in the flexible-wall permeameter, as it means that the cell pressure does not need to be adjusted throughout the test to maintain constant net confining pressure.

The water starts to flow out from the soil specimen when the matric suction is applied. At each matric suction, measure the volume of the water outflow from the soil specimen with time. The measurements should be done at regular intervals to obtain the nonlinear shape of the water outflow volume with time. The test at each matric suction can stop when the water outflow becomes negligibly small. In ASTM D7644-10, negligibly small water outflow is defined when the water outflow volume in 1 hour is less than 1% of the water outflow volume in the previous hour. The next matric suction is then applied and the measurement of water outflow with time is then repeated. This process is carried out until the last matric suction.

At the highest applied matric suction, the matric suction can be decreased in the reverse manner and volume of water inflow can be measured with time to obtain the unsaturated permeability for the wetting path. In the wetting test, it is possible to reduce the matric suction to zero to measure the saturated permeability.

12.4.2.6 Finishing

At the end of the permeability test, the air and water pressures are reduced to zero in quick steps and then the water pressure line is locked to prevent the soil specimen from swelling by imbibing the water. For the flexible-wall permeameter, the cell pressure will also have to be reduced in tandem with the air pressure. The final net confining pressure at zero air pressure is only reduced to zero after the water pressure line is locked.

The soil specimen can then be removed. The dimensions, weight and final water content of the specimen are then determined.

12.5 CALCULATIONS AND DATA INTERPRETATION

12.5.1 Instantaneous profile method

The instantaneous profile method applies mass continuity to transient profiles of suction and water content to determine the permeability of unsaturated soils. It does not matter which initial soil condition and boundary conditions were applied in the soil column test to obtain the transient profiles, the calculation to obtain the unsaturated permeability is the same. The profiles of suction and water content with time for the four different soil column tests are shown schematically in Figure 12.5.

The instantaneous profile method can also be applied when only either the matric suction profile or the volumetric water content is measured. However, the soil-water characteristic curve must be obtained separately. The missing matric suction profile or volumetric water content profile is then obtained using the soil-water characteristic curve.

The instantaneous profile method uses Darcy's law as given in Equation 12.1.

$$v = -k\frac{\partial H}{\partial z},$$

(12.1)

where
v = flow velocity
k = permeability
z = vertical coordinate (increasing in gravity direction)
H = total head for vertical flow (= $H_p + z$)
H_p = pressure head

The hydraulic gradient can be expressed using Equation 12.2.

$$-\frac{\partial H}{\partial z} = \frac{\partial H_p}{\partial z} + 1 = \frac{1}{\gamma_w}\frac{\partial u_w}{\partial z} + 1$$

(12.2)

Method	Suction profiles	Water content profiles
A1 (dry, infiltration)	Pore-water pressure	Volumetric water content
A2 (dry, imbibition)	Pore-water pressure	Volumetric water content

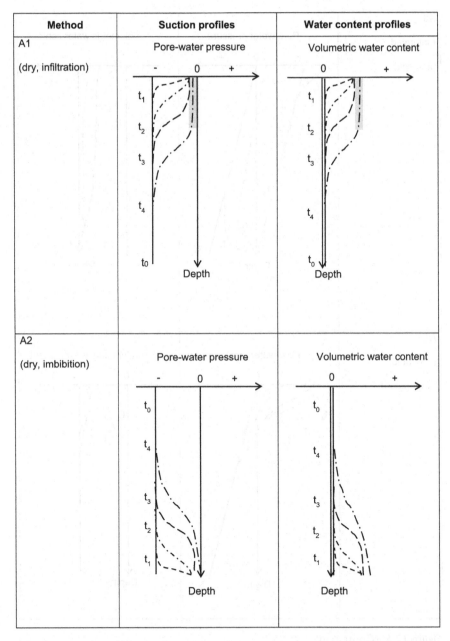

Figure 12.5 Schematic representation of suction and water content profiles in soil columns.

(Continued)

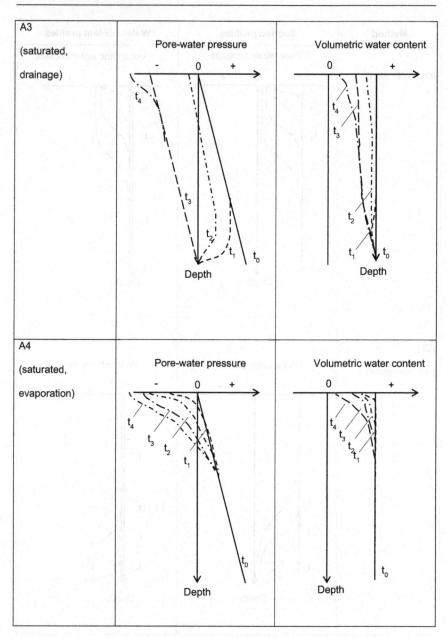

Figure 12.5 (Continued).

At any time t, Equation 12.2 can be expressed in finite difference form for two sensor depths z_m and z_{m+1} giving Equation 12.3.

$$-\left(\frac{\partial H}{\partial z}\right)_m = \frac{1}{\gamma_w}\left[\frac{(u_w)_m - (u_w)_{m+1}}{z_{m+1} - z_m}\right] + 1 \qquad (12.3)$$

The volume of water V_m passing depth m between time t and Δt can be obtained by integrating the volumetric water content θ downstream of depth m for the time interval Δt (Equation 12.4).

$$V_m = A\left[\sum_{i=m}^{n}\left(\theta_i^t - \theta_i^{t+\Delta t}\right)\left(z_{i+1} - z_i\right)\right] \qquad (12.4)$$

where
i = integration number representing sensor number
n = maximum number of sensors
A = cross-sectional area of soil column

Hence, the permeability $(k_w)_m$ at depth m can be obtained using Equation 12.5.

$$(k_w)_m = \frac{V_m}{A\Delta t}\left(\frac{1}{-\left(\partial H/\partial z\right)_m}\right) \qquad (12.5)$$

The value of $(k_w)_m$ is associated with the average suction of $[(u_w)_m + (u_w)_{m+1}]/2$ and the volumetric water content of θ_m.

The above calculations are repeated at each of the measurement points and for each time to define the permeability function. For infiltration (Method A1), it may be possible to reach a steady-state (unit hydraulic gradient) condition as indicated by the shaded zone in the suction profile and volumetric water content profile. If this is observed, the permeability at the suction and volumetric water content is given by the infiltration rate (Fredlund et al. 2012).

12.5.2 Multistep outflow method

The original multistep outflow method was developed by Gardner (1956). Gardner (1956) method is adopted in ASTM D7664-10 for data interpretation. The test results from the rigid-wall and flexible-wall permeameters are interpreted in the same way. To analyse the test results, it is assumed that the specimen is homogeneous and non-deformable, flow is one-dimensional, gravitational flow is negligible, unsaturated permeability k_w is constant and suction is linearly related to the volumetric water content for each suction increment, and impedance of the ceramic disk is negligible. The equation used to analyse test results from the multistep outflow method is Fick's second law of diffusion shown as Equation 12.6. There are different approaches to arriving at Equation 12.6 (Benson and Gribb 1997).

$$\frac{\partial\left(u_a - u_w\right)}{\partial t} = D\frac{\partial^2\left(u_a - u_w\right)}{\partial z^2} \tag{12.6}$$

where D is diffusivity given by Equation 12.7.

$$D = k_w \frac{\partial\left(u_a - u_w\right)}{\partial\theta} \tag{12.7}$$

The solution to Equation 12.6 is given by Gardner (1956) and is reproduced in Equation 12.8.

$$\ln\left(\frac{V_\infty - V}{V_\infty}\right) = \ln\left(\frac{8}{\pi^2}\right) - \left(\frac{D\pi^2}{4h_s^2}\right)t \tag{12.8}$$

where
V = outflow volume at time t
V_∞ = final outflow volume for suction increment
h_s = height of soil specimen

The diffusivity D for each suction increment can be obtained by plotting $\ln\left(\frac{V_\infty - V}{V_\infty}\right)$ versus t. A best fit line is drawn through the plot, and the slope of the line is given by $\left(\frac{D\pi^2}{4h_s^2}\right)$. Once D is obtained, k_w can be obtained using

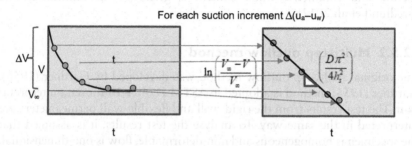

Figure 12.6 Illustration of obtaining D from multistep outflow test for one suction increment.

Equation 12.7. The steps are illustrated in Figure 12.6. Using Figure 12.6, k_w is given by Equation 12.9.

$$k_w = D \frac{\dfrac{\Delta V}{V_t}}{\Delta(u_a - u_w)} \tag{12.9}$$

where V_t is the total volume of the soil specimen. Repeating the calculation for each of the suction increments, the permeability function is obtained.

12.6 SUMMARY

The accuracy of the transient-state methods in obtaining the permeability function for unsaturated soils is not as good as the steady-state methods. For the instantaneous profile method, the range of suction over which permeability is determined is lower than in the multistep outflow method. Preparing a homogeneous soil column may not be easy, and the instruments (suction and water content sensors) may interfere with the flow process. For the large-diameter soil column, preferential flow and fingering processes may be observed and flow may not be one-dimensional. Vapour flow processes may also occur, and there is inaccuracy in measuring suction and water content as the changes may be small. There is also no control over the stress state. For the multistep outflow method, there is better control of the quality of the soil specimen, as it is small. The original multistep outflow method was designed to be used with the test to obtain the soil-water characteristic curve, thus it is an expedient process where both the soil-water characteristic curve and permeability function can be obtained at the same time. The accuracy of determining the permeability is obtained with smaller suction increments. However, this will increase the test duration and increases the likelihood of diffused air appearing in the water reservoir beneath the HAE ceramic disk and interferes with the water outflow/inflow volume measurement. In Gardner's (1956) method of calculating permeability, plate impedance has been neglected. Many have tried to improve the outflow method (Miller and Elrick 1958; Rijtema 1959; Kunze and Kirkham 1962; Kool et al. 1985; van Dam et al. 1994; Crescimanno and Iovino 1995; Hwang and Powers 2001; Figueras et al. 2006; Wayllace and Lu 2012; Shao et al. 2017).

REFERENCES

Benson, C. H. & Gribb, M. M. (1997). Measuring unsaturated hydraulic conductivity in the laboratory and field. In Houston, S. & Fredlund, D. G. (Eds.), *Unsaturated soil engineering practice*, Geotechnical Special Publication No. 68. New York: ASCE, pp 113–168.

Caicedo, B. (2017). Physical modelling of freezing and thawing of unsaturated soils. *Géotechnique*, 67(2): 106–126. http://dx.doi.org/10.1680/jgeot.15.P.098.

Crescimanno, G. & Iovino, M. (1995). Parameter estimation by inverse method based on one-step and multi-step outflow experiments. *Geoderma*, 68(4): 257–277.

Figueras, J., Gribb, M. M., & McNamara, J. P. (2006). *Design of a new automated multi-step outflow test apparatus, presented at the AGU fall meeting*. San Francisco, CA: American Geophysical Union, www.agu.org

Fredlund, D. G. & Houston, S. L. (2013). Interpretation of soil-water characteristic curves when volume change occurs as soil suction is changed. In *Proceedings of the First Pan-Am Conference on Unsaturated Soils*, UNSAT 2013, Advances in Unsaturated Soils, Editor: Bernardo Caicedo, February 20–22, pp. 15–31.

Fredlund, D. G., Rahardjo, H., & Fredlund, M. D. (2012). *Unsaturated soil mechanics in engineering practice*. New York: John Wiley & Sons, Inc., 926 pages (ISBN 978-1-118-13359-0).

Fujimaki, H. & Inoue, M. (2003). A transient evaporation method for determining soil hydraulic properties at low pressure. *Vadose Zone Journal*, 2: 400–408. https://doi.org/10.2136/vzj2003.4000a

Garnier, P., Rieu, M., Boivin, P., Vauclin, M., & Baveye, P. (1997). Determining the hydraulic properties of a swelling soil from a transient evaporation experiment. *Soil Science Society of America Journal*, 61: 1555–1563. https://doi.org/10.2136/sssaj1997.03615995006100060003x

Goh, S. G., Rahardjo, H., & Leong, E. C. (2015). Modification of triaxial apparatus for permeability measurement of unsaturated soils. *Soils and Foundations*, 55(1): 63–73. http://dx.doi.org/10.1016/j.sandf.2014.12.005.

Hopmans, J. W., Šimůnek, J., Romano, N., & Durner, W. (2002). Simultaneous determination of water transmission and retention properties: Inverse methods. In J. H. Dane & G. C. Topp (Eds.), *Methods of soil analysis*. Part 4. SSSA Book Ser. 5. Madison, WI: SSSA, pp. 963–1008.

Hwang, S. & Powers, S. E. (2001). Using a multi-step outflow technique to estimate unsaturated hydraulic conductivity function for quartz sands with variable interfacial properties, AGU fall meeting, San Francisco, CA: American Geophysical Union, www.agu.org

Kool, J. B., Parker, J. C., & Van Genuchten, M. T. (1985). Determining soil hydraulic properties from one-step outflow experiments by parameter estimation: I. Theory and numerical studies. *Soil Science Society Of America Journal*, 49(6): 1348–1354.

Kool, J. B., Parker, J. C., & Van Genuchten, M. Th. (1987). Parameter estimation for unsaturated flow and transport models – A review. *Journal of Hydrology*, 91: 255–293.

Krisdani, H., Rahardjo, H., & Leong, E. C. (2009). Use of instantaneous profile and statistical methods to determine permeability functions of unsaturated soils. *Canadian Geotechnical Journal*, 46(7): 869–874.

Kunze, R. J. & Kirkham, D. (1962). Simplified accounting for membrane impedance in capillary conductivity determinations. *Soil Science Society of America Journal*, 26(5): 421–426. https://doi.org/10.2136/sssaj1962.03615995002600050006x

Leij, F. J. & van Genuchten, M. Th. (1999). Characterization and measurement of the hydraulic properties of unsaturated porous media. In M. Th. van Genuchten et al. (Eds.), *Characterization and measurement of the hydraulic properties of unsaturated porous media: Proceedings of the International Workshop*. 22–24 October 1997. Riverside, CA: U.S. Salinity Lab., pp. 1–12.

McCartney, J. S. & Zornberg, J. G. (2010). Centrifuge permeameter for unsaturated soils II: Results and analysis. *Journal of Geotechnical and Geoenvironmental Engineering*, 136(8): 1064–1076.

Miller, E. E. & Elrick, D. E. (1958). Dynamic determination of capillary conductivity extended for non-negligible membrane impedance. *Soil Science Society of America Journal*, 22(6): 483–486. https://doi.org/10.2136/sssaj1958.03615995002200060002x

Minasny, B. & Field, D. F. J. (2005). Estimating soil hydraulic properties and their uncertainty: The use of stochastic simulation in the inverse modelling of the evaporation method. *Geoderma*, 126: 277–290. https://doi.org/10.1016/j.geoderma.2004.09.015

Parks, J., Stewart, M., & McCartney, J. S. (2012). Validation of a centrifuge permeameter for investigation of transient infiltration and drainage flow processes in unsaturated soils. *Geotechnical Testing Journal*, 35(1): 182–192.

Rijtema, P. E. (1959). Calculation of capillary conductivity from pressure plate outflow data with nonnegligible membrane impedance. *Journal of Agriculture Science*, 4: 209–215.

Samingan, A. S., Leong, E. C., & Rahardjo, H. (2003). A flexible wall permeameter for measurements of water and air coefficients of permeability of residual soils. *Canadian Geotechnical Journal*, June, 40(3): 559–574.

Schelle, H., Iden, S. C., & Durner, W. (2011). Combined transient method for determining soil hydraulic properties in a wide pressure head range. *Soil Science Society of America Journal*, 75: 1681–1693. https://doi.org/10.2136/sssaj2010.0374

Schindler, U., Durner, W., von Unold, G., & Müller, L. (2010). Evaporation method for measuring unsaturated hydraulic properties of soils: Extending the measurement range. *Soil Science Society of America Journal*, 74: 1071–1083. https://doi.org/10.2136/sssaj2008.0358

Shao, L., Wen, T., Guo, X., & Sun, X. (2017). A method for directly measuring the hydraulic conductivity of unsaturated soil. *Geotechnical Testing Journal*, 40(6): 907–916. https://doi.org/10.1520/GTJ20160197. ISSN 0149-6115

Tian, Z., Kool, D., Ren, T., Horton, R., & Heitman, J. L. (2019). Approaches for estimating unsaturated soil hydraulic conductivities at various bulk densities with the extended Mualem-van Genuchten model. *Journal of Hydrology*, 572: 719–731. https://doi.org/10.1016/j.jhydrol.2019.03.027

van Dam, J. C., Stricker, J. N. M., & Droogers, P. (1994). Inverse method to determine soil hydraulic functions from multistep outflow experiments. *Soil Science Society of America Journal*, 58: 647–652. https://doi.org/10.2136/sssaj1994.03615995005800030002x

Wayllace, A. & Lu, N. (2012). A transient water release and imbibitions method for rapidly measuring wetting and drying soil water retention and hydraulic conductivity functions. *Geotechnical Testing Journal*, 35(1): 1–15.

Wendroth, O., Ehlers, W., Hopmans, J. W., Kage, H., Halbertsma, J., & Wösten, J. H. M. (1993). Reevaluation of the evaporation method for determining hydraulic functions in unsaturated soils. *Soil Science Society of America Journal*, 57: 1436–1443. https://doi.org/10.2136/sssaj1993.03615995005700060007x

Wildenschild, D., Hopmans, J. W., & Šimůnek, J. (2001). Flow rate dependence of soil hydraulic characteristics. *Soil Science Society of America Journal*, 65: 35–48. https://doi.org/10.2136/sssaj2001.65135x

Wind, G. P. (1968). Capillary conductivity data estimated by a simple method. pp. 181–191. In Water in the unsaturated zone: Proc. Symp., Wageningen, the Netherlands. June 1966. Vol. 1. Gentbrugge, Belgium: International Association of Scientific Hydrology.

Yang, H., Rahardjo, H., Wibawa, B., & Leong, E. C. (2004). A soil column apparatus for laboratory infiltration study. *Geotechnical Testing Journal*, 27(4): 1–8.

Zornberg, J. G. & McCartney, J. S. (2010). Centrifuge permeameter for unsaturated Soils I: Theoretical basis and experimental developments. *Journal of Geotechnical and Geoenvironmental Engineering*, 136(8): 1051–1063.

FURTHER READING

Hopmans, J. W., Šimůnek, J., Romano, N., & Durner, W. (2002). 3.6.2. Inverse methods. In J. H. Dane & G. Clarke Topp (Eds.), *Methods of soil analysis.* https://doi.org/10.2136/sssabookser5.4.c40

Chapter 13

Oedometer test

13.1 BACKGROUND

The oedometer test (ASTM D2435/D2435M 2011; BS EN ISO 17892-5 2017) is an incremental loading compression test that is commonly carried out to determine the one-dimensional consolidation properties of saturated soils. The test can also determine the secondary compression (creep) of saturated soils by extending the test duration under a constant load and also the swelling potential of expansive soils or collapse potential of collapsible soil (ASTM D3877 2008; ASTM D4546 2008; and ASTM D5333 2003). While the oedometer test is commonly used for saturated soil, it can also be used for unsaturated soil. Three tests for unsaturated soils that can be performed using the oedometer apparatus are

1. unsaturated one-dimensional compression test,
2. wetting test, and
3. swelling/collapse test.

The test procedures for unsaturated soils using the oedometer apparatus have not been standardised for the first two tests. However, the test procedures for swelling and collapse of soils have been standardised in ASTM D3877 (2008), ASTM D4546 (2008) and ASTM D5333 (2003). In this chapter, the test procedures for the first two tests will be elaborated while the test procedures for the swelling and collapse of soils will only be described for unsaturated soils.

Unlike the saturated oedometer test, which is only dependent on effective stress, the unsaturated oedometer test is dependent on net normal stress and matric suction/water content. As a result, it requires more effort to determine the parameters to define the unsaturated compression curve. Thus, it is important to understand the field condition such as

1. the expected net normal stress that applies at the site, and
2. the initial water content/matric suction and its variation at the site.

13.2 RELATED STANDARDS

1. ASTM D3877. 2008. Standard Test Methods for One-Dimensional Expansion, Shrinkage, and Uplift Pressure of Soil-Lime Mixtures.
2. ASTM D4546. 2008. Standard Test Methods for One-Dimensional Swell or Collapse of Cohesive Soils.
3. ASTM D4829. 2011. Standard Test Method for Expansion Index for Soils.
4. ASTM D5333. 2003. Standard Test Method for Measurement of Collapse Potential of Soils.
5. ASTM D4186. 2006. Standard Test Method for One-Dimensional Consolidation Properties of Saturated Cohesive Soils Using Controlled-Strain Loading.
6. ASTM D2435/D2435M. 2011. Standard Test Methods for One-Dimensional Consolidation Properties of Soils Using Incremental Loading.
7. BS EN ISO 17892-5. 2017. Geotechnical Investigation and Testing. Laboratory Testing of Soil. Part 5: Incremental Loading Oedometer Test.
8. BS 1377-5 (1990) Methods of Test for Soil for Civil Engineering Purposes – Part 5: Compressibility, Permeability and Durability Tests, BSI.

13.3 THEORY

13.3.1 Settlement and heave

One major consideration in geotechnical engineering design is the estimation of settlement (downward movement of soil relative to the initial elevation) and heave (upward movement of soil relative to the initial elevation). There are five sources of settlement or heave that need to be considered:

1. Soil distortion
2. Primary compression (consolidation)
3. Secondary compression (creep)
4. Wetting-induced volume change (swelling and collapse)
5. Drying-induced volume change (shrinkage)

Soil distortion is a condition where the soil particles move due to loading. Figure 13.1 illustrates settlement-inducing soil distortion. Consider an initially horizontal ground level indicated by a straight line subjected to a load. Once the load has been applied, the soil is displaced sideways without causing volume change. The difference in elevation between the initial ground

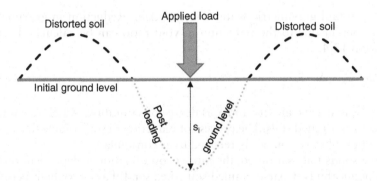

Figure 13.1 Settlement due to soil distortion.

level and the post-loading ground level is referred to as settlement/heave due to soil distortion (s_i).

Soil distortion occurs instantaneously and thus is considered immediate settlement. The analysis can be done by using elastic theory, and modulus is taken as undrained modulus (E_u). Modulus E_u can be obtained by using an undrained triaxial test, which is explained in Chapter 15.

Primary compression is a reduction in the volume of voids in a soil. For saturated soil, it is known as consolidation, as it involves the drainage of pore water with time. When a soil is subjected to a stress increase $\Delta\sigma$, its pore-water pressure u_w will increase (Δu_w). Under one-dimensional loading:

$$\Delta u_w = B\Delta\sigma \tag{13.1}$$

where B is the Skempton pore pressure parameter. For saturated soils, $B = 1$, and B is less than 1 for an unsaturated soil.

When the soil is subjected to an additional load of $\Delta\sigma$, σ is calculated using Equation 13.2.

$$\sigma = \sigma_0 + \Delta\sigma \tag{13.2}$$

where σ_0 is the initial overburden pressure. Excess pore-water pressure (u_e) which is equal to $\Delta\sigma$ is generated and u_w is equal to

$$u_w = u_0 + u_e \tag{13.3}$$

where u_0 is the initial pore-water pressure and is usually hydrostatic pressure. As σ equals to $\sigma_0 + \Delta\sigma$ all the time, σ' increases as u_e gradually dissipates over time as water flows out from the voids and causes reduction of the void volume, which is commonly represented as reduction of void ratio (Δe) and gravimetric water content (Δw). Reduction in gravimetric water content due to the reduction in void ratio can be calculated using Equation 13.4.

$$\Delta w \cdot G_s = \Delta e \cdot S_r \qquad\qquad\qquad (13.4)$$

where G_s is specific gravity and S_r is degree of saturation, which remains at 100% for saturated soils. The process of reduction in soil volume due to the dissipation of u_e is commonly referred to as consolidation.

If the soil is fully saturated, the rate of consolidation is dependent on the soil's permeability. Coarse-grained soils (i.e., sand) have very high permeability, and thus consolidation takes place rapidly (in terms of minutes to days). Thus, the compression settlement of coarse-grained soils is considered immediate settlement. For fine-grained soils (i.e., silts and clay), consolidation may take months to years, and thus is referred to as time-dependent settlement or consolidation settlement. If the soil is unsaturated ($S_r < 100\%$), compression of the voids filled with air will occur first and no water will flow out from the soil pores, and thus water content remains constant. Compression/decompression that does not cause any change in water content is referred to as the constant water content (CWC) condition. In the CWC condition, the pore-water pressure response depends on the initial degree of saturation, which is described as follows (Wijaya and Leong 2016; Wijaya 2017):

1. If the final degree of saturation remains less than 100%, no excess pore-water pressure will be generated, and no water will enter or leave the soil element. However, the pore-water pressure will increase (although it remains negative) as degree of saturation increases due to the reduction in void ratio under the CWC condition. Such behaviour is known as mechanical wetting.
2. If the final degree of saturation is 100%, excess pore-water pressure will be generated. Once the degree of saturation reaches 100%, the soil will behave like a saturated soil undergoing consolidation.

Unsaturated soil loaded under the CWC condition may experience wetting-induced volume change. However, it is difficult to separate pure compression behaviour from the wetting-induced volume change (Alonso et al. 1990), especially when there is no measurement of pore-water pressure during the test.

Secondary compression or creep is a change in volume under constant σ' without any change in u_w (Holtz and Kovacs 1981; Holtz et al. 2011). Secondary compression takes place over a very long time. It is unclear if secondary compression occurs at the same time as primary consolidation or only after primary consolidation ends. Secondary compression is usually negligible for most soils and is only significant for peaty or organic soils. The secondary compression index can be determined using an oedometer test by extending the test duration for a constant load.

Some soils may either swell or collapse on wetting. Swelling is defined as an increase in the soil volume due to wetting and may occur in both saturated and unsaturated soils. Collapse is usually associated with unsaturated soils on wetting. Soils that experience swelling will also experience shrinkage on drying (drying-induced volume change). The shrinkage curve described in Chapter 5 can be used to estimate the amount of swelling/shrinkage with respect to the change in water content. If soil movement is constrained for a swelling soil (i.e., constant volume), swelling pressure will be generated. Swelling and shrinkage potentials differ according to soil types, which can be categorised as follows:

1. Highly expansive soils, e.g., soils that contain montmorillonite or illite
2. Moderately expansive soils: fine-grained soils
3. Non-expansive soils: coarse-grained soils

Highly expansive soils require a thorough investigation of their swelling and shrinkage properties. Swelling and shrinkage can be ignored for non-expansive coarse-grained soils such as gravel and sand. Some soil deposits that are aeolian, loessal, subaerial, mudflows, alluvial, residual or manmade fills experience significant collapse when wetted. Collapse is defined as an irrecoverable settlement due to rearrangement of particles that is caused by wetting, additional loading or both. Collapse due to wetting is also known as hydro-consolidation, Whether a soil experiences swelling or collapse after being wetted depends on the applied stress and stress history.

13.3.2 Compression, shrinkage and wetting-induced volume change

Compression, shrinkage and wetting- and drying-induced volume change of soil are stress path dependent (different stress paths may lead to different results) as illustrated in Figure 13.2a. Figure 13.2b shows CWC curves under different initial water contents, while Figure 13.1c shows the wetting path along the CWC curve under different inundation pressures.

13.3.3 Saturated compression test

In Figure 13.2a, the initial soil specimen condition is located at point 1 where the net normal stress is zero and the natural or in situ soil suction is equal to s_n. When the soil specimen is subjected to a saturated compression test, the soil is first inundated with water. Inundation is commonly carried under constant stress conditions, and the plane that represents constant stress is referred to as the constant stress (CP) plane. Inundation causes the soil suction to go to zero, and the soil volume will increase (swelling). As a result, point 1 will move to point 2s. When the soil becomes saturated, the pore-air

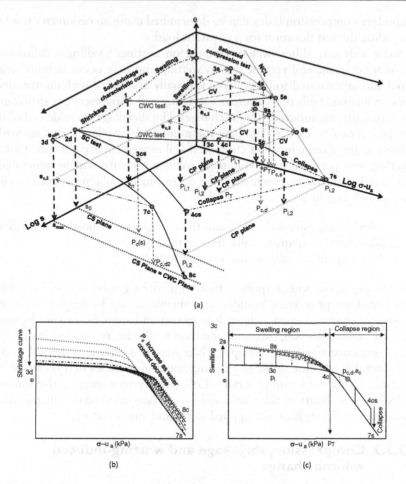

Figure 13.2 Schematic of unsaturated oedometer test in general. (a) 3D Schematic of compression shrinkage and wetting induced behaviour of unsaturated soil. (b) Constant water content compression curve under different initial water content. (c) Wetting induced volume change under different inundation pressure.

pressure (u_a) becomes pore-water pressure (u_w) and net normal stress $(\sigma-u_a)$ is equal to the effective stress $(\sigma-u_w)$. On further loading, the effective stress will increase and the soil specimen will move along the reloading curve until it reaches its saturated preconsolidation pressure (points 2s-3s-4s-5s), which is equal to $P_{c,s}$ with void ratio (e_c) equal to $e_{c,1}$. Further loading will cause the volume change to follow the normal compression line (NCL) which is indicated by points 5s-6s-7s. Figure 13.3 shows the saturated compression curve obtained from a kaolin specimen, KOdSat1 (Wijaya 2017) along with its typical parameters, such as swelling index (C_s), recompression index (C_r), compression index (C_c) and saturated preconsolidation pressure $(P_{c,s})$.

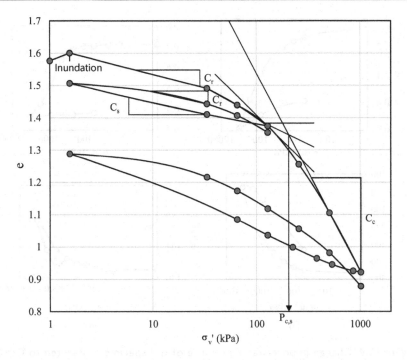

Figure 13.3 Saturated compression test of KOdSat1 from Wijaya (2017).

13.3.4 CWC compression test

If the soil is loaded without inundation (points 1-2c-5s), the loading is under CWC condition, and the test is referred to as the CWC test. Under this loading, the soil suction decreases due to mechanical wetting (the degree of saturation increases due to a decrease in void ratio under the CWC condition), as shown in Figure 13.4. If the soil suction becomes zero, further loading will cause the compression curve to merge with the NCL (points 5s-6s-7s).

If the soil specimen is dried (point 1 to point 2d), the water content decreases and soil suction increases. If the water content is low enough, loading the soil specimen under the CWC condition (points 2d-3c-4c-5c-6c) may cause the soil specimen to remain unsaturated beyond its apparent preconsolidation pressure or yield pressure $(P_{c,d})$ located at point 5c. Drying the soil specimen will cause the preconsolidation void ratio to decrease (similar to the soil specimen being subjected to a higher preconsolidation pressure and then unloaded) and e_c becomes $e_{c,2}$ while the yield pressure increases from $P_{c,s}$ to $P_{c,d}$.

Drying the soil specimen will not cause further shrinkage if the soil specimen has reached the zero-shrinkage line. The void ratio on the zero-shrinkage line is referred to as e_{min}. At this point, soil suction has no effect on soil structure. If points 2d and 3d are located on the zero-shrinkage line, any

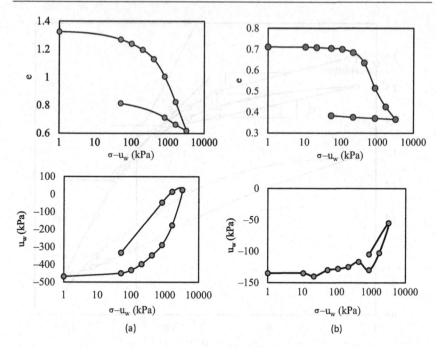

Figure 13.4 Change in pore-water pressure of soil specimen subjected to CWC test. (a) Kaolin specimen Kw30s468 (Wijaya 2017). (b) Compacted silty clay 7-10-H (Jotisankasa et al. 2007).

further drying will no longer reduce the prior compression void ratio (e_0) and will no longer reduce e_c. If the soil specimen is loaded under CWC condition (points 3d-7c-8c) the soil specimen will yield at $P_{c,d2}$, which is higher than $P_{c,d}$ despite $P_{c,d}$ and $P_{c,d2}$ having the same preconsolidation void ratio $e_{c,2}$. The reason is due to the presence of water, which acts to reduce friction between the soil particles. When there is less water at the interface between particles, it requires a higher pressure to reach the same amount of particle rearrangement.

13.3.5 Wetting-induced swelling/swelling pressure

If the soil is inundated at point 2c, the soil suction will reduce under constant net normal stress conditions. As swelling does not affect the preconsolidation void ratio (similar to unloading, which does not change the preconsolidation pressure), the soil specimen will swell to the saturated compression curve (point 3s) that is obtained if the soil specimen is inundated at point 1. However, if the soil specimen is inundated under constant volume conditions (i.e., the soil specimen is prevented from swelling), swelling pressure (P_s) will be generated that can be obtained by first projecting point 2c to the saturation plane under the constant void ratio (point 3s') and then drawing a horizontal line that intersects the saturated compression curve (point 4s). The effective stress required to reach point 4s is the swelling pressure. The value of e_c remains at $e_{c,1}$ regardless of if the soil is inundated at point 1, point 2c or not inundated, as inundation is equivalent to unloading, which does not affect the most compact structure that the soil has experienced before.

If the soil specimen is inundated at point 3c, the soil specimen will swell to point 8s, which is located on the second saturated reloading line (points 8s-9s-6s) that lies below the first saturated reloading line (points 2s-3s-4s-5s). The second saturated reloading line will merge into the NCL at point 6s with e_c equals to $e_{c,2}$, which is equivalent to CWC loading (points 2d-3c-4c-5c).

Point 4c is the point on the unsaturated compression curve (points 2d-3c-4c-5c-6c) which has the same void ratio as the saturated compression curve (points 8s-9s-6s-7s). At this point, inundation will not cause any swelling or swelling pressure. Net normal stress located at this point is referred to as the transition pressure P_T (Wijaya and Leong 2015).

Figure 13.5 shows the swelling test result on a kaolin specimen dried to 5% water content (25) and then subjected to stress path A-B-C-D-E. The soil was reloaded and then unloaded prior to wetting to remove any disturbance on the soil specimen. On inundation, the soil swells from point E and reaches point F. Any further loading will cause the specimen to follow the saturated path F-G-H-J compression line to be located at the NCL. The unsaturated compression line (UCL) is obtained from the specimen which was air dried to 1% water content (Kw1) and is observed to be parallel to NCL.

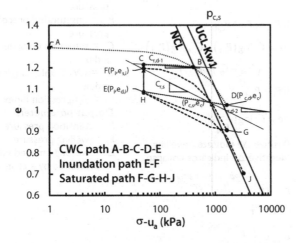

Figure 13.5 Wetting-induced swelling on specimen w5 (from Wijaya and Leong 2015).

13.3.6 Wetting-induced collapse

If the soil specimen is inundated at net normal stress higher than P_T (i.e., point 6c) with P_i equal to $P_{i,2}$, the soil specimen will undergo irreversible volumetric compression, which is referred to as collapse or hydroconsolidation from point 6c to point 7s (located at the same net normal stress). Collapse occurs as the unsaturated compression curve (points 2d-3c-4c-5c-6c) is located above the saturated compression curve (points 8s-9s-6s-7s). Table 13.1 shows the relationship between swelling, collapse and swelling pressure that can be obtained from the CWC test (Wijaya and Leong 2015). Figure 13.6 shows a kaolin specimen subjected to the CWC stress path A-B-C-D-E-D and then wetted when it is on the UCL. It shows that the soil volume decreases from point D to F and falls onto the NCL following the saturated path F-G.

13.3.7 Constant-suction compression test

For the constant-suction (CS) test, the soil suction is maintained constant given by the unsaturated compression curve (points 2d-3cs-4cs). As a result, the yield pressure is a function of soil suction $P_c(s)$. It is very difficult to

Table 13.1 Relationship between compressibility wetting-induced volume change/pressure of soils in CWC test (Wijaya and Leong 2015)

Relationship	Equation	Parameters
Swelling/ collapse volume change	If $P_i < P_{c,s}$ $$\Delta e = C_{r,s} \log\left(\frac{P_{c,s}}{P_i}\right) - C_{r,d} \log\left(\frac{P_{c,d}}{P_i}\right)$$ If $P_{c,s} \le P_i \le P_{c,d}$ $$\Delta e = C_c \log\left(P_{c,s}\right) - C_{r,d} \log\left(P_{c,d}\right) + \left(C_{r,d} - C_c\right)\log\left(P_i\right)$$ If $P_i > P_{c,d}$ $$\Delta e = C_c \log\left(\frac{P_{c,s}}{P_{c,d}}\right)$$ Positive Δe indicates swelling Negative Δe indicates collapse	**Input parameters** P_i = inundation pressure $P_{c,s}$ = saturated preconsolidation pressure $P_{c,d}$ = unsaturated preconsolidation pressure $C_{r,s}$ = saturated unloading/reloading index $C_{r,d}$ = CWC unloading/reloading index C_c = compression index **Output parameters** P_T = transition pressure P_s = swelling pressure Δe = wetting-induced volume change represented by a change in the void ratio
Transition pressure	$$P_T = 10^{\left[\frac{C_{r,s}\log(P_{c,s}) - C_{r,d}\log(P_{c,d})}{(C_{r,s} - C_{r,d})}\right]}$$	
Swelling pressure	If $P_i < P_T$ $$P_s = P_{c,s}\left(\frac{P_i}{P_{c,d}}\right)^{C_{r,d}/C_{r,s}}$$	

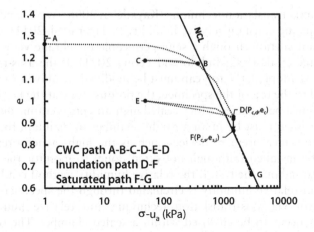

Figure 13.6 Wetting-induced collapse on specimen w10 (from Wijaya and Leong 2015).

maintain constant suction as compression causes instantaneous mechanical wetting. Thus, the test must be carried out at a very slow loading rate to allow the soil to equilibrate at the constant suction.

If the water content of the soil is very low, say at point 3d, the degree of saturation hardly changes, and thus mechanical wetting is insignificant. At such low water content (or high suction), the soil-water characteristic curves (w-SWCC and S-SWCC) merge into a single line as indicated in Chapter 10. Thus, the CWC test located at very low water contents is almost equivalent to the CS test.

13.3.8 Configuration of oedometer tests

Oedometer tests can be differentiated into:

1. Saturated oedometer test (BS EN ISO 17892-5 2017; ASTM D2435/ D2435M 2011).
2. Suction-control oedometer test (SC test).
3. Constant water content oedometer test (CWC test).

The saturated oedometer test is a compression test (under incremental loading), where the soil specimen is assumed to remain saturated throughout the test ($S_r = 100\%$). The saturation is achieved by inundating the soil specimen. It has been standardised in ASTM D4186 (2006) and BS EN ISO 17892-5 (2017) and mentioned in many books (e.g., Head 2011). Thus, it will not be further elaborated in this book.

The SC test requires extensive modification of the oedometer set-up as it requires suction control to be incorporated. Suction is controlled usually by using the axis-translation technique (Kayadelen 2008; Aversa and Nicotera 2002), applying constant relative humidity (Salager et al. 2011) and applying constant suction through a semi-permeable membrane via a salt solution (Delage et al. 1992; Bulolo and Leong 2021). If the axis-translation technique is adopted, the top cap must be modified to allow air pressure to be applied to the top of the specimen, the porous stone at the bottom of the specimen must be replaced with a sealed high-air entry ceramic disk and the whole specimen must be in an air-tight chamber to maintain the air pressure. Suction measurement devices such as the high-capacity tensiometer can also be installed (although not often done) to monitor the specimen suction throughout the test. If the relative humidity method is adopted, air at a certain relative humidity is circulated through the specimen. For suction control via axis-translation technique and relative humidity, the oedometer needs to be enclosed within a sealed chamber. The oedometer set-up where a constant suction is applied to the soil specimen through a semi-permeable membrane via a salt solution is commonly referred to as an osmotic oedometer.

In the SC test, the suction is maintained constant throughout the compression test in order to obtain the constant-suction compression curve.

As any stress change in unsaturated soil may cause an instantaneous change in the void ratio, only a small stress increment can be used, and suction equilibrium must be achieved during each loading/unloading stage. As a result, the SC test requires a very long test duration to be finished.

The procedure of the SC test is highly dependent on the technology used to control suction. Considering the cost and testing duration required in the SC test, the application of the SC test remains very much in the research laboratory and is difficult to implement in practice.

The CWC test does not require any modification to the oedometer apparatus. However, there is no control of soil suction or pore-water pressure measurement (due to latency in pore-water pressure response to loading). As soil suction is not controlled, the CWC compression curve is obtained instead of the constant-suction compression curve. Thus, the interpretation requires an alternative soil state, such as water content or degree of saturation.

As noted earlier, when the water content of the soil is very low such that void ratio, degree of saturation and suction are essentially constant during the test and the test approximates a CS test, the CWC test on such a soil specimen is similar to the SC test, and the compression curves can be treated as similar. Table 13.3 summarises some observations on the compression behaviour of unsaturated soils under CWC and SC tests.

If the CWC test incorporates suction measurement devices such as a high-capacity tensiometer (HCT), it is suggested to refer the test as CWC-P. One possible set-up of an oedometer with a HCT is shown in Figure 13.7. The advantages and disadvantages of each test are summarised in Table 13.2.

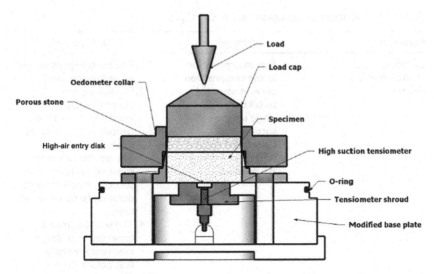

Figure 13.7 Modified base for CWC-P (from Wijaya and Leong 2016).

13.4 TEST METHODS

When testing unsaturated soil, a saturated soil test is also needed to provide a reference. In double oedometer test (Jennings and Knight 1957), two identical soil specimens are tested: one under saturated condition and the other under unsaturated condition (CWC test or SC test). In this section, general descriptions of oedometer test procedures for both saturated and unsaturated soils are given.

13.4.1 Saturated oedometer test

The test method for the saturated oedometer test has been standardised in BS EN ISO 17892-5 (2017), ASTM D2435/D2435M (2011), BS 1377-5 (1990), and it is briefly summarised next. More details can be found in Head (2011). There are three testing stages in the saturated oedometer test:

1. Specimen preparation stage
2. Inundation stage
3. Loading/unloading stage

13.4.1.1 Specimen preparation stage

The specimen preparation stage involves the following:

Table 13.2 Advantages and disadvantages of CWC procedures

Procedures	Advantages	Disadvantages
Suction-control oedometer test (SC test)	1. Easy interpretation as the compression curve is assumed to fall under a single unique soil suction. 2. Controlled suction wetting and drying are possible.	1. The most expensive type of oedometer test. 2. Maintaining the constant-suction condition is extremely difficult and technology dependent. 3. The equilibrium time required to maintain soil suction caused the testing duration to be extremely long. 4. The soil specimen is expected to undergo mechanical wetting first before the soil suction is recovered to the soil suction of interest. As a result, the specimen experiences complicated wetting and drying path.
Constant water content oedometer test with pore-water pressure measurement (CWC-P)	1. Relatively low cost in incorporating the HCT. 2. Pore-water pressure reading is available throughout the test. 3. Test duration is relatively short as it is controlled by the equilibrium time between HCT and soil specimen (in a range of 20 minutes to 1 hour per test).	1. HCT has to be sufficiently pressurised. The pressurising duration depends on the high-capacity tensiometer and the pressure magnitude. 2. Correlating the compressibility with soil suction remains difficult.
Constant water content oedometer test without pore-water pressure measurement (CWC)	1. The lowest cost, as there is no need to modify the conventional oedometer test. 2. Has the shortest testing duration.	1. Correlating soil compressibility with soil suction is not possible.

Table 13.3 Summary of unsaturated compression behaviour modified from Wijaya (2017)

Constant variable plane	Comment
Constant water content	1. Difficult to separate the pure compression behaviour from the mechanical collapse behaviour (Alonso et al. 1990). 2. Drying the soils will increase preconsolidation pressure and the preconsolidation void ratio (Leong et al. 2013; Jotisankasa 2005; Jotisankasa et al. 2007; Wijaya and Leong 2015; Wijaya and Leong 2016; Wijaya 2017). 3. Drying will not cause any further change in the preconsolidation void ratio once the soil has reached the zero-shrinkage line (Wijaya 2017) 4. The UCL can either merge or not merge with the saturated NCL, but the slope of the UCL seems to increase (less stiff) rather than decrease (Jotisankasa 2005; Burton et al. 2014). 5. The UCL is constant under the constant degree of saturation condition and will merge into the saturated compression line as the soil becomes saturated due to mechanical wetting (Wijaya and Leong 2015; Wijaya 2017). 6. Matric suction of soil will decrease when total stress is applied due to mechanical wetting (Wijaya 2017; Wijaya and Leong 2016; Jotisankasa et al. 2007) 7. Matric suctions at the top and bottom of the specimen may differ by around 5%–15% for a thin specimen of about 20 mm (Jotisankasa 2005). 8. The preconsolidation pressure is not only due to the past maximum compaction stress but also due to the inter-particle bonding which is caused by matric suction (Munoz et al. 2011). 9. Swelling does not change the preconsolidation void ratio (Wijaya 2017; Wijaya and Leong 2015) 10. The suction-degree of saturation relationship in the unloading-reloading path is reversible. However, the matric suction which is recorded prior to unloading is higher than the matric suction which is recorded at the end of the unloading-reloading cycle (Tarantino and De Col 2008). 11. Change in matric suction due to the change in net normal stress under the CWC condition becomes less significant as the soil becomes drier (Rahardjo 1990; Jotisankasa et al. 2007). 12. The slope of the unloading-reloading line appears to be independent of matric suction (Munoz et al. 2011). 13. The slope of the unloading-reloading line under the CWC condition is much smaller compared to the slope of the unloading-reloading line under the saturated condition.

(Continued)

Table 13.3 (Continued)

Constant variable plane	Comment
Constant matric suction	1. An increase in constant suction will increase the preconsolidation pressure (Alonso et al. 1990; Lloret et al. 2003).
	2. There are a number of uncertainties in the UCL of the unsaturated soils, and all of them are supported by experimental observations. The slope of the UCL may either reduce with suction (Alonso et al. 1990; Pham 2005), increase with suction and then merge with the saturated NCL (Toll 1990; Alonso 1993; Honda 2000; Sivakumar and Wheeler 2000; Futai et al. 2002; Silva et al. 2002; Wheeler and Sivakumar 1995) or be unaffected by matric suction (Larson et al. 1980; O'Sullivan 1992; Smith et al. 1997; Arvidsson and Keller 2004; Imhoff et al. 2004).
	3. When the degree of saturation equals 1, the UCL merges with NCL regardless of the matric suction of the soils (Mun and McCartney 2015).
	4. The slope of the unloading-reloading line appears to be independent of matric suction (Alonso et al. 1990; Wheeler and Sivakumar 1995; Munoz et al. 2011; Hoyos et al. 2012).
	5. Soils which are dried and then wetted to the desired matric suction are more susceptible to yield rather than soils which are only dried to the desired matric suction (Sivakumar et al. 2006).

1. *Set up the oedometer apparatus.*

   The oedometer must be set up such that it does not cause a change in water content or cause the specimen to swell. Thus, the procedures depend on soil type as follows:
   - For dry expansive soil, dry porous disk and dry filter paper must be used (ASTM D2435/D2435M 2011).
   - For initially unsaturated soil, damp porous disk and filter paper may be used (ASTM D2435/D2435M 2011).
   - For saturated soil with low affinity for water, saturated disk and saturated filter paper may be used (ASTM D2435/D2435M 2011).

2. *Place the specimen inside the oedometer apparatus.*

   If the specimen is placed into the oedometer apparatus and the test is not started immediately, a loose-fitting plastic or rubber membrane is used to cover the oedometer to minimise water content change due to evaporation (ASTM D2435/D2435M 2011).

3. *Apply the seating load on the specimen*

   ASTM D2435/D2435M (2011) recommends a nominal seating load of about 5 kPa, but a smaller seating load can be used if it is suspected that the soil will consolidate significantly under the nominal seating load. BS EN ISO 17892-5 (2017) recommends a seating load not exceeding 3 kPa, while BS 1377-5 (1990) recommends a seating load not exceeding 2 kPa.

13.4.1.2 Inundation stage

Inundation is carried out such that the specimen is not allowed to swell excessively prior to being loaded beyond its preconsolidation pressure (ASTM D2435/D2435M 2011). Inundation can be carried out as follows:

1. If the soil is saturated under field condition, ASTM, BS and BS EN ISO suggest proceeding to the loading stage immediately after the application of seating load to prevent the specimen from swelling.
2. If a specific condition is to be achieved, ASTM D2435/D2435M (2011) suggests applying a sufficiently large load (large enough to prevent the soil from swelling above its initial height) and inundating the specimen under the applied load. Record the inundation stress (vertical stress at which the soil is inundated) and the axial deformation during the inundation.

13.4.2 Loading/unloading stage

The test procedures are briefly summarised as follows:

1. Load is applied to the soil specimen under K_0 condition (the specimen is confined in an oedometer ring).
2. In the oedometer test, pore-water pressure measurement is generally not carried out (ASTM D2435/D2435M 2011). To determine the end of consolidation without measuring the pore-water pressure, several methods can be used:
   a. ASTM D2435/D2435M (2011) Method A and BS 1377-5 (1990): Use a standard load increment duration of 24 hours. Take at least two time-versus-deformation readings at two load increments, one of which must be taken after the preconsolidation pressure. The test duration may be extended if the time-versus-deformation reading during the 24 hours is suspected to be insufficient.
   b. ASTM D2435/D2435M (2011) Method B and BS EN ISO 17892-5 (2017): Use the settlement versus time curve to ensure that primary consolidation is completed, which can be shorter or longer than 24 hours.
3. Settlement is plotted against square-root time or log time to obtain t_{50} (time to 50% degree of consolidation) or t_{90} (time to 90% degree of consolidation), respectively, to determine the coefficient of

consolidation (c_v) and the permeability of the saturated soils (k_s) at the average void ratio before and after load increment.

4. Next, load increment is then applied, and steps 1 to 3 are repeated until the last load increment. Generally, a load increment ratio of one is used. Different numbers of load increments are suggested. BS 1377-5 (1990) recommends four to six load increments. BS EN ISO 17892-5 (2017) recommends at least seven load increments, while ASTM D2435/D2435M (2011) recommends that the number of load increments should be such that the following can be obtained:

   a. Three points which define a straight line in a log stress space
   b. Three points which define a concave up curve when plotted in log stress space
   c. Stress level which is eight times the estimated preconsolidation pressure

Convenient load increments to use are 6, 12, 25, 50, 100, 200, 400, 800, 1,600 and 3,200 kPa.

5. An unload-reload is performed at the end of the last load increment if required. For unloading, the load is removed in two load decrements and reloaded back to the maximum load. For the unload-reload, it is sufficient to wait until the vertical dial gauge stops moving at each load decrement or increment which is expected to occur within a short time.

6. After the oedometer test, a compression curve of void ratio versus load is plotted. The compression index (C_c), saturated reloading index (C_{rs}), saturated unloading index $(C_{u,s})$, saturated unloading-reloading index $(C_{ur,s})$ and saturated preconsolidation pressure $(P_{c,s})$ can be determined from the compression curve.

13.4.3 Unsaturated oedometer test

For the oedometer test on unsaturated soil (either CWC or SC test), there is an additional stage at the beginning compared to the oedometer test for saturated soil, which is the suction application stage.

The suction application stage is to bring the soil specimen to the initially desired soil suction. This stage can be carried out by air drying the sample or using an axis-translation apparatus such as the pressure plate. The initial suction application stage is carried out prior to cookie cutting the sample into the oedometer ring. In the SC test, the initial suction can also be applied to the specimen in the oedometer apparatus. However, shrinkage of the specimen may occur, and the specimen may shrink away from the sides of the oedometer ring. The advantage of the SC test over the CWC test is that the initial suction of the specimen can be ensured prior to testing. The procedures for the oedometer test on unsaturated soil are similar to the oedometer test for saturated soil.

13.4.4 CWC and CWC-P tests

CWC and CWC-P tests have similar testing procedures except that suction is monitored during the CWC-P test. The advantage of the CWC-P test is that it is able to provide suction versus void ratio (e) relationship. However, it is not possible to carry out the CWC-P test for soil suction higher than the maximum capacity of the HCT (typically 1,500 kPa). The CWC test does not have this limitation. Further information on HCT is given in Chapter 8.

It is important to note that in the CWC test, the water content of the soil must be maintained constant throughout the test. Thus, the porous stone and the filter paper must be dry (ASTM D4546-08, 2008), as wet filter paper and the wet porous disk might change the soil's water content. However, a dry porous disk is affected by evaporation, especially when the test takes a very long time. One method that can be used to reduce evaporation is by using a plastic disk at the top and bottom of the specimen.

13.4.4.1 CWC loading

Test procedures:

1. Increase/decrease the net normal stress applied to the soil specimen.
2. Wait until pore-water pressure reaches equilibrium. When pore-water pressure remains negative on increase of net normal stress, 20 minutes is sufficient. It is not recommended to wait for a long duration due to the possibility of water loss due to evaporation. However, when pore-water pressure becomes positive on increase of net normal stress 24 hours wait time is required, following the loading duration for saturated soils to ensure that the excess pore-water pressure has dissipated.
3. Record the final compression (and hence, void ratio) and the final soil suction after equilibrium.
4. Repeat steps 1 to 3.

Parameters to be reported: σ_i, e_i, s_i, w_i, S_{ri}

The primary compression of unsaturated soil under the CWC condition is very short as the permeability of air is much higher than the permeability of water. Therefore, the change in negative-pore-water pressure is due to the change in void ratio, and this is designated as mechanical wetting/drying. However, HCT may have a delayed response time, and therefore it is important to wait until HCT reaches equilibrium in order to obtain the correct pore-water pressure.

13.4.4.2 Wetting under constant net normal stress

Test procedures:

1. Inundate the soil specimen by filling the consolidometer chamber with water. The water must be either distilled water or water with similar chemical composition as the in situ pore water.
2. Observe the change of soil volume and pore-water pressure until pore-water pressure reaches 0 kPa and there is negligible change in soil volume. Usually, 48 hours duration is sufficient for the pore-water pressure to reach 0 kPa.
3. Record the amount of swelling Δe when pore-water pressure becomes zero.

Parameters to be reported: Δe, soil-swelling characteristic curve, swelling versus time graph

The advantage of CWC-P is that it is possible to obtain a continuous relationship between the void ratio and matric suction under the swelling condition in a relatively short duration compare to the SC test. Therefore, additional in situ swelling or shrinkage can be estimated based on the change in the in situ soil suction.

In CWC-P, either top or bottom is usually undrained due to the placement of HCT. Thus, it is not recommended to saturate the soil specimen from the bottom using a pressure line, as it may introduce additional water pressure which can lift the soil specimen.

13.4.4.3 Wetting under constant volume

Test procedures:

1. Inundate the soil specimen in the oedometer chamber with water. The water must be either distilled water or water with similar chemical composition as the in situ pore water.
2. Whenever the soil starts to swell, add additional net normal stress to maintain the volume of the soil specimen.
3. Once the soil suction has reached equilibrium, record the change in net normal stress and soil suction. Usually, 48 hours duration is sufficient for the soil to reach equilibrium.

Parameters to be reported: Swelling pressure, suction versus net normal stress curve

Based on this stage, it is possible to obtain a continuous relationship between the change in net normal stress and soil suction. Thus, in situ swelling pressure can be determined based on the change of the in situ soil suction. However, as it is impossible to maintain no volume change at all during this stage, care must be taken when interpolating the test result.

13.4.5 Stages in the CWC oedometer test

The CWC oedometer test is the simplest and only requires a conventional oedometer test. The loading duration is shorter than the saturated oedometer test, as the permeability of air is much higher than the permeability of water and compression is almost simultaneous on loading. However, as it does not give any information regarding matric suction, it is difficult to use CWC test results in models developed for the SC test or CWC-P test. The easiest way to use CWC test results is to duplicate the in situ loading stage and interpolate the CWC test results to estimate the in situ volume change. Another way is to use another state variable such as water content or degree of saturation as the model parameters.

The testing stages in CWC can be summarised as follow:

1. Specimen preparation
2. CWC loading/unloading
3. Wetting under constant net normal stress
4. Wetting under constant volume

13.4.5.1 Specimen preparation

Test procedures:

1. Set up the specimen with air-dried porous stones and air-dried filter papers in the oedometer.
2. Use a plastic sheet, moist paper towel or aluminium foil to cover the soil specimen to minimise evaporation.
3. Apply seating stress to the unsaturated soil specimen.

13.4.5.2 CWC loading/unloading

Test procedures:

1. Increase/decrease the net normal stress applied to the soil specimen.
2. Record the change in height of the specimen and calculate the change in degree of saturation assuming that the water content remains constant during the loading stage. If the degree of saturation reaches 100%, recalculate the water content.

3. Use a loading duration of either 5, 10 or 20 minutes. It is important to observe the change in void ratio with time for the first few loadings in order to ensure that 20 minutes is sufficient. Subsequent loading can be carried out when there is negligible change in void ratio. Loading duration should not be excessive to avoid water loss from the soil specimen due to evaporation.

Parameters to be reported: σ_i, e_i, w_i, S_{ri}

13.4.5.3 Wetting under constant net normal stress

Test procedures:

1. Inundate the soil specimen by filling the oedometer chamber with water or via a water pressure line connected to the bottom of the specimen. Use either distilled water or water with the same chemical composition as the in situ pore water. Saturation using a water pressure line from the bottom is preferred as air can be displaced from the bottom of the specimen upwards such that all the porous stone and soil specimens can become fully saturated. However, care must be taken to ensure that the water pressure does not uplift the soil specimen.
2. Observe the volume change until there is negligible change in soil volume. Usually, 48 hours is sufficient for the soil to reach equilibrium.
3. Record the amount of swelling Δe.

Parameters to be reported: Δe, swelling versus time graph
This stage provides the maximum swelling potential of the soil.

13.4.5.4 Wetting under constant volume

Test procedures:

1. Inundate the soil specimen by filling the oedometer chamber with water or by using a water pressure line connected to the bottom of the specimen. Either distilled water or water with the same chemical composition as the in situ pore water is used. Using a water pressure line to saturate the soil specimen is preferred, as air can be displaced from the bottom of the soil specimen upwards to ensure that the porous stone and soil specimen becomes fully saturated. However, care must be taken such that the water pressure does not lift the soil specimen.
2. Whenever the soil starts to swell, add additional net normal stress to maintain the volume of the soil specimen.
3. Record the change in net normal stress once the soil has reached equilibrium. Usually, 48 is sufficient for the soil to reach equilibrium.

Parameters to be reported: Swelling pressure

This stage provides the maximum swelling pressure which can be exerted by the soil.

13.4.5.5 Saturated loading/unloading stage

Test procedures:

1. Apply the load either increase (loading) or decrease (unloading) once the wetting stage is completed.
2. Maintain each loading stage for up to 24 hours or until there is negligible change in volume and pore-water pressure.

13.4.6 CWC and CWC-P oedometer test procedures

CWC test has been in use for a very long time due to its simplicity and some of the testing procedures have been standardised in ASTM:

1. ASTM D3877 (2008) for one-dimensional expansion, shrinkage and uplift pressure of soil-lime mixtures.
2. ASTM D4546 (2008) for one-dimensional swell or collapse of cohesive soil.
3. ASTM D5333 (2003) for measurement of collapse potential of soils.
4. ASTM D4829 (2011) for obtaining the expansion index of soils.

The ASTM standards mainly focus on swelling and collapse behaviours. However, the volume change of unsaturated soil is sometimes complicated as swelling and collapse are also affected by the change in net normal stress and suction or water content. For more robust testing procedures, Wijaya (2017) recommended a testing procedure to investigate the volume change of unsaturated soil due to the change in net normal stress and change in water content.

13.4.6.1 ASTM D3877-08 for one-dimensional expansion, shrinkage and uplift pressure of soil-lime mixtures

Lime is known to reduce the swelling potential of soils. However, different types of soil may require a different amount of lime to give optimum results in terms of cost and reduction in the amount of swelling. The soil-lime mixture must be prepared in accordance with ASTM D3551 using a mechanical mixer. The soil-lime mixture is made as follows:

1. Sieve the soil particles such that the particle size is less than 4.75 mm (Sieve no. 4).
2. Air or oven dry the soil for around 24 hours. If air drying or oven drying is suspected to affect the results, the soil should be cured at or near its final moisture content.

3. Determine the water content of the air-dried specimen.
4. Determine the weight of the soil which is required to provide the oven-dry weight of soil for the desired number of test specimens.
5. Determine the amount of lime based on the oven-dry weight of soil required for the final mixture.
6. Place both the dried soil and the lime inside a bowl.
7. Mix dried soil and lime for either about 1 minute or until the mixture appears uniform in colour.
8. Weigh or measure the required weight of water to produce the desired water content. The initial amount of water in the air-dried soil must be considered. Considering the possibility of evaporation, it is suggested to use the amount of water such that the final water content is around 1% + the desired water content.
9. Add the water into the soil mixture in a thin stream or a fine spray while mixing the soil with a mixer.
10. Mix the soil for two and a half minutes after all the water has been added.
11. Stop the mixer and then scrape all parts to return any caked mixture into the mixing bowl.
12. Repeat steps 10 and 11.
13. Blend the mixture briefly with a trowel or spatula and form a lightly compacted mound at the bottom of the bowl.
14. Cover the bowl immediately to minimise the evaporation of moisture and let the soil-lime-water mixture reach equilibrium for around 1 hour.
15. Weigh the oedometer ring.
16. Assembled the oedometer along with the oedometer ring, oedometer collar and load cap.
17. Measure the position of the load cap r_1 by using a dial gauge or linear variable differential transformer (LVDT). The dial gauge or LVDT must be positioned at the same place when the dial gauge or LVDT will be used during the test.
18. Assemble the oedometer ring and the extension collar. An extension collar is used such that a loose soil mixture can be placed.
19. Compact the specimen in the oedometer ring to the desired wet unit weight by using a suitable compaction hammer. The minimum height of the specimen shall be 19 mm and 6 mm greater than the depth of the ring gauge.
20. Remove the extension collar and trim the excess soil from the top of the soil specimen. It is preferred to make the specimen height equal to the oedometer ring height for ease of the initial height measurement.
21. Determine the water content of the soil-lime mixture using the trimmed soil.
22. Weigh the specimen with the ring.
23. Cover the exposed surface of the specimen using a glass plate, held with clamps until the specimen is placed in the oedometer.
24. Compute the initial wet density of the specimen.

25. If the initial wet density of the specimen is not obtained, discard the specimen and redo the specimen preparation.
26. If there is a need to cure the soil-lime specimen, it should be done once the desired wet density has been obtained.
27. Place the soil specimen inside the oedometer and use a seating load of approximately 2.5 kPa.
28. Re-measure the position of the load cap r_2 by using the dial gauge or LVDT. The difference between r_2 and r_1 should give the height of the specimen. Alternatively, if the soil specimen has the same height as the oedometer ring, the height of the specimen can be taken as the height of the oedometer ring.

Once the soil mixture has been prepared, the soil mixture can then be loaded. There are three procedures available for estimating the swelling pressure of soil-lime mixtures:

1. Loaded and expanded
2. Expanded then loaded
3. Individual load expansion

The testing procedures for loaded and expanded are given as follows:

1. Wetting under constant volume conditions: Record the maximum load.
2. Unloading stage: Reduce the load to ½, ¼ and $1/_8$ of the maximum load and finally to the seating load. Maintain each load for either 24 hours or longer such that there is no change in the void ratio. Record the change in the void ratio (or height) for each unloading step.
3. Determine the water content of the soil specimen.

The testing procedures for expanded and then loaded are given as follows:

1. Wetting under constant net normal stress: Allow the soil specimen to expand for at least 48 hours or until there is negligible change in the void ratio.
2. Loading stage: If the loaded and expanded test procedure has been conducted, the applied load is equal to $1/_8$, ¼ and ½ of the maximum load. The subsequent loading stage can be done using a load increment ratio (LIR) of 1. If no loaded and expanded test procedure has been carried, loading follows LIR of 1 as usually carried out in an oedometer test until the load reaches the capacity of the oedometer apparatus, usually either 3,200 or 6,400 kPa.
3. Unloading stage: Unload until the seating load. The LIR of 1 may be used for a normal oedometer test.
4. Determine the water content of the specimen.

The test procedures for individual load expansion are given as follows:

1. CWC loading stage: Increment the applied load until the desired net normal stress.
2. Wetting under constant net normal stress: Wet the soil for either 48 hours or until there is no more change in soil volume.
3. Unloading stage: Decrement the load using a load decrement ratio (LDR) of ½ until the seating load.
4. Determine the water content of the specimen.

The "loaded and expanded" procedure will give different results compared to the "expanded then loaded" procedure. While the "loaded and expanded" procedure is believed to give a more reliable result, it requires greater proficiency especially in controlling the load to be applied during wetting under the constant volume condition. Therefore, carrying out both test procedures is better for comparison. The "individual load expansion" procedure is carried out when the in situ soil will be subjected to load prior to wetting. Thus, the applied load must simulate the in situ condition.

13.4.6.2 ASTM D4546-08 for one-dimensional swell or collapse of cohesive soil

ASTM D4546-08 describes the procedure to measure wetting-induced behaviour under different net normal stresses and the post-wetting behaviour of natural or compacted soils. There are three methods:

1. Method A: Wetting-after-loading test on multiple specimens
2. Method B: Wetting-after-loading test on a single specimen
3. Method C: Loading-after-wetting test

The specimen preparation is identical for all three test methods described as follows:

1. Cookie cut the soil specimen using the oedometer ring.
2. Trim and ensure that the specimen height is less or equal to the height of the oedometer ring.
3. If the soil specimen's height is less than the height of the oedometer ring, measure the height of the specimen at three or four places.
4. Use air-dried porous stones and dry filter paper (or no filter paper).
5. Apply a seating load of 1 kPa.
6. Set the displacement indicator.

The three test methods are summarised in Table 13.4.

Table 13.4 ASTM D4546-08 procedures

Stage	Method A	Method B	Method C
Specimen preparation	Prepare as many specimens as needed following the previously mentioned procedure to be wetted under different net normal stress. The wetting pressure must be selected such that it covers the stress range that may be experienced in situ starting from before construction until future possible load applied after the construction is finished. All of the specimens can be tested at the same time or in sequence.	Prepare a single specimen to be tested.	Prepare as many soil specimens as needed following the previously mentioned procedures to be wetted under different net normal stresses. The wetting pressure must be selected such that it covers the stress range that may be experienced in situ starting from before construction until the future possible load applied after the construction is finished. All of the soil specimens can be tested at the same time or in sequence.
CWC loading stage	Increment the net normal stress at LIR of 1 until it reaches the wetting pressure.	CWC loading stage: Increment the net normal stress using LIR of 1 until it reaches the in situ net normal stress at its sampling depth.	CWC loading stage: Increase the net normal stress using LIR of 1 until it reaches the wetting pressure.
CWC unloading stage	—	Decrement the net normal stress using LDR of ½ until it reaches the seating pressure.	—

(Continued)

Table 13.4 (Continued)

Stage	Method A	Method B	Method C
CWC reloading stage	—	Increase the net normal stress up to the in situ net normal stress corresponding to the prior or post-construction condition, whichever is of interest.	—
Wetting under constant net normal stage	Inundate the soil specimen with water for either 24 to 72 hours, or until there is negligible change in the volume of the specimen. Water should be either distilled water or water with similar chemical composition as the primary water source of wetting in the field.		
Saturated loading stage	—	—	Increment the applied load using LIR of 1 up to either 3,200 or 6,400 kPa. Each loading stage is maintained for either 24 hours until there is no excess pore-water pressure, or there is negligible change in volume.
Saturated unloading stage	—	—	Decrement the applied load using LDR of ½ until its seating pressure. Each unloading stage is maintained for either 24 hours or until there is negligible change in volume and pore-water pressure.
Drying stage	Remove all water inside the oedometer using suction devices without unloading the specimen, and then use filter paper to soak up all free water in the oedometer.	—	—
CWC unloading stage	Remove all the vertical load applied to the soil specimen at once.		
Specimen removal	Remove the soil specimen and use dry filter paper to remove any free water on the surface of the soil specimen.		
Water content determination	Determine the water content of the soil specimen.		

13.4.6.3 ASTM D5333-03 for measurement of collapse potential of soils

ASTM D5333-03 determines the collapse potential of soil under two conditions:

1. At net normal stress corresponding to the desired in situ net normal stress (prior to or post-construction condition), I_c
2. At net normal stress equal to 200 kPa where the collapse potential is referred to as collapse index, I_e

The specimen which is tested can be either a remoulded, compacted or undisturbed specimen. The test procedures are given as follows:

1. Specimen preparation stage: Apply seating load of 5 kPa.
2. CWC loading stage: Increment the applied load using LIR of 1 with 1 hour loading duration until it reaches the desired load (either 200 kPa to obtain I_e, or another load to obtain I_c).
3. Wetting under constant net normal stress stage: Inundate the specimen with water for either 24 hours or until there is no more change in volume. Use distilled-deionised water to determine I_e and use water from the in situ wetting source for I_c. Water must flow upwards from the bottom of the soil specimen to avoid air being trapped inside the soil specimen.
4. Saturated loading stage: Increment the applied load using LIR of 1 with 24 hours loading duration until the applied load reaches either 3,200 kPa or 6,400 kPa.
5. Saturated unloading stage: Decrement the applied load using LDR of ½ with 24 hours loading duration until the applied load reaches seating pressure.
6. Take out the soil specimen from the oedometer and determine the final water content.

13.4.6.4 Wijaya (2017) CWC oedometer test procedure

Wijaya (2017) proposed a method to estimate both swelling and collapse by using at least two specimens. The first specimen is subjected to a saturated compression test while the second specimen is subjected to a CWC oedometer test. The stage procedures for the CWC oedometer test are given as follows:

1. Initial drying stage: Air dry the soil specimen until it reaches almost zero water content to obtain the shrinkage curve. Measure the dimensions of the soil specimen.
2. Specimen preparation stage: Cookie cut the air-dried soil specimen using the oedometer ring. If the soil becomes too brittle, use a plastic

film to wrap the perimeter of the specimen in order to provide a small confinement during the cutting process. Assemble the soil specimen into the oedometer apparatus.

3. CWC loading stage: Increase the applied load until it reaches the effective overburden stress. Maintain each loading for 20 minutes.
4. CWC unloading stage: Decrease the applied load using LDR of ½ until it reaches the seating pressure.
5. CWC loading stage: Increase the applied load using LIR of 1 until it reaches 6,400 kPa.
6. CWC unloading stage: Decrease the applied load using LDR of 1 until it reaches the seating pressure.
7. Remove the soil specimen from the oedometer and determine the final water content of the specimen.

As the soil specimen has been air dried until almost zero water content, the soil will no longer be affected by evaporation. By combining the parameters from the saturated and unsaturated compression curve, it is possible to estimate the volume change of the unsaturated soil which is subjected to a complex stress path.

13.4.7 SC Oedometer test

13.4.7.1 Constant-suction loading/unloading stage

Stage procedure:

1. Increase/decrease net normal stress σ_i.
2. Wait for soil suction to be in equilibrium with the applied suction.
3. Record void ratio after the soil specimen has reached equilibrium e_i.
4. Repeat steps 1 and 2 until the end of the constant-suction loading/unloading stage.

The constant-suction loading/unloading stage is carried out to obtain a constant-suction compression curve and to investigate the wetting-induced volume change under specific suction (Vázquez et al. 2013; Cuisinier and Masrouri 2005). It is important to note that during the loading stage, there will be mechanical wetting which induces swelling/collapse of the specimen, while unloading the specimen may cause mechanical drying which may cause the soil specimen to be stiffer.

As the collapse phenomenon is irreversible, the compression curve which is affected by the collapse phenomenon may not represent the compression curve under the constant-suction condition. Thus, to reduce the effect of mechanical wetting and drying, it is recommended to use small stress increments. Nevertheless, very small stress increments may cause the test duration to be excessively long (Cuisinier and Masrouri 2005).

13.4.7.2 Suction decrease (SD) stage

Stage procedure:

1. Record the net normal stress at SD stage σ_i.
2. Decrease the applied suction from s_0 to s_i.
3. Wait until soil suction and the applied suction are in equilibrium.
4. Record the change in void ratio Δe after equilibrium.
5. Either repeat steps 1 to 3 or end this stage.

SD stage will cause either swelling or collapse. Collapse settlement is dependent on the applied load direction and thus will occur only in the vertical direction. The parameter $K_{s,I}$ and $C_{w,I}$ are given by Equations 13.5 and 13.6, respectively.

$$\kappa_{s,i} = \frac{e(s_i) - e(s_0)}{\ln\left(\dfrac{s_0}{s_i}\right)} \tag{13.5}$$

$$C_{w,i} = \frac{e(s_i) - e(s_0)}{\log\left(\dfrac{s_0}{s_i}\right)} \tag{13.6}$$

For the SD stage, since s_i is smaller than s_0, $\kappa_{s,i}$ and $C_{w,i}$ are positive when the soil specimen is swelling and negative when the soil specimen is collapsing. The two parameters are equivalent and are a function of σ_i.

While it is possible to carry out the suction increase (SI) stage in the oedometer test, it is not recommended, as increasing soil suction may cause volume change to occur not only in the vertical direction but also in the radial direction, and may not be homogeneous. Volume change in the radial direction may cause non-k_0 conditions and cause difficulties in the interpretation. The SI stage is only recommended for the SC triaxial test.

REFERENCES

Alonso, E. E. (1993). *Unsaturated soil: Recent developments and application – constitutive model of unsaturated soils*. Barcelona: Civil Engineering European Courses Programme of Continuing Education.

Alonso, E. E., Gens, A., & Josa, A. (1990). A constitutive model for partially saturated soil. *Géotechnique*, 40(3): 405–430.

ASTM D2435/D2435m (2011). *Standard test methods for one-dimensional consolidation properties of soils using incremental loading*. West Conshohocken, PA: ASTM International.

ASTM D3877 (2008). *Standard test methods for one-dimensional expansion, shrinkage, and uplift pressure of soil-lime mixtures.* West Conshohocken, PA: ASTM International.

ASTM D4546 (2008). *Standard test methods for one-dimensional swell or collapse of cohesive soils.* West Conshohocken, PA: ASTM International.

ASTM D4829 (2011). *Standard test method for expansion index for soils.* West Conshohocken, PA: ASTM Standard.

ASTM D5333 (2003). *Standard test method for measurement of collapse potential of soils.* West Conshohocken, PA: ASTM International.

Arvidsson, J. & Keller, T. (2004). Soil precompression stress: I. A survey of Swedish arable soils. *Soil and Tillage Research*, 77(1): 85–95.

Aversa, S. & Nicotera, M. V. (2002). A triaxial and oedometer apparatus for testing unsaturated soils. *ASTM Compass*, 25(1): 13.

BS 1377-5 (1990). *Methods of test for soil for civil engineering purposes – part 5: Compressibility, permeability and durability tests.* UK: BSI.

BS EN ISO 17892-5 (2017). *Geotechnical investigation and testing. Laboratory testing of soil. part 5: incremental loading oedometer test.* UK: BSI.

Burton, G. J., Sheng, D., & Airey, D. (2014). Experimental study on volumetric behaviour of Maryland clay and the role of degree of saturation. *Canadian Geotechnical Journal*, 51(12): 1449–1455.

Cuisinier, O. & Masrouri, F. (2005). Hydromechanical behaviour of a compacted swelling soil over a wide suction range. *Engineering Geology*, 81(3): 204–212.

Delage, P., Suraj De Silva, G. P. R., & Vicol, T. (1992). Suction controlled testing of non saturated soils with an osmotic consolidometer. In *7th International Conference Expansive Soils*, Dallas, pp. 206–211.

Futai, M. M., De Almeida, M. S. S., Conciani, W., & Silva Filho, F. C. (2002). Experimental and theoretical evaluation of plate load test in collapsible soil. In *Proceedings 3rd International Conference on Unsaturated Soils (UNSAT 2002)*. Swets and Zeitlinger, Lisse, Recife, Brazil, pp. 721–726.

Head, K. H. (2011). *Manual of soil laboratory testing volume 2: Permeability.* Shear Strength and Compressibility Test.Scotland, UK: Dunbeath Mill.

Holtz, R. D. & Kovacs, W. D. (1981). *An introduction to geotechnical engineering.* Englewood Cliffs: Prentice–Hall.

Holtz, R. D., Kovacs, W. D., & Sheahan, T. C. (2011). *An introduction to geotechnical engineering*, 2nd ed. Upper Saddle River, NJ: Pearson Education, Inc.

Honda, M. (2000). *Research on the prediction methods for the mechanical behaviours of unsaturated soils.* Kobe, Japan: Kobe University,vol. Ph.D.

Hoyos, L., Pérez-Ruiz, D., & Puppala, A. (2012). Modeling unsaturated soil response under suction-controlled true triaxial stress paths. *International Journal of Geomechanics*, 12(3): 292–308.

Imhoff, S., Da Silva, A. P., & Fallow, D. (2004). Susceptibility to compaction, load support capacity, and soil compressibility of Hapludox. *Soil Science Society of America Journal*, 68(1): 17–24.

Jennings, J. E. & Knight, K. 1957. The additional settlement of foundations due to a collapse of structure of sandy subsoils on wetting. *Proceeding of the Fourth International Conference on Soil Mechanics and Foundation Engineering*, Vol. 1, pp. 316–319.

Jotisankasa, A. (2005). Collapse behaviour of a compacted silty clay. In *Civil and environmental engineering.* London: Imperial College, vol. Ph.D, p. 422.

Jotisankasa, A., Ridley, A., & Coop, M. (2007). Collapse behavior of compacted silty clay in suction-monitored oedometer apparatus. *Journal of Geotechnical and Geoenvironmental Engineering*, 133(7): 867–877.

Kayadelen, C. (2008). The consolidation characteristics of an unsaturated compacted soil. *Environmental Geology*, 54: 325–334.

Larson, W. E., Gupta, S. C., & Useche, R. A. (1980). Compression of agricultural soils from eight soil orders1. *Soil Science Society of America Journal*, 44(3): 450–457.

Leong, E. C., Widiastuti, S., & Rahardjo, H. (2013). Estimating wetting-induced settlement of compacted soils using oedometer test. *Geotechnical Engineering Journal of the SEAGS & AGSSEA*, 44(1): 26–33.

Lloret, A., Villar, M. V., Sa´Nchez, M., Gens, A., Pintado, X., & Alonso, E. E. (2003). Mechanical behaviour of heavily compacted bentonite under high suction changes. *Géotechnique*, 53(1): 27–40.

Mun, W. & McCartney, J. S. (2015). Compression mechanisms of unsaturated clay under high stresses. *Canadian Geotechnical Journal*, 52(12): 2099–2112.

Munoz, J., Castelblanco Delage, P., Pereira, J. M., & Cui, Y. J. (2011). Some aspect of the compression and collapse behaviour of an unsaturated natural loess. *Geotechnique Letters*, 1: 17–22.

O'Sullivan, M. F. (1992). Uniaxial compaction effects on soil physical properties in relation to soil type and cultivation. *Soil and Tillage Research*, 24(3): 257–269.

Pham, H. Q. (2005). A volume-mass constitutive model for unsaturated soils. In *Civil and Geological Engineering*. Saskatoon: University of Saskatchewan, vol. Doctor of Philosopy, p. 532.

Rahardjo, H. (1990). The study of undrained and drained behaviour of unsaturated soil. In *Department of Civil Engineering*. Saskatoon: University of Saskatchewan, vol. Ph.D.

Salager, S., Rizzi, M., & Laloui, L. 2011. An innovative device for determining the soil water retention curve under high suction at different temperatures. *Acta Geotechnica*, 6(3): 135–142. doi: 10.1007/s11440-011-0141-8

Silva, F. C. F., Almeida, M. S. S., & Futai, M. M. (2002) Critical state modelling of unsaturated soils. In *Proceeding of the Third International Conference on Unsaturated Soil*, Recife, Brazil, pp. 133–137.

Sivakumar, V., Tan, W. C., Murray, E. J., & McKinley, J. D. (2006). Wetting, drying and compression characteristics of compacted clay. *Géotechnique*, 56: 57–62.

Sivakumar, V. & Wheeler, S. J. (2000) Influence of compaction procedure on the mechanical behaviour of an unsaturated compacted clay. Part 1: Wetting and istropic compression. *Géotechnique*, 50(4): 359–368.

Smith, C. W., Johnston, M. A., & Lorentz, S. (1997). Assessing the compaction susceptibility of South African forestry soils. II. Soil properties affecting compactibility and compressibility. *Soil and Tillage Research*, 43(3–4): 335–354.

Tarantino, A. & De Col, E. (2008). Compaction behaviour of clay. *Géotechnique*, 58(3): 199–213.

Toll, D. G. (1990). A framework for unsaturated soil behaviour. *Géotechnique*, 40(1): 31–44.

Vázquez, M., Justo, J. L. & Durand, P. (2013). A simplified model for collapse using suction controlled tests. In *Proceedings of the 18th International Conference on Soil Mechanics and Geotechnical Engineering*, Paris, pp. 1203–1206.

Wheeler, S. J. & Sivakumar, V. (1995). An elasto-plastic critical state framework for unsaturated soil. *Géotechnique*, 45(1): 35–53.

Wijaya, M. (2017). Compression, shrinkage and wetting-induced volume change of unsaturated soils. In *Civil and environmental engineering (Ph.D.)* Singapore: Nanyang Technological University.

Wijaya, M. & Leong, E. C. (2015). Swelling and collapse of unsaturated soils due to inundation under one-dimensional loading. *Indian Geotechnical Journal*, 46: 239–251.

Wijaya, M. & Leong, E. C. (2016). Performance of high-capacity tensiometer in constant water content oedometer test. *International Journal of Geo-Engineering*, 7(1): 13.

FURTHER READING

Jotisankasa, A., Ridley, A., & Coop, M. (2007). Collapse ehavior of compacted silty clay in suction-monitored oedometer apparatus. *Journal of Geotechnical and Geoenvironmental Engineering*, 133(7): 867–877.

Wijaya, M. & Leong, E. C. (2015). Swelling and collapse of unsaturated soils due to inundation under one-dimensional loading. *Indian Geotechnical Journal*: 1–13.

Leong, E. C., Widiastuti, S., & Rahardjo, H. (2013). Estimating wetting-induced settlement of compacted soils using oedometer test. *Geotechnical Engineering Journal of the SEAGS & AGSSEA*, 44(1): 26–33.

Chapter 14

Constant rate of strain test

14.1 BACKGROUND

Soil deformation is rate dependent due to phenomena such as stress relaxation and creep (Bagheri et al. 2019b). Thus, the rate dependency of soil deformation starts to gain attention for both saturated and unsaturated soils (Kim and Leroueil 2001; Yin and Hicher 2008; Sorensen et al. 2010; Tong and Yin 2013; Yin et al. 2014; Rezania et al. 2017a; Rezania et al. 2017b; Bagheri et al. 2019b; Tanaka et al. 2014). One method to investigate the effect of strain rate on soil is by employing the constant rate of strain (CRS) test. In the CRS test, instead of applying an incremental load to the specimen, a constant rate of strain is applied. As the test can be completed faster than the oedometer test, it also serves as an alternative test method.

The challenge in the CRS test for saturated soils is to ensure that the excess pore-water pressure generated throughout the test is kept below a threshold value. Unlike the CRS test for saturated soils, the CRS test for unsaturated soils may not have this constraint depending on the initial degree of saturation of the soil. Many developments have been carried out on the use of the CRS test on unsaturated soil (with or without pore-water pressure measurement). In this chapter, the CRS test for unsaturated soils will be elaborated.

14.2 RELATED STANDARD

1. ASTM D4186/D4186m-12 (2012) Standard Test Method for One-Dimensional Consolidation Properties of Saturated Cohesive Soils Using Controlled-Strain Loading, ASTM International.

DOI: 10.1201/b22304-14

14.3 THEORY

14.3.1 Saturated CRS test (ASTM D4186/D4186M-12 2012)

The CRS test is an alternative to the oedometer test for testing saturated soils. Instead of applying an incremental loading and waiting for the excess pore-water pressure to dissipate, a CRS is imposed on the specimen such that the excess pore-water pressure does not exceed a certain threshold.

There are three obvious advantages of the CRS test compared to the oedometer test in testing saturated soils:

1. It provides a continuous compression curve and, hence, a continuous relationship between the void ratio (e) and permeability (k_w), coefficient of consolidation (c_v) instead of a discrete relationship based on the incremental loadings.
2. It is faster than incremental loading.
3. When a confined cell with a pressure system is used, it is possible to saturate the specimen by using a back pressure, and the saturation of the specimen can be checked by using the Skempton pore-pressure parameter B just like in the triaxial test.

The problem with the CRS test for saturated soil is that it generates excess pore-water pressure throughout the test. It is recommended that the pore-water ratio (R_u) is between 3% and 15%. Therefore, in order to calculate the applied effective stress, the pore-water pressure at the bottom of the soil specimen is monitored throughout the test.

The CRS apparatus is very similar to the oedometer apparatus except that the loading is strain controlled, and hence a compression machine or load actuator similar to that in a triaxial test is used. Some of the CRS apparatuses, such as those manufactured by GDS Instruments, UK, have the ability to do stress control as well and thus can replicate the oedometer test.

Figure 14.1 shows a schematic of a CRS apparatus. As the soil specimen is placed inside the CRS cell, it is possible to saturate the specimen by filling the CRS cell with water (instead of inundating the specimen like in the oedometer test) and applying a back pressure (cell pressure) to the specimen. The saturation of the specimen can be checked by increasing the cell pressure while observing the change in the pore-water pressure and calculating the Skempton B value, which is defined in Equation 14.1.

$$B = \frac{\Delta u_w}{\Delta \sigma_c} \tag{14.1}$$

where Δu_w is the change in the pore-water pressure (usually measured at the base), and $\Delta \sigma_c$ is the change in cell pressure.

Table 14.1 shows the equations to calculate the parameters in the CRS test. Some of the derived soil properties are based on the model used, either linear or non-linear. The linear model assumes that the soil has a constant

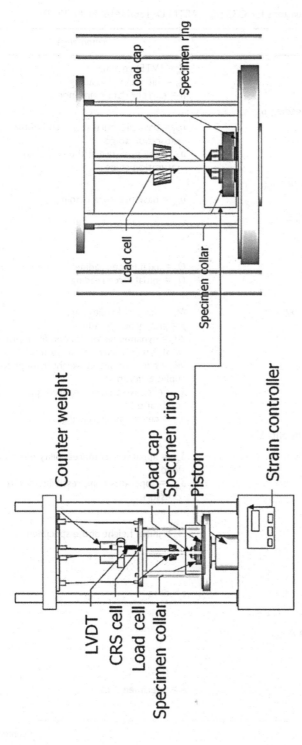

Figure 14.1 Standard CRS apparatus (from Wijaya 2017).

Table 14.1 List of equations for CRS test (ASTM D4186/D4186M-12 2012)

Axial deformation, Δd_n	Parameters
$\Delta d_n = d_n - d_0$	d_n = LVTD reading d_0 = initial LVDT reading n = current data point index
base pressure zero reading, $u_{b,0}$	
$u_{b,0} = u_{b,s} - \sigma_{c,s}$	$u_{b,s}$ = base pressure at the end of the saturation stage $\sigma_{c,s}$ = cell pressure at the end of the saturation stage
Corrected base pressure, $u_{m,n}$	
$u_{m,n} = u_{b,n} - u_{b,0}$	$u_{b,n}$ = base pressure reading
Base excess pressure, $\Delta u_{m,n}$	
$\Delta u_{m,n} = u_{m,n} - \sigma_{c,n}$	
Axial force	
$f_n = Q_n - Q_0$	Q_n = axial force reading Q_0 = initial force reading
Net axial force, $f_{a,n}$	
$f_{a,n} = f_n + M_L g - \Delta f_s + W_p - A_p \sigma_{c,n}$	M_L = mass of loading cap g = gravity acceleration Δf_s = dynamic piston friction from piston seal dynamic friction calibration W_p = effective piston weight from piston uplift calibration A_p = effective piston area from piston uplift calibration $\sigma_{c,n}$ = cell pressure reading
Change in specimen height, ΔH_n	
$\Delta H_n = \Delta d_n - \delta_{af,n} - \sigma_{ap,n}$	$\delta_{af,n}$ = apparatus compressibility due to axial load $\delta_{ap,n}$ = apparatus compressibility due to cell pressure
Axial strain, $\varepsilon_{a,n}$	
$\varepsilon_{a,n} = \dfrac{\Delta H_n}{H_0} 100$	H_0 = initial height of the specimen
Strain rate, $\dot{\varepsilon}$	
$\dot{\varepsilon} = \dfrac{\Delta H_{n+1} - \Delta H_{n-1}}{H_0} \cdot \dfrac{1}{t_{n+1} - t_{n-1}}$	t = time
Excess base pressure $\Delta u_{m,n}$	
$\Delta u_{m,n} = u_{m,n} - \sigma_{c,n}$	
Axial stress, $\sigma_{a,n}$	
$\sigma_{a,n} = \dfrac{f_{a,n}}{A}$	A = specimen area

(Continued)

Table 14.1 (Continued)

Axial deformation, Δd_n	Parameters

Steady-state factor, F_n

Linear equation:

$$F_n = \frac{(\sigma_{a,n} - \sigma_{a,l}) - (\Delta u_{m,n} - \Delta u_{m,l})}{(\sigma_{a,n} - \sigma_{a,l})}$$

l = index of data at the beginning of loading or unloading stage

Non-linear equation:

$$F_n = \frac{\left\{ \log\left[\sigma_{a,n} - (\Delta u_{m,n} - \Delta u_{m,l}) \right] \right\} - \log \sigma_{a,l}}{\log \sigma_{a,n} - \log \sigma_{a,l}}$$

Effective axial stress, $\sigma'_{a,n}$

Linear equation:

$$\sigma'_{a,n} = \left(\sigma_{a,n} - \frac{2}{3} \Delta u_{m,n} \right)$$

Non-linear equation:

$$\sigma_{a,n} = \left(\sigma_{a,n}^3 - 2\sigma_{a,n}^2 \cdot \Delta u_{m,n} + \sigma_{a,n}\Delta u_{m,n}^2 \right)^{1/3}$$

Permeability, k_n

Linear equation:

$$k_n = \frac{\dot{\varepsilon} H_n H_0 \gamma_w}{2\Delta u_{m,n}}$$

Non-linear equation:

$$k_n = -\frac{0.434\dot{\varepsilon} H_0 H_n \gamma_w}{2\sigma'_{a,n} \log\left(1 - \dfrac{\Delta u_{m,n}}{\sigma_{a,n}} \right)}$$

Coefficient of compressibility $m_{v,n}$

$$m_{v,n} = \frac{\varepsilon_{n+1} - \varepsilon_{n-1}}{\sigma'_{a,n+1} - \sigma_{a,n-1}}$$

Coefficient of consolidation, $c_{v,n}$

Linear equation:

$$c_{v,n} = \frac{k_n}{m_{v,n}\gamma_w}$$

Non-linear equation:

$$c_{v,n} = \frac{H_0 H_n \log\left(\dfrac{\sigma_{a,n+1}}{\sigma_{a,n-1}} \right)}{2(t_{n+1} - t_{n-1})\log\left(1 - \dfrac{\Delta u_{m,n}}{\sigma_{a,n}} \right)}$$

Pore-pressure ratio, $R_{u,n}$

$$R_{u,n} = \frac{\Delta u_{m,n}}{\sigma_{a,n}}$$

m_v, while the non-linear model assumes that the soil has a constant compressibility index (C_c) (ASTM D4186/D4186M-12 2012). The selection choice of linear or non-linear model depends on the specimen behaviour.

14.3.2 Unsaturated CRS test

Similar to the oedometer test, the CRS test can be performed on unsaturated soil as well through either a constant water content test or a constant suction test. For the constant water content CRS (CWC-CRS) test, the test can be performed either with or without pore-water pressure measurement (Wijaya 2017). In the CWC-CRS test with pore-water pressure measurement (Figure 14.2), the pore-water measurement is carried out at the bottom of the specimen (Wijaya 2017; Wijaya et al. 2014) or can be placed at the perimeter of the specimen (Bagheri et al. 2019b). For the constant suction CRS test, the modification can be made similar to the triaxial apparatus where the top cap is connected to the air pressure while the bottom porous disk is replaced with a ceramic disk. Additionally, a high-capacity tensiometer can be used to measure the soil suction at the perimeter of the specimen (Bagheri et al. 2019b).

The compression of unsaturated soil under the constant water content condition is almost instantaneous, as negligible excess pore-water pressure is generated during loading (Wijaya 2017). For a soil specimen that has a high initial degree of saturation, the soil specimen may become saturated during the test. Hence, the selected strain rate for such a specimen needs to follow the strain rate recommended for the CRS test for saturated soil such that the pore-water ratio (R_u) is between 3% and 15% and the steady-state factor (F_n) is bigger than 0.4 as recommended by ASTM D4186/D4186M-12 2012. It is therefore recommended that the CRS test be conducted for the saturated soil before conducting the CWC-CRS test to determine the appropriate strain rate.

The CWC-CRS test has an inundation stage where the soil specimen is flooded with water. Wijaya and Leong (2015) and Wijaya (2017) have shown that there is a transition pressure (P_T) in the unsaturated compression curve which causes the inundation stage and produces different phenomena due to the transition pressure which is located at the intersection between the saturated and unsaturated compression curves, as illustrated in Figure 14.3.

If the inundation stage is carried out at P_i (which is referred to as inundation pressure), when P_i is below P_T, the soil will swell and follow the swelling path (SP) when the soil volume is not constrained. As the soil is constrained in the CRS apparatus, it will follow the constant volume path (CVP), and the pressure will increase until the soil reaches the saturated compression curve (Figure 14.3a). The increase in pressure due to inundation under the constant volume condition is referred to as swelling pressure. Determining swelling pressure in the CRS apparatus is more convenient than using the oedometer apparatus, as it is effortless to maintain the constant volume

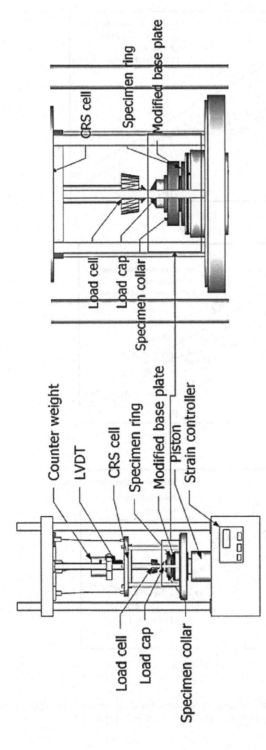

Figure 14.2 Modified CRS apparatus with pore-water pressure measurement (from Wijaya 2017).

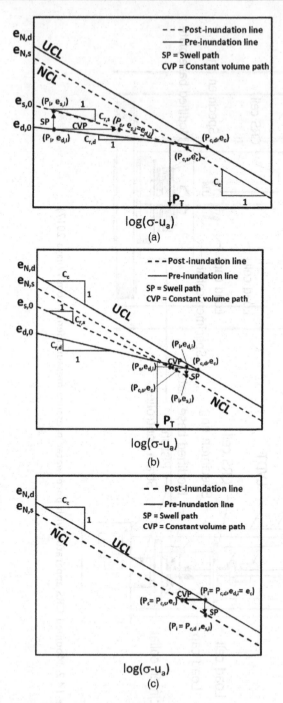

Figure 14.3 Wetting-induced behaviour of unsaturated soil (from Wijaya and Leong 2015). (a) $P_i < P_{c,s}$. (b) $P_{c,s} < P_i < P_{c,d}$. (c) $P_i = P_{c,d}$.

condition in the CRS apparatus. An example of an inundation stage which generates swelling pressure is shown in Figure 14.4.

However, if P_i is bigger than P_t, the soil will collapse. In the CRS apparatus, as the soil is constrained, and provided that the collapse is not significant enough to cause the load cap to no longer be in contact with the load cell, the soil state will move from the unsaturated compression curve to the saturated compression curve by following the CVP and thus reducing the net normal stress. The reduction in net normal stress associated with the collapse phenomenon is referred to as pressure relaxation. An example of an inundation stage which causes pressure relaxation is shown in Figure 14.5.

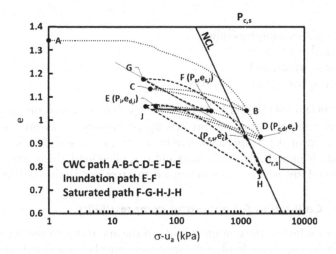

Figure 14.4 Swelling pressure due to inundation under transition pressure (from Wijaya and Leong 2015).

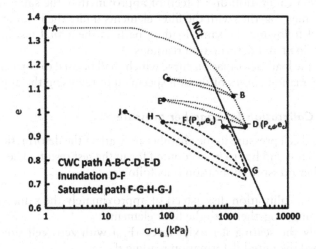

Figure 14.5 Pressure relaxation due to inundation above the transition pressure (from Wijaya and Leong 2015).

The normal compression line (NCL) determined in the two figures is obtained from a saturated oedometer test. It is shown that post inundation, the soil will follow the saturated path, and there is an agreement between the CRS test and the oedometer test results.

14.4 TEST METHODS

14.4.1 CRS apparatus calibration stages

For the proper conduct of the CRS test, the CRS apparatus requires calibration for

1. apparatus compressibility,
2. chamber pressure,
3. piston uplift, and
4. piston seal dynamic friction.

ASTM D4186/D4186M-12 (2012) requires that the correction be applied whenever the apparatus compressibility or chamber pressure may cause deformation of 0.1% specimen height. If the piston seal dynamic friction correction is up to 0.5% of the maximum applied load, the correction must be applied. However, it is recommended to carry out all the calibrations.

14.4.1.1 Calibration for apparatus compressibility

The contribution of the compressibility of the apparatus to the deformation reading when applying load to the specimen must be considered. The procedure for determining the apparatus compressibility is as follows:

1. Place a calibration disk (steel) of approximately the same size as the specimen together with the filter elements under the loading piston.
2. Load following the same loading path as the actual test and record both load and deformation readings.
3. Plot the load-deformation curve which shall be used to correct the load-deformation curves of the soil specimen tested using the apparatus.

14.4.1.2 Calibration for chamber pressure

The applied cell pressure in the chamber may affect the deformation reading in the CRS test and has to be accounted for in the test results. The procedure for chamber pressure calibration is as follows:

1. Place a calibration disk (steel) of approximately the same size as the specimen together with the filter elements.
2. Apply the seating net axial force ($F_{a,0}$) with zero cell pressure and record the initial deformation reading d_0.
3. Increase the cell pressure to $\sigma_{3,i}$.

4. Adjust the axial force back to $F_{a,0}$.
5. Record the axial deformation d_i and the cell pressure $\sigma_{3,i}$.
6. Repeat steps 2 to 5 for increasing cell pressures.
7. Establish a relationship between cell pressure and apparatus deformation.

This correction is subtracted from all deformation readings in a CRS test.

14.4.1.3 Calibration for piston uplift

Some CRS apparatus designs may cause an uplift of the piston due to the cell pressure and thus affect the force reading. Therefore, correction to the force reading due to the piston uplift might need to be carried out together with the calibration for the pressure chamber. The procedure for the piston uplift calibration is as follows:

1. Assemble the apparatus without a specimen.
2. Increase the cell pressure from zero to the maximum cell pressure.
3. Record the applied cell pressure and the force reading.
4. Create a plot between cell pressure and force reading.
5. The effective area of the piston (A_p) is given as the slope of the line between cell pressure and force reading and the effective piston weight (W_p) as the intercept of the line between cell pressure and force reading.

14.4.1.4 Calibration for piston seal dynamic friction

There is friction in the piston seal where the piston enters the CRS chamber, and this affects the axial force reading. Therefore, correction to the force reading needs to be carried out. The procedure for the calibration of piston seal dynamic friction is as follows:

1. Assemble the apparatus without a specimen.
2. Apply and record the targeted cell pressure σ_3.
3. Advance the piston at the rate used in the CRS test.
4. Compute the increment in axial force as the difference between the measured axial force and the uplift force.
5. Take the average of the increment in axial force as the dynamic seal friction force Δf_s.
6. Steps 2 to 5 are repeated for other cell pressures if required.

14.4.2 Specimen preparation

Specimen preparation for the CRS test is similar to the specimen preparation for the oedometer test. The specimen preparation method (ASTM D4186/D4186M-12 (2012)) is summarised as follows:

1. Record the weight (M_{ring}), inside diameter (D) and height (H_0) of the specimen ring. The diameter and height of the specimen can be taken as the inside diameter and height of the specimen ring, respectively.

2. The sample is first extruded into a short sample tube which is 2.5 mm larger than the CRS specimen diameter.

3. The sample is then extruded from the sampling tube. The extruding direction shall be the same as the direction that the soil enters the tube. If possible, the sample should be extruded vertically; otherwise, precaution must be taken to ensure that bending stress is not imposed on the sample due to gravity.

4. Either of the following can be done to reduce the specimen diameter into the specimen ring diameter:

    a) Trimming the specimen by using a trimming turntable until the specimen can fit into the specimen ring which has been coated with low-friction material (silicone/vacuum grease).

    b) Push in the specimen ring to the specimen. The specimen is trimmed to a gentle taper while the specimen ring is pushed by using an alignment guide until the specimen protrudes at the top of the ring.

5. Trim the top and bottom of the specimen to be flush with the specimen ring by using the following:

    a) Soft to medium soil: wire saw.

    b) Stiff soil: Straightedge with a sharpened cutting surface.

6. Any voids caused by trimming should be patched with the soil from the trimmings and noted in the test.

7. If the specimen height is below the height of the specimen ring, place the filter paper (or any filtering element) on top of the soil and do either of the following:

    a) Take at least four evenly spaced measurements of the specimen height by using a dial comparator or other measuring device which can minimise penetration into the soil during the measurement.

    b) Take the height of the confinement ring minus the recess spacer and the filtering element.

8. Clean the outside of the specimen ring and record the weight of both the specimen ring and the specimen ($M_{ring+soil}$) and determine the mass of soil (M_{soil}) as follows:

$$M_s = M_{ring+soil} - M_{ring} \tag{14.2}$$

9. Determine the initial water content (w_0) of the specimen using the trimmings.

14.4.3 Saturated CRS test

The saturated CRS test procedure is to follow ASTM D4186/D4186M-12 (2012). The stages in the saturated CRS test can be divided as follows:

1. Calibration and set up
2. Saturation
3. Loading
4. Unloading
5. Constant load

14.4.3.1 End calibration and set-up stage

The calibration stage for the CRS test involves the calibrations for apparatus compressibility:

1. Chamber pressure, piston uplift, and piston seal dynamic friction calibrations as detailed in ASTM D4186/D4186M-12 (2012). Once the calibration is completed, record the weight of the load cap and top porous stone.
2. Ensure that all tubes/systems which are used to apply cell pressure, back pressure and drainage system are completely filled with water. Record the zero readings of the cell pressure and base pressure, σ_0 and $u_{w,0}$, respectively.
   Place the porous stone as follows:
   a. For non-expansive saturated soil, use a saturated porous stone and place it into the CRS apparatus with the porous stone slot/base filled with water. Use a paper towel to remove the excess water.
   b. For non-expansive unsaturated soil, use a saturated porous stone and place it into the CRS apparatus with the porous stone slot/base under dry condition. Use a paper towel to remove any excess water from the porous stone.
   c. For expansive soil, use an air-dried porous stone and place it into the CRS apparatus with the porous stone slot/base under dry conditions.
      For all of the above procedures, the connection lines/tubes must be filled with water.
3. Prepare the specimen inside the specimen ring and place it into the CRS apparatus with filter elements at the top and bottom of the specimen (to avoid intrusion of material into the pores of the porous disk).
4. Record the zero-reading force F_0.
5. Adjust the axial loading device such that the load cap of the specimen is in contact with the piston or load cell.
6. Set the deformation measuring device and record the zero-reading d_0.
7. Applies a small seating pressure. The seating pressure must be enough to prevent swelling but not cause significant consolidation. Recommended seating pressure is about 10% of the in situ effective stress or 0.2% axial strain.
8. Open the valve to fill the CRS cell with water.

14.4.3.2 Saturation stage

The specimen can be saturated by using back pressure as follows:

1. Apply back pressure (cell pressure) to saturate the specimen while maintaining a constant seating pressure or constant specimen height. The back pressure is typically from 400 kPa to 1,000 kPa. There are two methods for applying back pressure:

   a.   Stepwise increment. A starting increment is typically equal to the seating pressure and then doubled (typically in the range of 35 kPa to 140 kPa). A typical first-time increment is 10 minutes.
   b.   Constant rate increment. Back pressure is progressively increased up to the maximum value in several hours.
        If constant seating pressure or constant specimen height cannot be maintained, the back pressure increment or rate must be reduced.

2. Check the saturation by applying an increment of cell pressure. If the pore-water pressure at the base increases almost instantaneously as the increase in cell pressure, the saturation can be considered successful.
3. Record the post-saturation seating force, cell pressure, pore-water pressure at the base, axial deformation and the time prior to the loading stage.
4. Close the drainage valve and check the pore-water pressure at the base. Ensure that the pore-water pressure at the base remains constant else there is a leak at the base.

14.4.3.3 Loading stage

Select a suitable strain rate such that the pore-water ratio R_u is between 3% and 15%. The appropriate strain rate can be achieved by increasing the strain rate and monitoring R_u until it reaches approximately 5% and then maintaining the constant strain rate. As the equation shown in Table 14.1 is based on $F_n > 0.4$, it is also important to select a strain rate such that F_n is bigger than 0.4; otherwise, the parameters calculated in Table 14.1 are approximate. Some recommended strain rates are as follows:

1. MH material: 10%/h
2. CL material: 1%/h
3. CH material: 0.1%/h

Once an appropriate strain rate for the typical specimen has been obtained, a nearly constant strain rate (the variation is not more than a factor of 5) throughout the test must be used.

14.4.3.4 Unloading stage

For the unloading stage, use half of the strain rate used for the loading stage. During the unloading stage, it is possible for the pore-water pressure at the base to become negative. Thus, cell pressure must be high enough such that the pore-water pressure at the base is greater than the saturation pressure.

14.4.3.5 Constant load stage

In the constant load stage, the stress is maintained until the excess pore-water pressure dissipates. Record the time, the elapsed duration, cell pressure, base pressure, axial load and axial displacement reading. The constant load stage is maintained for a specified time or until the base excess pressure is nearly 0 (about 1% of the total stress).

14.4.3.6 End-stage

1. At the end of the test, open the bottom drainage valve and decrease the cell pressure to atmospheric pressure.
2. Unload and record the final zero readings for the axial force, cell pressure and base pressure.
3. Drain the water from the cell and remove the specimen ring from the CRS apparatus.
4. Extrude the specimen and determine the final water content of the soil.

14.4.4 CWC-CRS test

The CWC-CRS test procedure developed by Wijaya (2017) can be divided into the following stages:

1. Suction initialisation
2. Calibration and set up
3. Loading
4. Unloading
5. Inundation
6. End

14.4.4.1 Suction initialisation stage

The CWC-CRS test can be done on a specimen with natural water content (and, hence, soil suction is measured) or soil suction can be predetermined either by using suction control or water content control. Water content control can be carried out by air drying the specimen to the desired water content and soil suction is measured (Wijaya 2017; Bagheri et al. 2019a; Bagheri et al. 2019b).

14.4.4.2 Calibration and set-up stage

The calibration of the CWC-CRS test apparatus is similar to that for the CRS except that no cell pressure is applied in the CWC-CRS test, so the calibration for the cell pressure and calibration for the piston uplift can be omitted.

If pore-water pressure at the base is measured, a high-capacity tensiometer must be used.

Placed a kaolin paste on the ceramic disk of the high-capacity tensiometer (HCT) to prevent evaporation of the HCT during set up and to ensure good contact condition during the loading stage.

1. Record the weight of the load cap and top porous stone.
2. Prepare the specimen inside the specimen ring and placed it into the CRS apparatus with dry filter elements at the top and bottom of the specimen.
3. Record the zero-reading force F_0.
4. Adjust the axial loading device such that the load cap of the specimen is in contact with the piston or load cell.
5. Set the deformation measuring device and record the zero-reading d_0.
6. Apply a small seating pressure. Recommended seating pressure is about 10% of the in situ effective stress (or total stress if the soil is unsaturated at the site) or 0.2%axial strain.
7. If pore-water pressure at the base is measured, let the HCT reading equilibrate and record its initial value.

14.4.4.3 Loading stage

The loading rate for the CWC-CRS test can be based on the strain rate used for the saturated CRS test. The strain rate can be increased if the unsaturated soil specimen will not reach a fully saturated condition during the test. If HCT is used to measure the pore-water pressure at the base, the strain rate should consider the response rate of the HCT such that the representative pore-water pressure reading is recorded.

14.4.4.4 Unloading stage

For the unloading stage, use half of the strain rate used for the loading stage.

14.4.4.5 Inundation stage

The inundation stage is only carried out if there is a need to investigate the phenomenon related to swelling (swelling pressure) and collapse (relaxation pressure). If pore-water pressure measurement is carried out

1. soil can be inundated by filling the chamber with water (without applying any cell pressure), and
2. the pore-water measurement should be checked to ensure that the pore-water pressure has reached zero.

14.4.4.6 End-stage

1. Unload and record the final zero readings for the axial force and the matric suction of the soil specimen.
2. Extrude the specimen and determine the final water content of the soil.

REFERENCES

ASTM D4186/D4186m-12 (2012). *Standard test method for one-dimensional consolidation properties of saturated cohesive soils using controlled-strain loading.* West Conshohocken, PA: ASTM International.

Bagheri, M., Mousavi Nezhad, M., & Rezania, M. (2019a). A CRS oedometer cell for unsaturated and non-isothermal tests. *Geotechnical Testing Journal, 43*(1): 20–37.

Bagheri, M., Rezania, M., & Nezhad, M. M. (2019b). Rate dependency and stress relaxation of unsaturated clays. *International Journal of Geomechanics, 19*(12): 04019128.

Kim, Y. T. & Leroueil, S. (2001). Modeling the viscoplastic behaviour of clays during consolidation: Application to Berthierville clay in both laboratory and field conditions. *Canadian Geotechnical Journal, 38*(3): 484–497.

Rezania, M., Bagheri, M., Mousavi Nezhad, M., & Sivasithamparam, N. (2017a). Creep analysis of an earth embankment on soft soil deposit with and without PVD improvement. *Geotextiles and Geomembranes, 45*(5): 537–547.

Rezania, M., Nezhad, M. M., Zanganeh, H., Castro, J., & Sivasithamparam, N. (2017b). Modeling pile setup in natural clay deposit considering soil anisotropy, structure, and creep effects: Case study. *International Journal of Geomechanics, 17*(3): 04016075.

Sorensen, K. K., Baudet, B. A., & Simpson, B. (2010). Influence of strain rate and acceleration on the behaviour of reconstituted clays at small strains. *Géotechnique, 60*(10): 751–763.

Tanaka, H., Tsutsumi, A., & Ohashi, T. (2014). Unloading behavior of clays measured by CRS test. *Soils and Foundations, 54*(2): 81–93.

Tong, F. & Yin, J.-H. (2013). Experimental and constitutive modeling of relaxation behaviors of three clayey soils. *Journal of Geotechnical and Geoenvironmental Engineering, 139*(11): 1973–1981.

Wijaya, M. (2017). Compression, shrinkage and wetting-induced volume change of unsaturated soils. In *Civil and environmental engineering (Ph.D.).* Singapore: Nanyang Technological University.

Wijaya, M. & Leong, E. C. (2015). Swelling and collapse of unsaturated soils due to inundation under one-dimensional loading. *Indian Geotechnical Journal, 46:* 239–251.

Wijaya, M., Leong, E. C., & Rahardjo, H. (2014). CRS compression tests of an unsaturated kaolin with pore-water pressure measurement In N. Khalili, A. R. Russell, & A. Khoshghalb (Eds.), *Proceedings of unsaturated soils: Research & applications*. Australia: CRC Press Taylor & Francis, vol. 2, pp. 1639–1643.

Yin, Z.-Y. & Hicher, P.-Y. (2008) Identifying parameters controlling soil delayed behaviour from laboratory and in situ pressuremeter testing. *International Journal for Numerical and Analytical Methods in Geomechanics*, 32(12): 1515–1535.

Yin, Z.-Y., Zhu, Q.-Y., Yin, J.-H., & Ni, Q. (2014). Stress relaxation coefficient and formulation for soft soils. *Geotechnique Letters*, 4(1): 45–51.

FURTHER READING

Bagheri, M., Mousavi Nezhad, M., & Rezania, M. (2019a). A CRS oedometer cell for unsaturated and non-isothermal tests. *Geotechnical Testing Journal*, 43(1): 20–37.

Chapter 15

Triaxial test (volume change)

15.1 BACKGROUND

The compressibility of soil is commonly investigated using an oedometer or constant rate of strain (CRS) test. However, such a test is only applicable when the soil in the field can be assumed to be under one-dimensional loading (k_0) condition and no lateral displacement is allowed. However, when the load is applied over a finite area, the soil will also deform laterally, and loading is no longer under k_0 condition. Moreover, water is generally restricted to only flow in the vertical direction under k_0 loading in an oedometer or CRS test. In order to overcome the limitation of k_0-loading tests, compressibility of soil can be determined using a triaxial apparatus. The soil specimen in a triaxial apparatus can be subjected to either an isotropic, anisotropic or k_0-loading condition. The other parameters determined from the triaxial test are shear strength and permeability, which are dealt with in the other chapters.

15.2 RELATED STANDARDS

1. BS 1377-6 (1990) Methods of Test for Soil for Civil Engineering Purposes – Part 6: Consolidation and Permeability Tests in Hydraulic Cells and with Pore Pressure Measurement, BSI.
2. BS 1377-7 (1990) Methods of Test for Soils for Civil Engineering Purposes – Part 7: Shear Strength Tests (Total Stress), BSI.
3. BS 1377-8 (1990) Methods of Test for Soils for Civil Engineering Purposes – Part 8: Shear Strength Tests (Effective Stress), BSI.
4. BS EN ISO 17892-7 (2018) Geotechnical Investigation and Testing – Laboratory Testing of Soil Part 7: Unconfined Compression Test (ISO 17892-7:2017), BSI.

DOI: 10.1201/b22304-15

5. BS EN ISO 17892-8 (2018) Geotechnical Investigation and Testing – Laboratory Testing of Soil Part 8: Unconsolidated-Undrained Triaxial Test, BSI.
6. BS EN ISO 17892-9 (2018) Geotechnical Investigation and Testing – Laboratory Testing of Soil Part 9: Consolidated Triaxial Compression Tests on Water-Saturated Soils, BSI.
7. ASTM D2166/2166m-16 (2016) Standard Test Method for Unconfined Compressive Strength of Cohesive Soil.
8. ASTM D2850-15 (2015) Standard Test Method for Unconsolidated-Undrained Triaxial Compression Test on Cohesive Soils.
9. ASTM D4767-11 (2011) Standard Test Method for Consolidated Undrained Triaxial Compression Test for Cohesive Soils.
10. ASTM D7181-11 (2011) Standard Test Method for Consolidated Drained Triaxial Compression Test for Cohesive Soils.

15.3 THEORY

15.3.1 Stress and strain of soils

To determine the stress-strain behaviour of soil, it is important to understand the loading condition which is applicable to the soil. The stress-strain relationship of soil is not only related to the initial stress state and magnitude of the applied load but also affected by the degree of saturation of the soil, the soil permeability, the rate of load application and duration after which the load is applied.

If the soil is saturated, the permeability of the soil dictates the loading behaviour, as it governs the rate of pore-water pressure dissipation. If the permeability of the soil is low (fine-grained soils such as silt and clay), the applied load will generate excess pore-water pressure equal to the applied stress which needs to dissipate at a rate governed by the soil's permeability. Thus, there is no change in the effective stress at the start of the loading. This is referred to as the undrained condition and is applicable for short-term analysis. If the permeability of the soil is high (coarse-grained soils such as sand and gravel), there is sufficient time for all the excess pore-water pressure to dissipate, and the loading condition is referred to as drained. In fine-grained soils, the drained condition is applicable for long-term analysis. In coarse-grained soils, the undrained condition can prevail under a transient loading condition such as earthquake, dynamic or blast loading.

If the soil is unsaturated, there are both air and water phases present in the voids of the soil. The pore-water pressure in the soil is negative (also referred to as soil suction) and the engineering behaviour of the soil is no longer governed by the conventional effective stress principle. In this case, the volumetric response of the soil to the applied load is separated into effects due to the net normal stress and matric suction (Fredlund et al. 2012). Commonly, two types of loading conditions are applicable to unsaturated soil:

1. Constant water content loading
   When load is applied to an unsaturated soil, the volume change (and hence void ratio) is immediate, as air is expelled with the compression of the air voids. With the reduction in air voids, the degree of saturation of the soil increases and matric suction reduces. This is termed mechanical wetting (degree of saturation change due to the change in void ratio under the constant water content condition) by Wijaya and Leong (2017). However, if the initial degree of saturation of the soil is high, the soil may approach full saturation on load application and excess pore-water pressure will be generated as the soil now behaves as a fully saturated soil.
2. Constant suction loading
   The matric suction is maintained constant while the soil is loaded. It is very difficult to maintain matric suction at a constant value, as the water content of the soil needs to equilibrate at the matric suction for the new void ratio of the soil. Hence, the loading rate has to be slow so that there is sufficient time for the water content to come to equilibrium at the matric suction for the new void ratio of the soil. The rate at which water can move in and out of the soil to equilibrate depends on the method employed to maintain matric suction constant. If the hanging column or axis-translation method is used, the loading rate is dependent on the lower permeability of the soil or the high-air entry (HAE) ceramic disk. If the vapour equilibrium method is used, it depends on the rate of vapour diffusion.

In the analysis of the stress-strain behaviour of soil, the soil is treated as a continuum. The stresses at a point in a continuum can be represented as stresses acting on a small element of the soil. The principal stresses (σ_1, σ_2, σ_3) in the three orthogonal directions can be used to describe the stresses of the soil at the point. In most cases, the vertical stress (σ_z) is the major principal stress (σ_1), while horizontal stresses (σ_x and σ_y) are the intermediate and minor principal stresses (σ_2 and σ_3). Pore-water pressure (σ_w) and pore-air pressure (u_a) act equally in all directions.

It is convenient to use stress invariant and strain invariant counterparts, which are independent of the coordinate system to describe the stresses and strains acting on a soil element. Commonly used stress and strain invariants are as follows:

1. Mean stress (p), which can be either mean total stress (p_t), mean effective stress (p'), or net mean stress (p_m). The strain invariant associated with mean stress is volumetric strain (ε_v).
2. Deviatoric stress (q) and deviatoric strain (ε_q).

The stress and strain invariants for both saturated and unsaturated soils are shown in Table 15.1. Mean total stress (p_t) and mean effective stress (p') are commonly used in saturated soil while net mean stress (p_m) is commonly used in unsaturated soils.

Stress and strain are related by a proportionality constant known as modulus. Depending on the loading condition (i.e., drained, undrained or

Table 15.1 Stress and strain invariants

Mean total stress, pt	Parameters
General condition $$p_t = \frac{\sigma_1 + \sigma_2 + \sigma_3}{3}; p_t = \frac{\sigma_x + \sigma_y + \sigma_z}{3}$$ **Axisymmetric condition ($\sigma_2 = \sigma_3$)** $$p_t = \frac{\sigma_1 + 2\sigma_3}{3}$$ **Mean effective stress, p′** General condition $$p' = p_t - u_w$$ k_0 condition $(\sigma'_x = \sigma'_y = k_0.\sigma'_z; \varepsilon_x = \varepsilon_y = 0)$ $$p' = \sigma'_z\left(\frac{1 + 2k_0}{3}\right)$$ **Net mean stress, p** $$p = p_t - u_a$$ **Volumetric strain, ε_p** $$\varepsilon_p = \varepsilon_1 + \varepsilon_2 + \varepsilon_3 = \varepsilon_x + \varepsilon_y + \varepsilon_z$$ **Deviatoric stress** General condition $$q = \frac{1}{\sqrt{2}}\left[(\sigma_1 - \sigma_2)^2 + (\sigma_2 - \sigma_3)^2 + (\sigma_3 - \sigma_1)^2\right]^{1/2}$$ $$q = \frac{1}{\sqrt{2}}\left\{\left[(\sigma_{xx} - \sigma_{yy})^2 + (\sigma_{yy} - \sigma_{zz})^2 + (\sigma_{zz} - \sigma_{xx})^2\right]\right.$$ $$\left. + 6\tau_{xy}^2 + 6\tau_{yz}^2 + 6\tau_{zx}^2\right\}^{1/2}$$ Axisymmetric condition $$q = \sigma_1 - \sigma_3$$ k_0 condition $$q = \sigma'_z(1 - k_0)$$ **Deviatoric strain** $$\varepsilon_q = \frac{\sqrt{2}}{3}\left[(\varepsilon_1 - \varepsilon_2)^2 + (\varepsilon_2 - \varepsilon_3)^2 + (\varepsilon_3 - \varepsilon_1)^2\right]^{1/2}$$	P_t = mean total stress σ_1 = maximum principal stress σ_2 = intermediate principal stress σ_3 = minimum principal stress σ_x = total stress at x direction σ_y = total stress at y direction σ_z = total stress at z direction p' = mean effective stress p = net means tress ε_p = volumetric strain ε_1 = strain at maximum principal stress direction ε_2 = strain at intermediate principal stress direction ε_3 = strain at minimum principal stress direction ε_x = strain at x direction ε_y = strain at y direction q = deviatoric stress ε_q = deviatoric strain u_a = pore-air pressure u_w = pore-water pressure k_0 = lateral at rest pressure coefficient

unsaturated), there are five moduli which are commonly used to describe stress-strain relationship, namely:

1. Bulk modulus (K): related to the change in volumetric strain (ε_v) due to the change in mean stress (p).
2. Shear modulus (G): related to the change in deviatoric strain (ε_q) due to the change in deviatoric stress (q).
3. Young's modulus (E): related to the change in one directional strain (axial or lateral) due to the change in one directional stress (axial or lateral).
4. Constrained modulus (E_c): related to the change in axial strain under the constrained condition (where lateral strain is not allowed) due to the change in axial stress. Hence, axial strain is equal to the volumetric strain.
5. Modulus of soil structure with respect to change in matric suction (H): related to the change in volumetric strain due to the change in matric suction.

Table 15.2 summarises different stress-strain relationships which are commonly used in geotechnical engineering under drained, undrained and unsaturated conditions. For soil loaded under the drained condition, the correlation between effective/drained modulus $(E', E_c', G, \text{ and } K')$ is easily established provided that the effective Poisson's ratio (v') is known, as shown in Table 15.3. Poisson's ratio (v) generally defined as shown in Equation 15.1.

$$v = \frac{\varepsilon_x}{\varepsilon_z} \tag{15.1}$$

where ε_x is the strain at the lateral direction, while ε_z is the strain at the axial direction. Based on Table 15.3, only two of E', G, K' and v' are needed to obtain the other parameters.

For soil loaded under the undrained condition, there is no volume change. Hence, the Poisson's ratio under the undrained condition (v_u) is equal to 0.5. The correlation between drained and undrained moduli can be found in Table 15.4. For a soil under k_0 loading condition (i.e. oedometer test), it is possible to correlate the effective drained modulus (E') with the constrained modulus (E_c') provided that Poisson's ratio (v') is known as shown in Table 15.4.

Unfortunately, Poisson's ratio (v') is rarely determined. Table 15.5 shows some typical values of Poisson's ratio (v'), effective Young's modulus (E')

and shear modulus G for various soil types. Alternatively, it is quite common to assume 0.3 for Poisson's ratio (ν').

For unsaturated soil, the modulus is a function of either matric suction, water content or both. As there are several soil models developed for unsaturated soil, the interpretation of the unsaturated soil modulus depends on the soil model and the test procedures.

Table 15.2 Stress and strain relationship in soils

Type of strain	Drained	Undrained	Unsaturated
Volumetric strain, ε_p	$\Delta\varepsilon_p = \dfrac{\Delta p'}{K'}$	$\Delta\varepsilon_p = 0$	$\Delta\varepsilon_p = \dfrac{\Delta p}{K} + 3\dfrac{\Delta s}{H}$
Deviatoric strain, ε_q	$\Delta\varepsilon_q = \dfrac{\Delta q}{3G}$	$\Delta\varepsilon_q = \dfrac{\Delta q}{3G}$	$\Delta\varepsilon_q = \dfrac{\Delta q}{3G}$
Axial strain, ε_z	$\Delta\varepsilon_z = \dfrac{\Delta\sigma'_z}{E'_z}$	$\Delta\varepsilon_z = \dfrac{\Delta\sigma_z}{E_{u,z}}$	$\Delta\varepsilon_z = \dfrac{\Delta(\sigma_z - u_a)}{E_z}$
Lateral strain, $\varepsilon_x, \varepsilon_y, \varepsilon_r$	$\Delta\varepsilon_x = \dfrac{\Delta\sigma'_x}{E'_x}$	$\Delta\varepsilon_x = \dfrac{\Delta\sigma_x}{E_{u,x}}$	$\Delta\varepsilon_x = \dfrac{\Delta(\sigma_x - u_a)}{E_x}$
Axial k_0 (constrained) strain, ε_{z0}	$\Delta\varepsilon_{z,0} = \dfrac{\Delta\sigma'_z}{E'_c}$	$\Delta\varepsilon_{z,0} = 0$	$\Delta\varepsilon_{z,0} = \dfrac{\Delta(\sigma_z - u_a)}{E_c}$

Stress:
$\Delta p'$ = change in mean effective stress
Δp = change in net mean stress
Δq = change in deviatoric stress
$\Delta\sigma$ = change in total stress which can be in x, y or z directions
$\Delta\sigma'$ = change in effective stress which can be in x, y or z directions
$\Delta(\sigma - u_a)$ = change in net normal stress which can be in x, y or z directions
u_a = air pressure
Δs = change in soil suction
Strain:
$\Delta\varepsilon_p$ = change in volumetric strain
$\Delta\varepsilon_q$ = change in deviatoric strain
$\Delta\varepsilon_z$ = change in axial strain
$\Delta\varepsilon$ = change of strain which can be in x, y or z directions
$\Delta\varepsilon_{z,0}$ = change in axial strain under k_0 condition ($\varepsilon_x = \varepsilon_y = \varepsilon_r = 0$)
Modulus:
K' = effective bulk modulus
H = modulus of soil structure with respect to change in soil suction
G = shear modulus
E' = Effective Young's modulus
E_u = Undrained modulus
E = Young's modulus
E'_c = Constrained (oedometer) modulus under the drained condition
E_c = Constrained (oedometer) modulus

Table 15.3 Various expressions of effective (drained) modulus

Effective Bulk modulus, K'	Effective Young's modulus, E'	Shear modulus, G	Poisson's ratio, υ'
$K' = \dfrac{E'}{3(1-2\upsilon')}$	$E' = 3(1-2\upsilon')K'$	$G = \dfrac{E'}{2(1+\upsilon')}$	$\upsilon' = 0.5\left(1-\dfrac{E'}{3K'}\right)$
$K' = \dfrac{E'G}{3(3G-E')}$	$E' = \dfrac{9K'G}{3K'+G}$	$G = \dfrac{3K'E'}{9K'-E'}$	$\upsilon' = \dfrac{E}{2G}-1$
$K' = \dfrac{2G(1+\upsilon')}{3(1-2\upsilon')}$	$E' = 2G(1-\upsilon')$	$G = \dfrac{3K'(1-2\upsilon')}{2(1+\upsilon')}$	$\upsilon' = \dfrac{3K'-2G}{2G+6K'}$

Table 15.4 Modulus under special condition

Undrained condition	Parameters
Undrained Poisson's ratio, υ_u $\upsilon_u = 0.5$ Undrained Young's modulus, E_u $E_u = \dfrac{1.5E'}{1+\upsilon'}$ Shear modulus, G $G = \dfrac{E_u}{2(1+\upsilon_u)}$ K_0 (Constrained) condition Constrained modulus, E'_c $E'_c = E'\dfrac{(1-v')}{(1+v')(1-2v')}$ $E'_c = \dfrac{1}{m_v}$	υ_u = undrained Poisson's ratio υ' = effective Poisson's ratio E_u = undrained Young's modulus E' = effective Young's modulus G = shear modulus E'_c = constrained modulus m_v = modulus of volume compressibility (usually obtained from an oedometer test)

Table 15.5 Typical values for effective Poisson's ratio (Budhu 2010)

Soil type	Description	υ'	E' (MPa)	G (MPa)
Clay	Soft	0.35–0.40	1–15	0.5–5
	Medium	0.30–0.35	15–30	5–15
	Stiff	0.20–0.30	30–100	15.40
Sand	Loose	0.15–0.25	10–20	5–10
	Medium	0.25–0.30	20–40	10–15
	Dense	0.25–0.35	40–80	15–35

15.3.2 Compression curve based on stress and strain invariants

It is also possible to describe the compression curve of soil using mean effective stress (i.e., e vs ln (p') plot) similar to the one-dimensional compression curve for effective vertical stress (i.e., e vs log (σ'_v) plot). The parameters obtained from the $e - \ln (p')$ and from $e - \log (\sigma'_v)$ are similar. Take the following, for instance (Budhu 2010):

1. The maximum effective vertical stress that has been applied to the soil is referred to as preconsolidation pressure (σ_{vc}) and an equivalent stress invariant for the preconsolidation pressure is mean effective preconsolidation pressure (p'_c), which is the maximum mean effective stress that has been applied to the soil specimen.
2. The ratio between σ_{vc} and overburden pressure (σ_{v0}) is referred to as the over-consolidation ratio (OCR $= \sigma_{vc}/\sigma_{v0}$) while the ratio between p'_c and effective mean overburden pressure p'_0 is referred to as the preconsolidation ratio $(R_0 = p'_c/p'_0)$.
3. The slope of the normal compression line (NCL) in the $e - \log (\sigma'_v)$ plot is referred to as C_c while it is referred to as λ in the $e - \ln (p')$ plot, where

$$\lambda = \frac{C_c}{2.3}.$$ (15.2)

4. The slope of the unloading-reloading line (URL) in the $e - \log (\sigma'_v)$ plot is referred to as C_s (for unloading) or C_r (for reloading), while it is referred to as κ in the e-ln(p') plot, where

$$\kappa = \frac{C_r}{2.3}.$$ (15.3)

When the soil is unsaturated, the compressibility parameters (i.e., λ and κ) are now a function of matric suction or water content. It has been observed that the intersection between elastic and plastic ranges of the compression curve (in the log scale or ln scale) is increased when the soil is unsaturated (Alonso et al. 1990; Wheeler and Sivakumar 1995; Wijaya 2017; Wijaya and Leong 2015). However, as the intersection no longer represents the maximum past effective stress that has been applied to the soil, the terms "apparent preconsolidation pressure" $\sigma_{c,a}$ for vertical stress and "yield stress" $p_{c,a}$ for net mean stress are used, and both are a function of either matric suction or water content.

15.4 TRIAXIAL TEST FOR SATURATED AND UNSATURATED SOIL

A study on the stress-strain relationship of soil is usually carried out using a triaxial test. The triaxial test is a more comprehensive test as it allows control of the lateral pressure by using a fluid (i.e., water or air). Figure 15.1 shows a schematic of the triaxial apparatus. A brief description of the triaxial apparatus is as follows:

1. Triaxial cell, which contains the fluid where cell pressure is applied to the soil specimen.
2. Soil specimen, which sits on a bottom pedestal and is loaded through a top load cap.
3. Rubber membrane, which encloses the soil specimen to separate the fluid in the triaxial cell from the fluid within the soil specimen. Filter elements are placed at the top and bottom of the specimen.
4. Drainage line to allow drainage of fluid to and from the soil specimen to be controlled and a back pressure line to allow water pressure to be applied to saturate the soil specimen.
5. Pore-pressure transducer to measure the pore-water pressure of the soil specimen.
6. Dial gauge or linear-variable differential transformer (LVDT) to measure axial displacement.

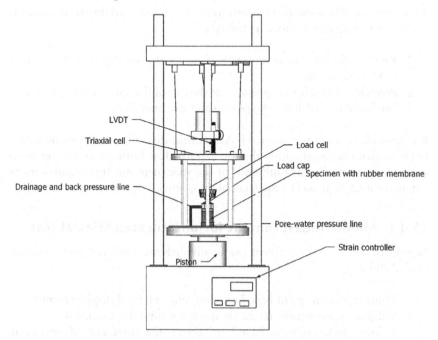

Figure 15.1 Conventional triaxial apparatus.

7. Load cell to record the deviatoric load applied to the soil specimen via the loading piston.
8. Strain controller to control the speed of loading applied to soil specimen.

Several features of the triaxial test are as follows:

1. Allows the test of large specimens. BS recommends specimen diameter between 38 mm to 100 mm (the size of the largest soil particle must be less than one-fifth of the specimen diameter) with length-to-diameter ratio of around 2 while ASTM recommends a minimum diameter of 33 mm with length-to-diameter ratio between 2 and 2.5 (the largest soil particle must be less than one-sixth of the specimen diameter).
2. Allows application of back pressure to saturate and maintain saturation of the specimen during the test.
3. Allows continuous measurement of displacement, stress, volume change and pore-water pressure.
4. Allows control of cell pressure.
5. Allows various combinations of loading to be applied (i.e., isotropic consolidation test, permeability test, shearing test, k_0 consolidation test).
6. Allows multi-stage test to be conducted (one specimen tested under different cell pressures) to obtain multiple data.

Minimum modification to the conventional triaxial apparatus for unsaturated soil testing can be done as follows:

1. Replace the porous stone at the base of the specimen with a sealed HAE ceramic disk.
2. Provide control for air pressure at the top of the specimen while drainage is controlled from the pore-water pressure line.

It is possible to have a combined HAE ceramic disk and porous stone sealed in the bottom pedestal and in the top cap to allow both air and water pressures to be controlled at both ends of the specimen. But this requires more extensive modifications to the triaxial apparatus.

15.4.1 Volume measurement for unsaturated triaxial test

Generally, volume measurement can be divided into three categories (Leong et al. 2004):

1. Volume measurement based on confining cell fluid displacement
2. Volume measurement based on specimen fluid displacement
3. Volume measurement based on direct measurement of specimen dimension

For unsaturated soil, the change in the volume of water in the specimen is not equal to the change in the volume of the specimen. Hence, only methods 1 and 3 are applicable to measure the volume change of unsaturated soil. However, method 2 can be used to measure the change in the volume of water inside the soil specimen which can then be used to calculate gravimetric and volumetric water content of the soil specimen.

15.4.1.1 Volume measurement based on confining cell fluid displacement

Volume measurement based on confining cell fluid displacement is one of the most common techniques in unsaturated triaxial testing. In this method, the change in soil volume is assumed to be equal to the change in the volume of fluid inside the confining cell. However, as the amount of fluid in the confining cell increases, the error in the volume measurement will increase as well. Hence, a double cell is commonly employed.

Bishop and Donald (1961) used a double cell with mercury at the bottom of the inner cell surrounding the soil specimen. The rest of the inner cell is filled with water and connected to the outer cell so that the pressure between inner cell and outer cell is equal. A stainless-steel ball floater is placed on top of the mercury surface to serve as an indicator for mercury level change which represents the change in volume of the soil specimen. Mercury is no longer allowed in the laboratories of many countries due to health risks.

Toyota et al. (2001) and Aversa and Nicotera (2002) used pressurised air to provide confining pressure above an inner cell filled with water as shown in Figure 15.2a. However, pressurised air may cause safety concerns.

Wheeler (1986) and Sivakumar (1993) used a double-wall triaxial cell (Figure 15.2b) instead of an inner cell. The water in the inner and outer cells is separated. The volume change of the specimen is measured based on the change in the fluid volume within the inner cell.

When volume measurement is conducted by measuring the water column head, water in the column head is subjected to evaporation in long-duration tests (Leong et al. 2004). Hence, a thin layer of paraffin (Ng et al. 2002; Sivakumar 1993) or silicon oil (Aversa and Nicotera 2002) is commonly added to the water column used to reduce the effects of evaporation.

A slightly different double-wall cell from the apparatuses described earlier consists of a triaxial cell (inner cell) enclosed by a larger cell (Wheeler 1988). In this double-wall cell, the inner triaxial cell is subjected to equal water pressures from inside and outside, and, hence, any extension of the tie rods of the inner triaxial cell which affects the volume measurement is eliminated. A double-cell triaxial cell requires calibration to quantify the creep of the inner cell. For example, Mancuso et al. (2012) applied a pressure of 600 kPa to both cells for weeks to quantify the creep of the inner cell.

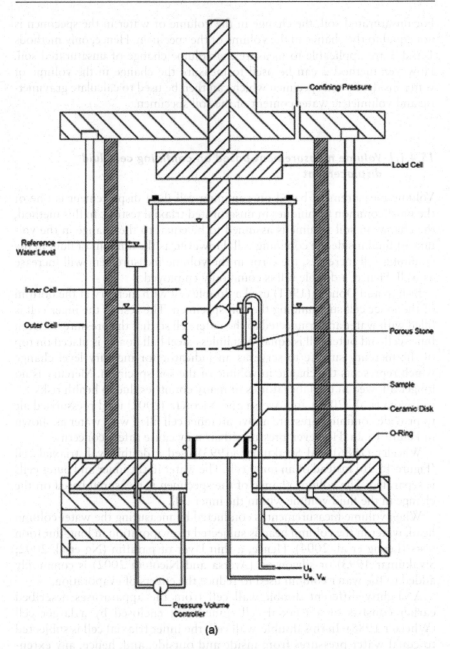

Figure 15.2 Suction-control triaxial apparatuses (from Leong et al. 2004).
(a) Double cell triaxial apparatus (Toyota et al. 2001).

(Continued)

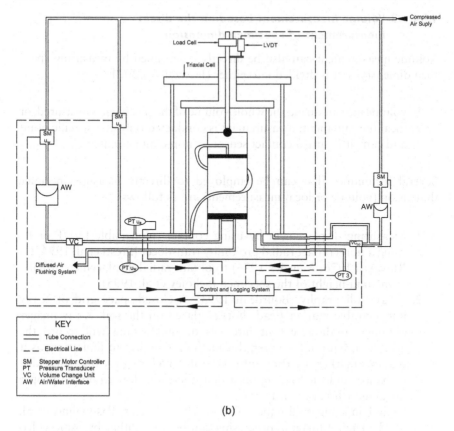

(b)

Figure 15.2 (Continued) (b) Double walled triaxial apparatus (Sivakumar 1993).

15.4.1.2 Volume measurement based on specimen fluid displacement

Volume measurement based on specimen fluid displacement is typically used in saturated triaxial testing. As the soil is assumed to be saturated, the volume change of the specimen is equal to the volume of water coming out from the specimen. For an unsaturated soil specimen, the presence of air makes the pore compressible. Hence, the amount of water coming out from the specimen is not equal to the volume change of the specimen.

15.4.1.3 Volume measurement based on the direct measurement of specimen dimension

Volume measurement can also be directly determined by measuring specimen dimensions. It is divided into either (Jastrzębska 2021)

1. volume measurement: the whole volume of the specimen is measured, or
2. local measurement: measurement is conducted only at a specific point and can either use a contact sensor or non-contact sensor.

Several techniques that can be employed to directly measure specimen dimension (volume or local measurement) are as follows:

1. Measuring displacement by using contact submersible LVDT or by using a non-contact radial proximity transducer (Presti et al. 1994; El-Ruwayih 1976; Menzies 1976) to obtain radial displacement usually at the mid-height of the specimen (Scholey et al. 1995).
2. X-ray radiography (Bourdeau 1993). A closed space grid of X-ray impenetrable marker (lead shot) is placed on the soil. X-ray images obtained at discrete time intervals during loading will record the grid deformation. However, this method is subject to limitation, as it requires marker insertion into the soil, which might cause some disturbance, and the loading must be sufficiently slow to obtain proper imaging (Scholey et al. 1995).
3. Digital imaging techniques (Parker 1987; Macari-Pasqualino et al. 1993). Digital imaging processing can be done either by using video or camera, and image processing can then be used to obtain radial, axial and even volume deformation (Scholey et al. 1995).

15.5 TRIAXIAL TEST CALIBRATIONS

There are three typical corrections which are required prior to conducting a triaxial test:

1. Load cell uplift correction (applicable for non-submersible load cell)
2. Filter-paper correction (applicable when filter paper is used)
3. Membrane correction (applicable when a membrane is used)

The principle of the calibration is the same, but the procedures and equation differ slightly based on the standard followed. Thus, it is important to state which standard is used as reference for the calibration. The effect of calibration can be significant when the soil is very soft and negligible when the soil is stiff. However, it is recommended to always carry out the calibration prior to testing.

15.5.1 Load cell uplift correction

If the load measuring device is located outside the triaxial cell (non-submersible load cell), the fluid in the triaxial cell will produce an uplift force. The correction can be done by re-zeroing the load cell reading after the cell pressure is applied prior to contacting the soil specimen with the piston. For submersible load cell, no correction is needed.

15.5.2 Filter-paper correction

Vertical filter-paper strips around the perimeter of the soil specimen are commonly used in the triaxial test to accelerate the consolidation and drained shearing stages. It is compulsory to apply the filter-paper correction to the deviator stress when the filter-paper correction (q_{fp}) is higher than 5% of the recorded deviatoric stress (ASTM D4767-11 2011; ASTM D7181-11 2011). The filter-paper correction q_{fp} is given by Equation 15.4.

$$q_{fp} = \min\left(K_{fp} \frac{P_{fp}}{A_c}, 50\varepsilon_1 K_{fp} \frac{P_{fp}}{A_c} \right) \tag{15.4}$$

where K_{fp} is the load carried by filter-paper strips per unit length of perimeter covered by filter paper, which is typically 0.19 kN/m (ASTM D4767-11 2011; ASTM D7181-11 2011); P_{fp} is the perimeter covered by filter paper; and A_c is the cross-sectional area of the specimen after consolidation.

BS (BS 1377-8 1990) recommends the filter-paper correction to be applied to the deviator stress when the vertical strain exceeds 2%. The vertical side drain correction can be interpolated from Table 15.6.

Table 15.6 Vertical side drains correction (BS 1377-8 1990)

Specimen diameter (mm)	Drain correction, q_{fp} (kPa)
38	10
50	7
70	5
100	3.5
150	2.5

15.5.3 Membrane correction

The rubber membrane used for the triaxial specimen has several requirements (BS 1377-7 1990; BS 1377-8 1990; ASTM D7181-11 2011; ASTM D4767-11 2011):

1. The rubber membrane must be made of high-density latex.
2. The unstretched internal diameter of the rubber membrane shall not be less than 90% of the specimen diameter.
3. The unstretched internal diameter of the rubber membrane shall not be greater than the specimen diameter
4. The rubber membrane thickness shall not exceed 1% of the specimen diameter
5. The length of the rubber membrane shall be 50 mm greater than the specimen length.

For a 50-mm diameter specimen, a thickness of 0.2 mm is recommended and when the soil has angular particles, more than one rubber membrane can be used to avoid puncturing the rubber membrane by the soil particles (BS 1377-7 1990; BS 1377-8 1990).

According to the ASTM Standard (ASTM D2850-15 2015; ASTM D4767-11 2011; ASTM D7181-11 2011), correction for the rubber membrane must be applied when the correction due to the rubber membrane (q_m) is higher than 5% of the recorded deviatoric stress.

Based on ASTM and BS EN ISO, q_m can be determined using Equations 15.5 and 15.6 (BS EN ISO 17892-8 2018; ASTM D4767-11 2011; ASTM D2850-15 2015; ASTM D7181-11 2011).

$$q_m = \frac{4E_m t_m \varepsilon_1}{D_c} \qquad (15.5)$$

$$D_c = \sqrt{4A_c/\pi} \qquad (15.6)$$

where E_m is the modulus and t_m is the thickness of the rubber membrane. The E_m can be determined by hanging a circumferential strip membrane with a width (W_s) of 15mm. The membrane is hung by using a thin rod at the top of the membrane and another rod at the bottom of the membrane to apply a uniform load (F). The Young's modulus of the membrane (E_m) is determined as shown in Equation 15.7.

$$E_m = \frac{F/A}{\Delta L/L} \qquad (15.7a)$$

where

$$A_m = 2t_m W_s \qquad (15.7b)$$

and t_m is the thickness of the membrane. Typical E_m value for latex membrane is 1,400 kPa (ASTM D4767-11 2011; ASTM D7181-11 2011; BS EN ISO 17892-8 2018).

When volume measurement is available, BS EN ISO 17892-9 (2018) recommends that both total vertical stress (q_{m1}) and total horizontal stress (q_{m3}) must be corrected using Equations 15.8 and 15.9, respectively.

$$q_{m1} = \frac{4E_m t_m}{D_c}\left(\varepsilon_1 + \frac{\varepsilon_p}{3}\right) \tag{15.8}$$

$$q_{m3} = \frac{4E_m t_m}{D_c}\frac{\varepsilon_p}{3} \tag{15.9}$$

British Standards (BS 1377-7 1990; BS 1377-8 1990) recommend that q_m be calculated using Equation 15.10.

$$q_m = q_{m,38\text{-}0.2}\frac{38}{D}\frac{t_m}{0.2} \tag{15.10}$$

where $q_{m,38\text{-}0.2}$ is the membrane correction for soil specimen of diameter 38 mm and membrane of thickness 0.2 mm obtained from Figure 15.3, and D is the soil specimen diameter.

15.5.4 Triaxial test calculation

Once all the correction has been conducted, typical calculation to apply the correction in triaxial test is shown in Table 15.7. Comprehensive procedures for triaxial testing will be discussed in Chapter 17.

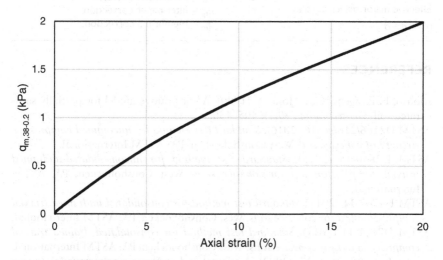

Figure 15.3 Membrane correction for 38 mm diameter specimens fitted with 0.2 mm thick membrane.

Table 15.7 Typical calculation in a triaxial test

Axial strain, ε_1	Parameters
$\varepsilon_1 = \dfrac{\Delta H}{H_c}$	ε_1 = axial strain $\Delta H\varepsilon$ = change in the specimen length/height during the shearing stage
Average cross-sectional area, A	H_c = specimen height after consolidation/specimen height prior to shear
During isotropic consolidation stage:	A = average cross-sectional area
$A = \dfrac{V}{H}$	V = volume of the specimen
	H = height of the specimen
$V = V_0 - \Delta V$	V_0 = initial volume of the specimen
$H = H_0 - \Delta H$	ΔV = change in volume of the specimen due to saturation stages and isotropic consolidation stages
During undrained shearing stage:	H_0 = initial height of the specimen
$A = \dfrac{A_c}{1 - \varepsilon_1}$	ΔH = change in height of the specimen due to saturation stages and isotropic consolidation stages
During the drained shearing stage:	
$A = \dfrac{V_c - \Delta V_\varepsilon}{H_c - \Delta H_\varepsilon}$	A_c = specimen area after consolidation stage/specimen area prior to shear
Uncorrected deviatoric stress, q_{uncor}	V_c = specimen volume after consolidation stage/specimen volume prior to shear
$q_{uncor} = \dfrac{P}{A}$	$\Delta V\varepsilon$ = change in specimen volume due to shearing
Corrected deviatoric stress, q	q_{uncor} = uncorrected deviatoric stress P = recorded load
$q = q_{uncor} - q_{fp} - q_m$	q = corrected deviatoric stress
Effective minor principal stress, σ'_3	σ_3 = total minor principal stress or cell pressure
$\sigma'_3 = \sigma_3 - u_w$	σ'_3 = effective minor principal stress u_w = pore-water pressure
Effective major principal stress, σ'_1	q_{fp} = filter-paper correction
$\sigma'_1 = q + \sigma'_3$	q_m = membrane correction

REFERENCES

Alonso, E. E., Gens, A., & Josa, A. (1990). A constitutive model for partially saturated soil. *Géotechnique,* 40(3): 405–430.

ASTM D2166/2166m-16 (2016). *Standard test method for unconfined compressive strength of cohesive soil.* West Conshohocken, PA: ASTM International.

ASTM D2850-15 (2015). *Standard test method for unconsolidated-undrained triaxial compression test on cohesive soils.* West Conshohocken, PA: ASTM International.

ASTM D4767-11 (2011) *Standard test method for consolidated undrained triaxial compression test for cohesive soils.* West Conshohocken, PA: ASTM International.

ASTM D7181-11 (2011). *Standard test method for consolidated drained triaxial compression test for cohesive soils.* West Conshohocken, PA: ASTM International.

Aversa, S. & Nicotera, M. (2002). A triaxial and oedometer apparatus for testing unsaturated soils. *Geotechnical Testing Journal,* 25(1): 3–15.

BS 1377-6 (1990). *Methods of test for soil for civil engineering purposes - part 6: consolidation and permeability tests in hydraulic cells and with pore pressure measurement.* UK: BSI.

BS 1377-7 (1990). *Methods of test for soils for civil engineering purposes-part 7:shear strength tests (total stress).* UK: BSI.

BS 1377-8 (1990). *Methods of test for soils for civil engineering purposes-part 8: shear strength tests (effective stress).* UK: BSI.

BS EN ISO 17892-7 (2018). *Geotechnical investigation and testing – laboratory testing of soil part 7: unconfined compression test (ISO 17892-7:2017).* UK: BSI.

BS EN ISO 17892-8 (2018). *Geotechnical investigation and testing – laboratory testing of soil part 8: Unconsolidated undrained triaxial test.* UK: BSI.

BS EN ISO 17892-9 (2018). *Geotechnical investigation and testing – laboratory testing of soil part 9: Consolidated triaxial compression tests on water saturated soils.* UK: BSI.

Bishop, A. W., & Donald, I. B. (1961). The experimental study of partly saturated soil in triaxial apparatus. In *Fifth International Conference on Soil Mechanics and Foundation Engineering*, vol. 1, pp. 13–21.

Bourdeau, P.L. (1993). Radiographic visualization in experimental soil mechanics. In *Proceedings of Engineering Foundation-National Science Foundation Conference on Digital Image Processing: Techniques and Applications in Civil Engineering*, ASCE, pp. 125–134.

Budhu, M. (2010)*Soil mechanics and foundations.* 3rd ed. Hoboken, NJ: Wiley.

El-Ruwayih, A. A. (1976). Design manufacture and performance of a lateral strain device. *Géotechnique*, 26(1): 215–216.

Fredlund, D. G., Rahardjo, H., & Fredlund, M. D. (2012). *Unsaturated soil mechanics in engineering practice.* Hoboken, New Jersey: John Wiley & Sons, Inc.

Jastrzębska, M. (2021). Modern displacement measuring systems used in geotechnical laboratories: Advantages and disadvantages. *Sensors*, 21(12): 4139.

Leong, E., Agus, S., & Rahardjo, H. (2004). Volume change measurement of soil specimen in triaxial test. *Geotechnical Testing Journal*, 27(1): 47–56.

Macari-Pasqualino, E. J., Costes, N. C., & Parker, J. K. (1993). Digital image techniques for volume change measurement in triaxial tests. In *Proceedings, Engineering Foundation – National Science Foundation Conference on Digital Image Processing: Techniques and Applications in Civil Engineering.* ASCE, New York, pp. 211–219.

Menzies, B. K. (1976). Discussion: Design manufacture and performance of a lateral strain device. *Géotechnique*, 26(3): 542–544.

Ng, C. W. W., Zhan, L. T., & Cui, Y. J. (2002). A new simple system for measuring volume changes in unsaturated soils. *Canadian Geotechnical Journal*, 39: 757–764.

Parker, J. K. (1987). Image processing and analysis for the mechanics of granular materials experiment. In *ASME Proceedings of 19th S.E. Symposium on System Theory.*, Nashville, TN, March 2. New York: ASME, pp. 536–541.

Piriyakul, K. & Haegeman, W. (2005). Automated K_0 consolidation in stress path cell. In *Proceedings of 16th International Conference on Soil Mechanics and Geotechnical Engineering*, IOS Press, pp. 575–578.

Presti, D. C. L., Pallara, O. V., Rainò, M., & Maniscalco, R. (1994). A computer controlled triaxial apparatus: Preliminary results. *Rivista Italiana di Geotecnica*, 28(1): 43–60.

Scholey, G. K., Frost, J. D., Lo Presti, D. C. F., & Jamiolkowski, M. (1995). A review of instrumentation for measuring small strains during triaxial testing of soil specimens. *Getechnical Testing Journal*, *18*(2): 137–156.

Sivakumar, V. (1993). *A critical state framework for unsaturated soil*. University of Sheffield, Sheffield, UK, vol. Ph.D.

Toyota, H., Sakai, N., & Nishimura, T. (2001). Effects of stress history due to unsaturation and drainage condition on shear properties of unsaturated cohesive soil. *Soils and Foundations*, *41*(1): 13–24.

Wheeler, S. J. (1986). *The stress-strain behaviour of soils containing gas bubbles*. Oxford University, Oxford, UK, vol. Ph.D.

Wheeler, S. J. (1988). The undrained shear strength of soils containing large gas bubbles. *Géotechnique*, *38*(3): 399–413.

Wheeler, S. J. & Sivakumar, V. (1995). An elasto-plastic critical state framework for unsaturated soil. *Géotechnique*, *45*(1): 35–53.

Wijaya, M. (2017). Compression, shrinkage and wetting-induced volume change of unsaturated soils. *Ph.D.* thesis. Singapore: Nanyang Technological University.

Wijaya, M. & Leong, E. C. (2015). Swelling and collapse of unsaturated soils due to inundation under one-dimensional loading. *Indian Geotechnical Journal*, 46: 1–13.

FURTHER READING

Laloui, L., H. Peron, F. Geiser, A. Rifa, & L. Vulliet. (2006). Advances in volume measurement in unsaturated soil triaxial tests. *Soils and Foundations*, *46*(3): 341–349.

Leong, E., Agus, S., & Rahardjo, H. (2004). Volume change measurement of soil specimen in triaxial test. *Geotechnical Testing Journal*, *27*(1): 47–56.

Menzies, B. K. (1976). Discussion: Design manufacture and performance of a lateral strain device. *Géotechnique*, *26*(3): 542–544.

Chapter 16

Direct shear test

16.1 BACKGROUND

The direct shear test is the oldest and simplest test to determine the shear strength of soils. A historical account of the development of the direct shear test is given in Head and Epps (2014). The commonly used direct shear apparatus nowadays was designed by Casagrande in 1932. Most commercially available direct shear apparatuses are displacement controlled, where displacement rates can vary from ten thousandths of a millimetre to a few millimetres per minute.

The direct shear test has been standardised in BS1377: Part 7:1990:4 and 5, BS EN ISO 17892-10, and ASTM D3080/D3080M-11. The test is usually conducted on dry sand or fully saturated clay. The test is deemed a drained test, as there is no provision for pore-water pressure control. The direct shear test has been adapted for testing unsaturated soils. There are two types of direct shear tests performed for unsaturated soils: constant suction or constant water content (CWC) tests. To date, more constant suction (CS) direct shear tests were conducted than CWC direct shear tests.

16.2 RELATED STANDARDS

1. ASTM D3080 /D3080M-11 Standard Test Method for Direct Shear Test of Soils Under Consolidated Drained Conditions.
2. BS 1377-7 Methods of Test for Soils for Civil Engineering Purposes – Part 7: Shear Strength Tests (Total Stress).
3. BS EN ISO 17892-10 Geotechnical Investigation and Testing – Laboratory testing of soil – Direct Shear Tests.

16.3 THEORY

The theory of interpreting the direct shear tests for unsaturated soils is similar to that for saturated soils. However, the interpretation of the test results in the literature can be separated into two categories: (1) those without and (2) those with consideration of unsaturated soil mechanics. In the first category, the tests are treated as total stress tests, and the peak, or the ultimate, shear stress is plotted against normal stress to obtain the so-called total stress peak or residual shear strength parameters (Figure 16.1). While the interpretation is correct for dry sand and saturated clays, the interpretation may be problematic for partially saturated sands and clays. Examples of interpretation in the first category are Lacerda (2010) and Xu et al. (2018). In the second category, there is a need to consider the type of direct shear test conducted. For the constant matric suction test, the interpretation is simpler. There are two axes to consider: a net normal stress axis and a matric suction axis. Usually, it is more common to plot the peak shear stress on a net normal stress axis. For the constant matric suction test, the peak shear stresses usually fall on a straight line with a slope of ϕ' and a y-axis intercept that is dependent on the matric suction (Figure 16.2). The y-axis intercept is known as the total cohesion. For the CWC test, the matric suction is not maintained constant. Thus, for the low degree of saturation soils, all the peak shear stresses usually fall on a straight line with a slope of ϕ', but for soil with a high degree of saturation, the peak shear stresses fall on parallel lines with a slope of ϕ', but the y-axis intercept value will continue to decrease as the net normal stress increases, resulting in a curve line that deviates from the straight line with a slope of ϕ' (Figure 16.3). This is because the degree of saturation of the specimen increases as it is being compressed, (mechanical wetting) resulting in a decrease in matric suction.

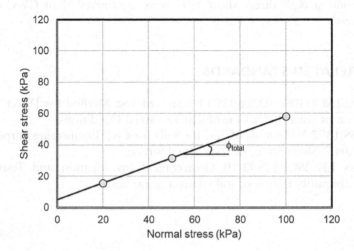

Figure 16.1 Total stress test.

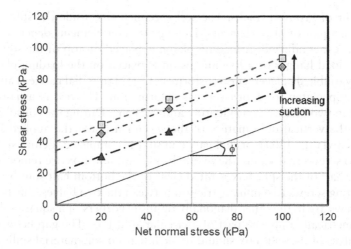

Figure 16.2 CS test.

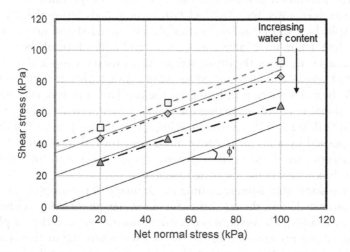

Figure 16.3 CWC test.

16.4 TEST METHODS

The apparatus for conducting a direct shear test on unsaturated soil is similar to the conventional direct shear apparatus. The modifications needed depend on the type of test to be conducted, i.e., CS or CWC test. It consists of a split box with top and bottom halves. The split box can be either square or circular in cross-section. The two halves of the split box are secured together with two screws at diagonally opposite corners for a square split box or diametrically opposite for a circular split box for specimen preparation. The split box sits on either rollers or ball bearings that are positioned in a pair of grooved tracks. The soil specimen sits inside the split box in between a pair of filter elements. The type of filter elements depends on the type of test conducted. A loading platen is placed on top of the soil specimen. A load hanger or a loading piston is placed on the loading platen. A linear variable displacement transformer (LVDT) or a dial gauge is attached to the top platen to record vertical deformation. The test is usually conducted in two phases: consolidation and shearing. During the consolidation phase, the vertical deformation of the specimen under the applied load is observed via the LVDT or dial gauge until it is completed. Before shearing, the two screws that secure the two halves of the split box are removed, and the top half of the split box is lifted clear of the bottom half split box with two lifting screws to eliminate friction between the two halves. The two lifting screws are in the other diagonally opposite corners of a square split box or diametrically opposite part of a circular split box. The gap between the two halves of the split box should be such that no soil material spills out of the split box. The gap is recommended in ASTM D3080/D3080M (2011) to be the maximum grain-size diameter of the soil specimen or 0.64 mm for fine-grained soils. BS 1377-7 (1990) recommends lifting the upper half of the split box so that no soil materials are extruded and give guidelines of less than 1 mm for coarse-grained soils and half a turn of the screw for fine-grained soils. Once the lifting screws have raised the top half of the box, the lifting screws can be retracted to avoid contact with the bottom half of the split box. This action of lifting up the top half split box may have the unintentional effect of subjecting the soil specimen to tensile stress (ASTM D3080/D3080M 2011). For the shearing phase, the bottom half of the split box is pushed by a motor to move along the grooved tracks at a constant displacement rate. A load cell or proving ring connected to the top half of the split box registers the shearing resistance.

There is some confusion regarding the drainage of the air phase. For testing of saturated soils, the condition for the water phase can be either undrained or drained. However, when testing unsaturated soil, the air phase is drained, i.e., air pressure is constant. In a CS test where the suction is applied using the axis-translation method, the pore-air pressure and the pore-water pressure are maintained constant, which means that both the air and water phases are drained. Air or water can leave or enter the soil specimen to maintain CS. This condition can be easily extrapolated to the case where the pore-air pressure is maintained at atmospheric pressure, and hence in a CS test, this would mean maintaining the pore-water pressure at some negative value. Thus, in a CWC test, the condition of the pore water is undrained while the condition of the pore air is drained.

The soil specimen can be intact/undisturbed or reconstituted. For intact/undisturbed, the soil specimen can be extruded directly or trimmed into the split box. If the soil specimen is trimmed, care needs to be excised, especially when the specimen is stiff to minimise sample disturbance. Depending on the intent of the test, a reconstituted soil specimen may be used sometimes.

Table 16.1 Specimen size requirement for direct shear test

Standard	Maximum particle size	Specimen height	Specimen diameter/ width
BS1377:Part 7:1990:4 and 5	≤1/10 specimen height or 20 mm for large shear box	≤ 20 mm, 25 mm, 150 mm	60 mm, 100 mm, 305 mm
BS EN ISO 17892-10	1/6 specimen height	≥ 20 mm or 6 × largest particle size	≥ 50 mm Width/height ≥ 2.5
ASTM D3080/ D3080M-11	1/10 specimen height	≥ 13 mm	≥ Max (10 times largest particle size, 50 mm) Width/height ≥ 2.1

The soil specimen can be compacted directly in the split box using static/ dynamic compaction or pluviation. The particle size, height and diameter/ width requirements of the soil specimen for the standards are summarised in Table 16.1.

16.4.1 CWC test

The direct shear apparatus can be used for the CWC test without major modification. In the CWC test, the pore air is drained, while the pore water is undrained. To interpret the CWC test, the matric suction of the specimen is needed. The matric suction can be measured before and after the direct shear test, or during the test. Measurement of the matric suction after the test will reflect the change in matric suction of the soil specimen after appli- cation of the normal load. Any matric suction measurement method can be adopted if the matric suction of the specimen is to be measured before and after the test. If the matric suction is measured during the test, a miniature tensiometer such as a small-tip tensiometer (limited to less than 100 kPa) or a high-capacity tensiometer (limited to less than 1,500 kPa) can be added to the direct shear apparatus. Set-ups using one or two tensiometers have been done, as shown in Figure 16.4. Regardless of when matric suction is being measured, the matric suction is representative only of the soil specimen and is not the matric suction at the shearing plane. Thus, it is not surprising that experiences in measuring matric suction in a direct shear test show that the matric suction remains relatively unchanged during shearing (Tarantino and Tombolato, 2005; Jotisankasa and Mairaing 2010; Tang et al. 2018).

In the CWC test, the rate of shearing is only important when the degree of saturation of the soil specimen is high after the application of the normal load. At a high degree of saturation (>90%), the soil specimen may experi- ence positive excess pore-water pressure similar to testing saturated soil. Hence, the peak shear stress in such a test may fall below the failure line for the saturated soil at the drained condition. This situation can be confirmed

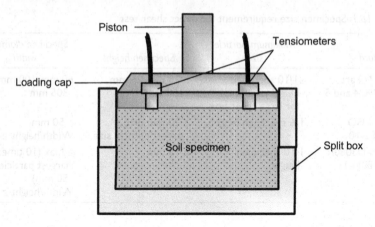

Figure 16.4 Schematic drawing of a direct shear box with tensiometers.

by applying the shearing rate determined for an equivalent saturated soil specimen to ensure that the shearing takes place under the fully drained condition to verify that the peak shear stress falls on the failure line for the saturated soil under the drained condition.

There are two common problems in the CWC direct shear test: (i) loss of water due to evaporation and (ii) loss of soil during shearing. The loss of water through evaporation can be minimised by selecting a high shearing rate if the degree of saturation of the soil specimen remains below 90% throughout the test or minimising the air space of the top half of the box using a latex membrane as done by Caruso and Tarantino (2004).

16.4.2 CS test

The direct shear apparatus has to be modified for the CS test. In the CS test, it is more common to control suction via the axis-translation technique. To apply the axis-translation technique, a high-air entry (HAE) ceramic disk is sealed at the base of the direct shear box with a water inlet line to apply the water pressure. The whole direct shear box is enclosed in a sealed chamber for application of the air pressure. Hence, the top loading element, motor and horizontal load measurement need to pass through airtight sleeves as they enter the sealed chamber. Hence the friction on the vertical and horizontal (shear) loads need to be accounted for if the load-measuring devices are outside the chamber, as shown in Figure 16.5a. Calibration for the frictional stresses can be done using load cells placed inside the chamber below the vertical loading element and between the horizontal piston and the shear box. This should be done for the range of air pressure and shearing rates that are applied to the soil specimen. If the load-measuring devices are all within the chamber, as shown in Figure 16.5b, then the frictional stresses do not affect the vertical and horizontal load readings. Melinda et al. (2004) used a similar set-up that is shown in Figure 16.5b.

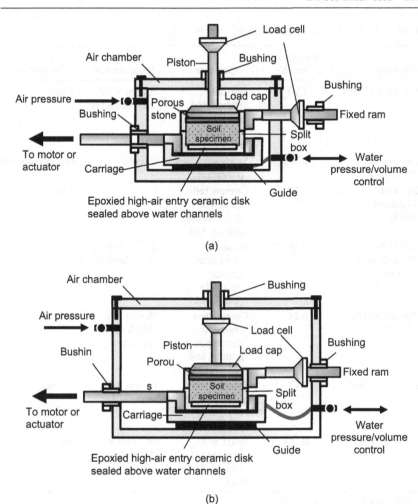

Figure 16.5 Modified direct shear apparatus for testing unsaturated soil (a) external load cells. (b) internal load cells.

For the CS test, the rate of shearing is important, as the soil specimen needs to maintain a constant suction condition during shearing. The shearing rate used should be the rate below which the shear strength remains constant (Gan 1986). This can only be ascertained by conducting several CS direct shear tests with different shear rates, which is not practical. Escario and Sáez (1986) and Gan (1986) conducted CS direct shear tests on statically and dynamically compacted clays to sands using shear rates of 0.075–0.612 mm/h. Gan et al. (1988) found that a shear displacement rate of below 0.792 mm/h is satisfactory for a compacted glacial till. Since then, shearing rates based loosely on these rates were used by others, as summarised in Table 16.2.

Table 16.2 Shearing rates used in CS direct shear tests

References	Shearing rate (mm/h)	Soil tested	Specimen size	Ceramic disk
Kim et al. (2010)	1.20	Statically compacted granitic residual soil (USCS: SP-SM)	20 mm × 60 mm diameter	5 bar
Sun and Xu (2007)	0.576	Sandy silt – wet pluviation	Circular	Not mentioned
Gan and Fredlund (1994)	0.3	Compacted completely decomposed granitic soil		5 bar
Han (1997), Melinda et al. (2004)	0.24	Statically compacted granitic residual soil (USCS: CL)	20 mm × 63.5 mm diameter	5 bar
Hossain and Yin (2010)	0.24	Compacted completely decomposed granitic soil (USCS: SM)	40 mm × 100.5 mm square	5 bar
Shimada (1998)	0.12	Fraser River sand (USCS: SP) – wet pluviation	16 mm × 70 mm diameter	> 1 bar
Ng et al. (2020)	0.114	Intact and compacted loess (USCS: CL)	24.4 mm × 50.8 mm diameter	NA

USCS: Unified soil classification system.

16.5 TEST PROCEDURES

Similar to direct shear tests on saturated soils, direct shear tests on unsaturated soils can be divided into two stages:

1. consolidation, and
2. shearing.

The general principles of the test procedures for CWC and CS direct shear tests are given in the following sections. For detailed test procedures, the reader is requested to check BS1377: Part 7:1990:4 and 5, BS EN ISO 17892-10 and ASTM D3080/D3080M-11.

16.5.1 CWC test

1. Preparation of apparatus

   The shear box, upper and lower perforated plates and base plate should be clean and free of moisture. Apply a thin layer of silicone grease to the mating surfaces of the split box. Assemble the two halves of the box with the clamping screws and check the internal dimensions of the box using Vernier calipers to an accuracy of 0.1 mm. Measure the mean depth of the split box from the top surface of the upper split box to the baseplate to 0.1 mm accuracy. Measure the thickness of the upper and lower porous grid plates to 0.1 mm accuracy. Place the lower porous plate into the bottom of the shear box, taking care that the grids are perpendicular to the direction of the shear. Place the assembled shear box on the carriageway and check that the shear box can run freely on the bearings. Check that the horizontal piston can move freely in its sleeve. Check that the lower half of the shear box rests firmly against the spacer block at the driving end of the carriage. Check that the horizontal piston can engage the lower half of the shear box and is able to drive the lower half of the shear box smoothly. The shear box is now ready to receive the soil specimen. If it is more convenient, the shear box can be taken out of the carriage for placement of the soil specimen.

2. Preparation of test specimen

   See Chapter 3 for preparation of the soil specimen. For the CWC test, the initial matric suction of the soil specimen can be imposed by equilibrating it in a pressure plate apparatus until its weight change is negligible or the matric suction of the specimen can be measured using a tensiometer (see Chapter 8), null-type axis-translation apparatus (see Section 8.4, Chapter 8), hygrometer (see Section 8.5, Chapter 8) or filter paper (see Chapter 9). For hygrometer, the measurement is performed on another small specimen, while for the filter paper method, the matric suction will be only known after the test (about 14 days later). Measure the weight of the soil specimen.

3. Assembly of apparatus

   Place the soil specimen into the shear box and place the upper porous plate on the specimen and determine the average specimen height. Place the top loading platen and place the shear box into the carriage. Adjust the top dial gauge/LVDT and move the horizontal piston so that they are all in contact.

4. Applying normal stress

   Place the normal load on the soil specimen and note the deformation using the vertical dial gauge reading (to 0.002 mm accuracy). The deformation is due to the compression of the air voids. As the water content is constant, the new degree of saturation is computed. If the degree of saturation exceeds 90%, then care must be taken to use a shearing rate based on consolidated drained (CD) direct shear test for saturated soils.

5. Lifting the top half of the box

   Once the soil specimen does not deform anymore under the vertical load, the lifting screws are turned into the top half of the split box until contact with the lower half split box can be felt and the clamping screws are removed. Rotate the lifting screws such that the upper half of the shear box is lifted by about 0.5 mm (BS1377: 1990) or the maximum grain-size diameter (ASTM D3080/3080M 2011). The lifting screws can then be removed from the shear box.

6. Final checks

   Check that contact is made at all contact points, the vertical dial gauge, the horizontal gauge (accuracy 0.01 mm) and the horizontal load device are all initialised to zero. Enclose the top half of the box using a latex membrane to minimise evaporation.

7. Shearing

   Turn on the motor and timer to start the test. If the readings are taken manually, the readings can be taken at intervals of 0.1 mm of the horizontal dial gauge reading. This interval can be adjusted based on the rate of change of the shear load readings. The recommendation in BS1377: 1990 Part 7 is that at least 20 readings be taken up to the maximum load. If a peak load is approaching, more readings can be taken to capture the peak load more accurately. Continue shearing beyond the peak or until the full travel of the shear box has been reached if there is no peak load.

8. Removal of load

   At the end of the test, turn off the motor. When the motor has completely stopped, reverse the motor until the drive unit returns to its initial position.

9. Removal of shear box

   Remove the vertical load, vertical dial gauge and the top loading cap and upper porous plate. Lift out the shear box from the carriage using the lifting lugs and place it on a bench.

10. Removal of specimen

    Remove the top grid plate and remove the soil specimen for final water content determination.

11. Repeat tests

    To obtain the test results for the next normal load, repeat steps 1 to 10. Typically, direct shear tests are conducted at three normal loads.

12. Calculations

    Calculations for the test results are similar to those found for saturated soil tests. Readers are referred to ASTM D3080/3080M (2011) and BS1377: 1990 Part 7.

16.5.2 CS test

1. Preparation of apparatus

   The preparation of the apparatus is similar to that of the CWC test with two exceptions. The shear box needs to be fully enclosed in an airtight chamber for application of air pressure in the axis-translation technique, and the bottom half of the shear box needs to be fitted with an HAE ceramic disk epoxied at the base to separate the air phase in the chamber from the water phase at the bottom of the test specimen. For the chamber, checks need to be done to ensure that there is no leakage of air when air pressure is applied. If the load-measuring devices are outside the chamber (Figure 16.5a), correction to the loads due to friction where the loading pistons (vertical and horizontal) enter the chamber must be applied to get the correct normal and shear stresses. In such cases, the effect of both air pressure and the shear rate on friction needs to be determined. For the case where both vertical and horizontal load-measuring devices are inside the chamber (Figure 16.5b), a check is made to ascertain that the readings are not affected by the range of air pressure applied during the text. The ceramic disk sealed into the base of the closed bottom half shear box needs to be flushed to ensure that the ceramic disk is fully saturated, and no air is trapped beneath the ceramic disk. For more efficient flushing, a spiral groove can be etched into the base of the closed bottom half shear box (Figure 16.6) such that water can be pushed in from one end of the groove, travel along the groove and exit at the other end to ensure that there is no dead space where air can be trapped. The saturation of the disk can be checked by applying air pressure in the chamber at the air-entry value of the ceramic disk and monitoring the outlet tube from the base of the shear box submerged in a beaker of water. If no air bubbles can be observed after a couple of hours, it can be assumed that the system is fully saturated. If the volume of water exiting from the test specimen needs to be measured accurately, it is recommended

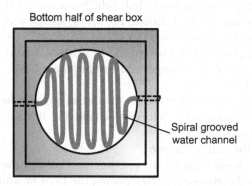

Bottom half of shear box

Spiral grooved water channel

Figure 16.6 Grooved water channels below an HAE ceramic disk.

to use a steel tubing instead of a plastic tubing, which is deformable under high air pressure.

2. Preparation of test specimen

Unlike the preparation of the test specimen for a constant water test where the initial suction of the test specimen is either measured or induced outside the direct shear apparatus, suction is imposed on the test specimen together with the normal stress in the consolidation stage. Hence, the test specimen before placement into the shear box can be in a fully saturated or partially saturated condition at unknown suction.

3. Assembly of apparatus

For the CS test, the shear box is always on the carriageway because of the bottom water pressure line. Lock the bottom water pressure line. The soil specimen is placed into the shear box followed by the upper perforated plate/porous stone on the specimen. Determine the average specimen height and place the top loading platen. The top dial gauge/ LVDT and horizontal dial gauge/LVDT are set up outside the chamber. The axial load and horizontal load can be measured outside or inside the chamber. Usually, the instrumentations for a CS test are electronic devices where the data is recorded automatically using a data acquisition system. The use of load cells makes it possible to place them inside the chamber due to their compact sizes compared to a proving ring.

4. Applying normal stress

Lower the piston to apply the normal load on the soil specimen and note the deformation. If the initial condition of the test specimen is saturated, obtain the consolidation properties of the test specimen as per for a direct shear test on saturated soil. Drainage is at the top surface of the soil specimen as the water pressure line is closed. If the initial condition of the test specimen is unsaturated, note the deformation and calculate the new void ratio.

5. Lifting the top half of the box

Once the soil specimen does not deform anymore under the vertical load, the lifting screws are turned into the top half of the split box until contact with the lower half split box can be felt and the clamping screws are removed. Rotate the lifting screws such that the upper half of the shear box is lifted by about 0.5 mm (BS1377: 1990) or the maximum grain-size diameter (ASTM D3080/3080M 2011). The lifting screws can then be removed from the shear box.

6. Final checks

Check that contact is made at all contact points, the vertical deformation, the horizontal deformation and the horizontal load are all initialised to zero.

7. Applying the matric suction

Close the chamber, apply air pressure and control the water pressure or vent to the atmosphere by opening the bottom water pressure line and setting a datum for the exiting water. The water that flows out is collected to check for the equilibrium condition of the test specimen with the applied suction and to calculate the water content of the test specimen. It is difficult to obtain an accurate estimate of the water content of the specimen, and usually, the water content of the test specimen at the end of the test is measured to verify.

8. Shearing

   Once matric suction has equilibrated, turn on the motor/actuator and timer to start the test. If the readings are automatically recorded, sampling of data can be done at smaller time intervals to obtain a continuous reading with time. The same stopping criterion as the constant water content test can be applied.

9. Removal of load

   At the end of the test, turn off the motor/actuator. When the motor/actuator has completely stopped, reverse the motor (retract the actuator) until the drive unit (horizontal piston) returns to its initial position.

10. Removal of matric suction

    The water pressure line is then locked. The air pressure and water pressure are reduced to atmospheric.

11. Removal of vertical load and test specimen

    The vertical load is reduced to zero and the chamber is opened quickly to remove the test specimen for water content determination. This step should be performed in as short a time as possible to reduce the possibility of the specimen drawing water from the saturated ceramic disk.

12. Repeat tests

    To obtain the test results for the next normal load, repeat steps 1 to 10. Typically, direct shear tests are conducted for three normal loads.

13. Calculations

    Calculations for the test results are similar to those found for saturated soil tests. Readers are referred to ASTM D3080/3080M (2011) and BS1377: -7 (1990).

16.6 INTERPRETATION OF TEST RESULTS

The interpretation of the direct shear tests is the same irrespective of if the test is conducted on saturated or unsaturated soil. Plots of shear stress versus horizontal displacement and vertical displacement versus horizontal displacement are required. For the direct shear test on unsaturated soil at the CWC condition, matric suction versus horizontal displacement is also plotted if matric suction is measured during the test. For the direct shear test on unsaturated soil at the CS condition, air and water pressures versus horizontal displacement and water volume versus horizontal displacement are also plotted.

For computation of shear and normal stresses in the direct shear test, there are several ways to report: (i) no area correction (ASTM D3080/

D3080M-11, BS EN ISO 17892-10), (ii) with area correction applied to both the shear and the normal stresses (Shwan and Smith 2015; Gallage and Uchimura 2016) and (iii) area correction applied to only the normal stress (Murthy 2002). The effect of area correction on shear and normal stresses is negligible for horizontal displacement less than 2 mm. However, there can be a difference of 10% in the shear and normal stresses at large horizontal displacement (Skuodis and Tamosiūnas 2014). In some direct shear tests, the shear stress does not peak or plateau but shows a slow increase with horizontal displacement. In such tests, the recommendation is to take the failure shear stress at a particular horizontal displacement, usually 15 mm (15% of the shear box length), or displacement limit of the shear box (ASTM D3080/D3080M-11, BS EN ISO 17892-10). The arbitrary use of a limit horizontal displacement to interpret the test results is not needed if the decreasing contact area is considered for both the normal stress and shear stress calculations and the "stress path" is plotted. The calculation of the corrected contact area for the square shear box is trivial. The corrected contact area, A_c, for the circular shear box, is given by Equation 16.1.

$$A_c = F\left(\frac{\pi D^2}{4}\right) \tag{16.1}$$

where
D = initial diameter of the soil specimen
F = correction factor F, given by Equation 16.2.

$$F = \frac{2}{\pi}\left\{\cos^{-1}\left(\frac{\delta}{D}\right) - \left(\frac{\delta}{D}\right)\sqrt{1-\left(\frac{\delta}{D}\right)^2}\right\} \tag{16.2}$$

where
δ = horizontal displacement and the arccosine in radians.

Using the corrected area, the stress path in the shear stress and normal stress can be plotted. Using data from a CS test reported by Gallage and Uchimura (2016), the shear stress versus horizontal displacement and stress path in the shear stress-net normal stress space without and with area correction is illustrated in Figure 16.7 together with the effective failure line. As can be seen in Figure 16.7, the stress path lies on the failure line with decreasing contact area. Thus, taking any point of the test on this failure line for the failure stress does not lead to any error but may be problematic if no area correction is made.

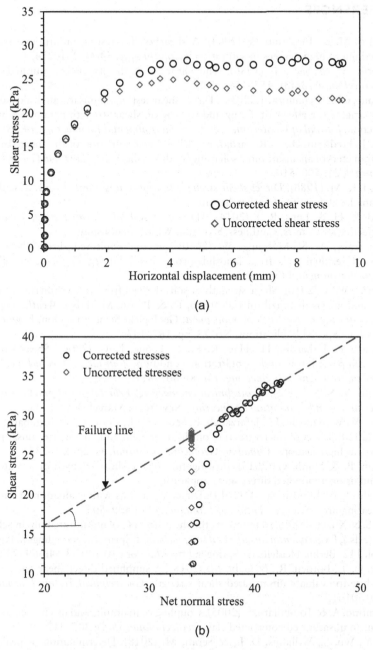

Figure 16.7 CS direct shear test results from Gallage and Uchimura (2016) without and with area correction. (a) Shear stress versus horziontal distance. (b) Shear stress versus net normal stress.

REFERENCES

Caruso, M. & Tarantino, A. (2004). A shearbox for testing unsaturated soils at medium to high degrees of saturation. *Géotechnique, 54*(4): 281–284.

Escario, V. & Sáez, J. (1986). The shear strength of partly saturated soils. *Géotechnique, 36*(3): 453–456.

Gallage, C. & Uchimura, T. (2016). Direct shear testing on unsaturated silty soils to investigate the effects of drying and wetting on shear strength parameters at low suction. *Journal of Geotechnical and Geoenvironmental Engineering, 142*(3): 1–9.

Gan, J., Fredlund, D., & Rahardjo, H. (1988). Determination of the shear strength parameters of an unsaturated soil using the direct shear test. *Canadian Geotechnical Journal, 25*: 500–510.

Gan, J. K. M. (1986). *Direct shear strength testing of unsaturated soils*. MSc thesis. Canada: University of Saskatchewan.

Head, K. H. & Epps, R. J. (2014). *Manual of soil laboratory testing. Volume 3, Effective stress tests*. Caithness, Scotland: Whittles Publishing.

Jotisankasa, A. & Mairaing, W. (2010). Suction-monitored direct shear testing of residual soils from landslide-prone areas. *Journal of Geotechnical and Geoenvironmental Engineering, 136*(3): 533–537.

Lacerda, W. A. (2010). Shear strength of soils derived from the weathering of granite and gneiss in Brazil. In Calcaterra, D. & Parise, M. (Eds.), *Weathering as a predisposing factor to slope movement*. Geological Society London, Engineering Geology Special Publications, No. 23, pp. 167–182.

Melinda, F., Rahardjo, H., Han, K. K., & Leong, E. C. (2004). Shear strength of compacted soil under infiltration condition. *Journal of Geotechnical and Geoenvironmental Engineering, 130*(8): 807–817.

Murthy, V. N. S. (2002). *Geotechnical engineering: Principles and practices of soil mechanics and foundation engineering*. New York: Marcel Dekker AG.

Ng, C. W. W., Sadeghi, H., Jafarzadeh, F., Sadeghi, M., Zhou, C., & Baghbanrezvan, S. (2020). Effect of microstructure on shear strength and dilatancy of unsaturated loess at high suctions. *Canadian Geotechnical Journal, 57*(2): 221–235.

Shwan, B. & Smith, C. (2015). Investigation of the shear strength of unsaturated sand using a modified direct shear apparatus.

Skuodis, Š. & Tamošiūnas, T. (2014). Direct shear tests with evaluation of variable shearing area. *Science - Future of Lithuania, 6*(5): 499–503.

Sun, S. & Xu, H. (2007). Determining the shear strength of unsaturated silt. In Schanz, T. (Eds.), *Experimental unsaturated soil mechanics*. Springer Proceedings in Physics, vol. 112. Berlin, Heidelberg: Springer. https://doi.org/10.1007/3-540-69873-6_19

Tang, C.-T., Borden, R., & Gabr, M. (2018). A simplified direct shear testing procedure to evaluate unsaturated shear strength. *Geotechnical Testing Journal, 41*: 20150161.

Tarantino, A. & Tombolato, S. (2005). Coupling of hydraulic and mechanical behaviour in unsaturated compacted clay. *Géotechnique, 55*(4): 307–317.

Xu, Y., Wu, S., Williams, D. J., & Serati, M. (2018). Determination of peak and ultimate shear strength parameters of compacted clay. *Engineering Geology, 243*: 160–167.

Chapter 17

Triaxial test (shear strength)

17.1 BACKGROUND

Soil shear strength plays a very important role in ensuring the safety of civil engineering structure. However, interpretation of soil shear strength remains challenging for geotechnical engineers, especially when the soil becomes unsaturated. In saturated soil mechanics, it is well-known that the soil shear strength is dependent on whether the soil is under a drained or undrained condition. However, when the soil becomes unsaturated, it is necessary to incorporate the effect of soil suction into the shear strength of the unsaturated soil. In this chapter, the shear strength of both saturated and unsaturated soil will be discussed. Through this chapter, it is expected that the reader will be able to:

1. Differentiate between saturated and unsaturated shear strength
2. Determine the differences between saturated and unsaturated triaxial tests
3. Determine the suitable test for unsaturated soils
4. Interpret the shear strength parameters for unsaturated soils

17.2 RELATED STANDARDS

1. BS 1377-6 (1990) Methods of Test for Soil for Civil Engineering Purposes – Part 6: Consolidation and Permeability Tests in Hydraulic Cells and with Pore Pressure Measurement, BSI.
2. BS 1377-7 (1990) Methods of Test for Soils for Civil Engineering Purposes – Part 7: Shear Strength Tests (Total Stress), BSI.
3. BS 1377-8 (1990) Methods of Test for Soils for Civil Engineering Purposes – Part 8: Shear Strength Tests (Effective Stress), BSI.
4. BS EN ISO 17892-7 (2018) Geotechnical Investigation and Testing – Laboratory Testing of Soil Part 7: Unconfined Compression Test (ISO 17892-7: 2017), BSI.

5. BS EN ISO 17892-8 (2018) Geotechnical Investigation and Testing – Laboratory Testing of Soil Part 8: Unconsolidated-Undrained Triaxial Test, BSI.
6. BS EN ISO 17892-9 (2018) Geotechnical Investigation and Testing – Laboratory Testing of Soil Part 9: Consolidated Triaxial Compression Tests on Water-Saturated Soils, BSI.
7. ASTM D2166/2166m-16 (2016) Standard Test Method for Unconfined Compressive Strength of Cohesive Soil.
8. ASTM D2850-15 (2015) Standard Test Method for Unconsolidated-Undrained Triaxial Compression Test on Cohesive Soils.
9. ASTM D4767-11 (2011) Standard Test Method for Consolidated Undrained Triaxial Compression Test for Cohesive Soils.
10. ASTM D7181-11 (2011) Standard Test Method for Consolidated Drained Triaxial Compression Test for Cohesive Soils.

17.3 THEORY

It is quite common to check the safety of geotechnical structures in short-term and long-term conditions. The definition of short-term and long-term is related to the condition of the excess pore-water pressure. When a saturated soil is loaded, excess pore-water pressure equal to the applied load is generated immediately. As a result, there is no change in the mean effective stress (p′). This condition is referred to as an undrained condition. An undrained condition is only applicable in the short term as the excess pore-water pressure takes time to dissipate. An undrained condition is also referred to as a short-term condition. Shear strength related to an undrained condition is referred to as undrained shear strength.

As the excess pore-water pressure dissipates, the mean effective stress of the soil will increase. The process of the excess pore-water pressure dissipating with time is referred to as consolidation. Once the excess pore-water pressure has fully dissipated (i.e., reaching equilibrium condition), the condition is the drained condition. As the drained condition requires a long time to be achieved, the condition after the full dissipation of excess pore-water pressure is also referred to as the long-term condition. Shear strength related to a drained condition is referred to as drained or effective shear strength.

For soil which has high permeability (i.e., coarse-grained soil), it takes only a few seconds for the excess pore-water pressure to build up and then dissipate. Thus, the undrained condition is only applicable for a very short duration and might be relevant only for special conditions (e.g., earthquake and liquefaction cases). For normal design purposes, only the drained condition is relevant for coarse-grained soil. Thus, conducting an undrained test for coarse-grained soil is invalid.

For soil which has low permeability (i.e., fine-grained soils), it might take a few days to years for the excess pore-water pressure to be fully dissipated. Thus, both undrained and drained conditions are relevant for fine-grained soils.

However, for the soil which is unsaturated, water does not fill up the void, and the pore-water pressure is negative. When the load is applied, the soil structure directly withstands the load under the constant water content condition and the settlement will be instantaneous until the soil is mechanically saturated (Wijaya 2017). The change in pore-water pressure is due to the change in the void volume which affects the size of the water meniscus. The shear strength of the unsaturated soil will be the drained shear strength and the additional shear strength due to the contribution of soil suction.

17.3.1 Mohr-Coulomb (MC) model for saturated and unsaturated soil

The MC model has been commonly used to describe the shear strength of saturated soils. In the MC model, the shear strength of saturated soil can be described using Equation 17.1.

$$\tau = c' + \sigma' \tan \phi' \tag{17.1}$$

where τ is the shear strength, c' is the effective cohesion, σ' is the effective stress and ϕ' is the effective friction angle. The effective cohesion could be due to (Budhu 2010)

1. intermolecular forces, or
2. cementation.

However, when the soil is unsaturated, the extended MC model is commonly expressed using Equation 17.2.

$$\tau = c' + (\sigma - u_a) \tan \phi' + c_t \tag{17.2}$$

where c_t is the cohesion due to the contribution of matric suction $(u_a - u_w)$, σ is the normal stress, u_a is the air pressure and $(\sigma - u_a)$ is the net normal stress. The term soil suction (s) can be used to replace matric suction, as the effect of soil suction on the soil shear strength is applicable to the entire range of suction (Fredlund et al. 2012).

The simplest form of c_t is when c_t is assumed to be a linear function of soil suction such as given by Fredlund et al.'s (1978) equation shown as Equation 17.3.

$$\tau = c' + (\sigma - u_a) \tan \phi' + s \cdot \tan \phi^b \tag{17.3}$$

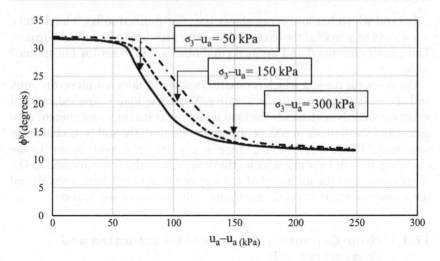

Figure 17.1 Change in ϕ^b as a function of soil suction under different net normal stress (Trinh et al. 2006).

where ϕ^b is the friction angle with respect to the change in soil suction. However, a lot of researchers have found that the contribution of soil suction is non-linear (Escario and Sáez 1986; Gan et al. 1988; Vanapalli et al. 1996; Rassam and Williams 1999) and the value of ϕ^b can be taken as ϕ' when the soil suction is below air-entry value (AEV) and then slowly diminishes as the soil suction increases (Goh et al. 2010; Tekinsoy et al. 2004). Figure 17.1 shows the change in the ϕ^b as the soil suction changes under different net normal stress. To accommodate such non-linear behaviour, many forms of equations were proposed to represent the contribution of soil suction to the shear strength of unsaturated soils. Some of the proposed equations are shown in Table 17.1.

17.3.2 Failure envelope of unsaturated soil based on the MC model

The definition of failure is important to define the failure envelope, which is then used to determine soil shear strength. Several definitions of failure are as follows:

1. Peak failure condition: Failure is taken as the highest shear stress that can be withstood by the soil.
2. Critical state condition: Failure is taken when continuous shearing occurred at constant shear stress to normal effective stress ratio and constant volume (Budhu 2010).
3. Residual condition: Failure is taken when the soil is sheared under large strain.

Table 17.1 Cohesion due to soil suction

Authors	Equation	Parameters
Fredlund et al. (1978)	$c_t = s \cdot \tan\phi^b$	c_t = Cohesion due to soil suction
Shen and Yu (1996)	$c_t = \left(\dfrac{s}{1 + s \cdot a}\right)\tan\phi'$ Or $c_t = \left(\dfrac{s}{\cot\alpha + \dfrac{s}{b}}\right)\tan\phi'$	s = Soil suction AEV = Air-entry value s_r = Residual soil suction Θ = Normalised volumetric water content θ_w = Volumetric water content θ_r = Residual volumetric water content θ_s = Saturated volumetric water content
Vanapalli et al. (1996)	$c_t = s \cdot \Theta^\kappa \tan\phi'$ $\Theta = \dfrac{\theta_w - \theta_r}{\theta_s - \theta_r}$ κ is taken as 1 when data is not available Garven and Vanapalli (2006): $\kappa = -0.0016.PI^2 + 0.0975.PI + 1$	PI = Plasticity index S_r = Degree of saturation ϕ' = Friction angle ϕ^b = Friction angle with respect to the change in soil suction a, b, α, β, κ, λ = Empirical parameters
Oberg and Sallfors (1997)	$c_t = s \cdot S_r \tan\phi'$	τ_{sr} = Residual shear strength due to soil suction
Bao et al. (1998)	$c_t = s \cdot \zeta \tan\phi'$ $\zeta = \dfrac{\log(s_r / s)}{\log(s_r / AEV)}$	P_{atm} = Atmospheric pressure (101 kPa)
Khalili and Khabbaz (1998)	$c_t = s\left(\dfrac{s}{AEV}\right)^{-0.55}\tan\phi'$	
Rassam and Cook (2002)	$c_t = s \cdot \tan\phi' - \varphi(s - AEV)^\beta$ $\varphi = \dfrac{s_r \tan\phi' - \tau_{sr}}{(s_r - ARV)^\beta}$ $\beta = \dfrac{\tan\phi'(s_r - ARV)}{s_r \tan\phi' - \tau_{sr}}$	
Tekinsoy et al. (2004)	$c_t = (AEV + P_{atm})\ln\left(\dfrac{s + P_{atm}}{P_{atm}}\right)\tan\phi'$	
Lee et al. (2005)	If $s \leq AEV$ $c_t = s\tan\phi'$ Else $c_t = \left[AEV + (s - AEV)\Theta^\kappa(1 + \lambda \cdot s)\right]\tan\phi$	

(Continued)

Table 17.1 (Continued)

Authors	Equation	Parameters
Vilar (2006)	$c_t = \left(\dfrac{s}{\cot\phi + \dfrac{s}{c_{ult} - c'}} \right) \tan\phi'$	
Goh et al. (2010)	If $s \le AEV$ $c_t = s \cdot \tan\phi'$ Else $c_t = \left[AEV + (s - AEV) b \cdot \Theta^\kappa \right] \tan\phi'$ $\kappa = \left[\log(s / AEV) \right]^a$	

The easiest way to determine the MC failure envelope is by using s-t MIT stress path where s' and t' are defined as shown in Equations 17.4 and 17.5, respectively.

$$s' = \frac{\sigma'_1 + \sigma'_3}{2} \tag{17.4}$$

$$t = \frac{\sigma'_1 - \sigma'_3}{2} = \frac{\sigma_1 - \sigma_3}{2} \tag{17.5}$$

By determining the value of s and t at failure under several different net normal stresses, and using a regression method, the correlation between s' and t can be made, as shown in Equation 17.6.

$$t = a + s' \tan\alpha \tag{17.6}$$

The effective stress parameter (c' and ϕ') can then be determined using Equations 17.7 and 17.8, respectively (BS 1377-8 1990).

$$\phi' = \arcsin(\tan\alpha) \tag{17.7}$$

$$c' = \frac{a}{\cos\phi'} \tag{17.8}$$

For unsaturated soil, the failure envelope is also a function of soil suction, as shown in Figure 17.2. To obtain the failure envelope under different matric suctions, either constant water content (CWC) or constant suction (CS)

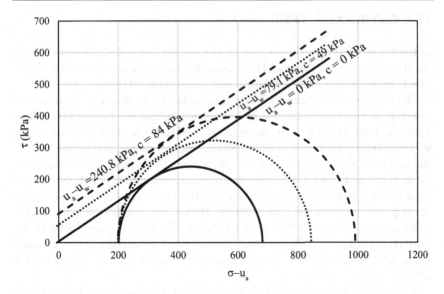

Figure 17.2 Failure envelope under different matric suction. (Trinh et al., 2006).

test can be performed. However, matric suction at failure condition (s_f) must be known. By assuming a constant ϕ' or α, the value of c_t can be easily determined by as using the s-t MIT stress path as well as using Equations 17.9 to 17.11.

$$s_n = \frac{\sigma_1 + \sigma_3}{2} - u_a \tag{17.9}$$

$$a_a(s_f) = t - s_n \tan \alpha \tag{17.10}$$

$$c_t(s_f) = \frac{a_a(s_f)}{\cos \phi'} - c' \tag{17.11}$$

Any equation in Table 17.1 can then be used to determine the relationship between s_f and c_t.

Table 17.2 Strain rate used in the CS triaxial test

Soil type	Strain rate	Reference
Siliceous sand	0.03 mm/min (0.034%/min)	Farouk et al. (2004)
Silt (ML)	0.167 mm/min (0.11%/min)	Burrage et al. (2012)
Residual soil	0.0009mm/min	Rahardjo et al. (1995)
Weathered granite (silty sand)	0.006 mm/min (0.006%/min)	Lee et al. (2005)

17.4 CS TRIAXIAL TEST

In the CS triaxial test, the suction of soil is controlled such that it remains constant throughout the test. The most common method to control soil suction is by using the axis-translation technique (Hilf 1956).

The CS triaxial test can be easily interpreted to establish c_t versus s relationship, as there is no change in soil suction before the test, at failure and after the test. However, to maintain the constant suction condition, the shearing speed must be extremely low. During shearing, the volume of the soil specimen decreases causing the soil suction to decrease due to mechanical wetting. As the air and water pressures are maintained constant during a CS test, the water in the soil specimen will flow out to re-establish its suction at the preset soil suction (Zhang 2016). Hence, the soil suction is in a quasi-constant state. To ensure that fluctuations in soil suction caused by mechanical wetting is kept small, the shearing rate must be carried at a very small strain rate. Table 17.2 shows the typical strain rate that has been used in the CS triaxial test.

While the contribution of the CS triaxial test is significant in helping us to understand the shear strength of unsaturated soil, such a small strain rate causes the test to require months to complete, and hence, it is difficult to be implemented in practice and more suitable for research work.

17.5 CWC TRIAXIAL TEST

The CWC triaxial test can be done easily by using the existing triaxial apparatus or with minor modification if suction measurement is required. Suction measurement can be carried out using a tensiometer (Marinho et al. 2016; Trinh et al. 2006).

When the soil is sheared under the CWC condition, the shearing can be done in less than 2 hours (Marinho et al. 2016) and is thus more practical for engineering application. However, as compression increases the degree of saturation of the specimen, the specimen suction decreases during the consolidation and shearing stages (Trinh et al. 2006; Marinho et al. 2016), as shown in Figure 17.3. It is recommended to use the shearing rate for saturated soil if the soil is expected to be saturated due to the compression during the test to avoid the build-up of excess pore-water pressure.

By knowing the suction and deviatoric stress at failure and combining both with effective shear strength parameters from a saturated triaxial test, it is possible to work out the relationship between c_t versus s.

The CWC triaxial test can be done either with pore-water pressure measurement or without pore-water pressure measurement. When pore-water pressure measurement is available, it is easier to determine soil suction at

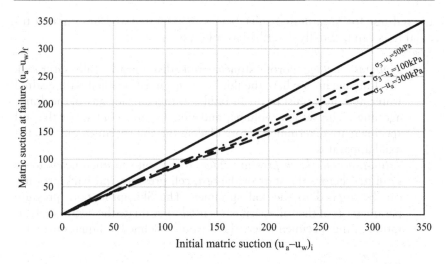

Figure 17.3 Change in suction during CWC triaxial test. (Trinh et al., 2006).

failure condition. However, when pore-water pressure measurement is not available, soil suction at failure can be determined as follows:

1. Before test (Farouk et al. 2004): Soil suction before the test is used as soil suction at failure. It will overestimate soil suction at failure, as the void ratio at failure is less than the void ratio before the test. It is recommended only when the degree of saturation is very low and, hence, compression will not cause a significant change in soil suction (Wijaya 2017).

2. After test: Soil suction after the test will underestimate soil suction at failure, as the void ratio at failure is higher than the void ratio after the test. If this procedure is to be carried out, it is recommended to stop the test when the failure condition has been achieved.

It is recommended to always measure the soil suction before and after the test in order to obtain the range of change in soil suction pre and post-triaxial test. However, if the range of soil suction is very large, it might be necessary to estimate the soil suction at failure.

17.6 TEST METHODS

The common triaxial tests for saturated soils are, namely, unconsolidated-undrained (UU), consolidated undrained (CU) and consolidated drained (CD) triaxial tests. Additional procedures are needed when conducting triaxial test for unsaturated soils. Stages in saturated and unsaturated triaxial tests can be briefly described as follows:

1. *Preparation stage.* Porous stone/ceramic disk must be fully saturated. All drainage lines where the fluid is water must be filled with deaired water, and there must be no air bubble trapped inside any of the drainage lines. The weight, volume, and initial water content of the soil specimen are determined before placing the soil specimen into the triaxial apparatus.

2. *Saturation with back pressure stage.* If the soil specimen needs to be saturated before the start of the test, cell pressure and back pressure can be applied to the soil specimen. The Skempton pore pressure parameter B of the soil specimen is used to evaluate the degree of saturation of the specimen where the B value is defined in Equation 17.12.

$$B = \frac{\Delta u_b}{\Delta \sigma_3} \tag{17.12}$$

where Δu_b is the change in the back pressure, and $\Delta \sigma_3$ is the change in the cell pressure. The acceptable B value for full saturation is typically 0.95 (BS 1377-6 1990).

3. *Suction increase/decrease stage.* Matric suction of the soil specimen is increased or decreased to the desired value under the constant net mean stress $(p - u_a)$ condition. Matric suction is increased by increasing the air pressure (u_a). It is important to note that as u_a changes, there is a change in $(p - u_a)$. Thus, cell pressure needs to be adjusted to maintain $(p - u_a)$ constant. If there is no deviatoric stress, $(p - u_a)$ is equal to $(\sigma_3 - u_a)$. Therefore, only cell pressure needs to be adjusted.

4. *Saturated isotropic consolidation stage.* The cell pressure (σ_3) is applied to the soil specimen after saturation. This stage is commonly referred to as the consolidation stage in the saturated triaxial test (CU or CD).

5. *Constant water content isotropic consolidation stage.* The cell pressure (σ_3) is applied to the specimen and the specimen is not saturated. Water is not allowed to flow out of the specimen. It is sometimes used as a method to mechanically saturate the specimen (decreasing the void ratio under the constant water content condition will increase the degree of saturation).

6. *Constant suction isotropic consolidation stage.* The cell pressure (σ_3) is applied under the constant matric suction condition. This stage requires control of the air pressure supply to the specimen, and the load must be applied in small increments to ensure that the constant suction condition can be maintained.

7. k_0 *consolidation stage.* The cell pressure (σ_3) is increased when the deviatoric stress (q) is increased such that there is zero lateral strain on the soil specimen. Excess pore-water pressure is maintained at zero.

8. *Undrained shearing stage.* Deviatoric stress (q) is applied to the soil specimen without allowing the water to flow out of or drained from the soil specimen. ASTM D2166/2166M-16 (2016), ASTM D2850-15 (2015) and ASTM D4767-11 (2011) recommend that the *undrained shearing stage* be continued until 15% strain or until deviatoric stress has dropped by 20% or when 5% of additional axial strain occurs after reaching the peak of deviatoric stress, whereas BS 1377-7 (1990) recommends stopping the *undrained shearing stage* when the axial strain has reached 20% or when the peak deviatoric stress has been achieved.

9. *Constant water content shearing stage.* Deviatoric stress (q) is applied when the soil specimen is not saturated.

10. *Drained shearing stage.* Deviatoric stress (q) is applied when the soil specimen has been saturated while allowing the water to flow out or drained from the specimen such that no excess pore-water pressure develops during the shearing stage.

11. *Constant suction shearing stage.* Deviatoric stress (q) is applied under the constant suction condition. The shearing rate must be slow enough to maintain the constant matric suction condition.

12. *Finishing stage.* Soil specimen is taken out from the triaxial apparatus. A sketch or image of the specimen is obtained. Final weight of the soil specimen is measured and the water content of the soil specimen is determined.

From the possible combination of the preceding stages, the triaxial test can be separated into the following:

1. Preparation stage
2. Saturation stage (*saturation with back pressure stage* and *constant water content isotropic consolidation stage*)
3. Consolidation stage (*saturated isotropic consolidation stage, constant water content isotropic consolidation stage* or *constant suction isotropic consolidation stage*)
4. Shearing stage (*undrained shearing stage, drained shearing stage, constant water content shearing stage, constant suction shearing stage*)
5. Finishing stage

By using different combinations of stages, several test procedures can be used to determine the consolidation properties of soils using the triaxial test. The tests are given below with their relevant standards when applicable:

1. Unconfined compression test (UC)
   a.   ASTM D2166/2166M-16 (2016)
   b.   BS 1377-7 (1990)
   c.   BS EN ISO 17892-7 (2018)

2. Unconsolidated-undrained test (UU)
   a. ASTM D2850-15 (2015)
   b. BS 1377-7 (1990)
   c. BS EN ISO 17892-8 (2018)
3. Consolidated undrained test (CU)
   a. ASTM D4767-11 (2011)
   b. BS 1377-8 (1990)
   c. BS EN ISO 17892-9 (2018)
4. Consolidated drained test (CD)
   a. ASTM D7181-11 (2011)
   b. BS 1377-8 (1990)
   c. BS EN ISO 17892-9 (2018)
5. Isotropic consolidation test (IC)
   a. BS 1377-6 (1990)
6. At rest (k_0) consolidation test (k_0C)
7. Consolidated constant suction shearing test (CCSS)
8. Isotropic constant suction consolidation test (ICSC)
9. Consolidated constant water content shearing test (CCWCS)
10. Isotropic constant water content consolidation test (ICWCC)

17.6.1 UC test

Test suitability and requirement:

1. Specific gravity (G_s) for basic soil properties calculation
2. Suitable for fine-grained soils or materials with low permeability where the undrained condition is relevant
3. Soil specimen must be fully saturated in order to do the *undrained shearing stage*. If the soil specimen is not fully saturated, the shearing stage is the *constant water content shearing stage*, and the result will be misleading

The UC test is the simplest test, as there is no need to saturate the specimen, no cell pressure is applied to the specimen and the test is carried out under the undrained condition. The test result is equivalent to the UU test, but the test results are more sensitive to soil disturbance. The UC test is suitable for fine-grained soils with low permeability. The test sequence is as follows:

1. *Preparation stage*
2. *Undrained shearing stage/constant water content shearing stage*
3. *Finishing stage*

Obtainable parameters:

1. Basic soil properties: void ratio, water content, density, degree of saturation
2. Main parameters: E_u, q_u, s_u

17.6.2 UU test

Test suitability and requirement:

1. Specific gravity (G_s) for basic soil properties calculation.
2. Suitable for fine-grained soils or materials with low permeability where the undrained condition is relevant
3. Soil specimen must be fully saturated prior to the *undrained shearing stage*. If the *constant water content IC stage* failed to saturate the specimen, the shearing stage is the *constant water content shearing stage*, and the result will be misleading
4. Control of cell pressure is required
5. Water pressure transducer can be used to optionally measure pore-water pressure during the test
6. Impermeable condition at the top and bottom of the specimen is preferred to ensure the undrained condition

When the soil specimen is taken out from the ground, there might be a small dilation (and thus, the soil can become unsaturated) due to the stress release during sampling. As there is no cell pressure in the UC test, the small expansion of the soil specimen might affect the test results. In the UU test, the cell pressure is applied in the constant water content IC stage. BS 1377-7 (1990) recommends that the cell pressure must be related to the total vertical in situ stress (σ_v) and must cover the total stress range that will be experienced by the soil in situ. BS 1377-7 (1990) recommends that cell pressure of $0.5\sigma_v$, σ_v and $2\sigma_v$ might be generally appropriate, while for over-consolidated clays, the minimum cell pressure must not be less than σ_v.

As water is not allowed to drain from the specimen, excess pore-water pressure will build up when the soil becomes saturated affecting the effective mean stress (p') of the soil specimen. The sequence of the test procedure is as follows:

1. *Preparation stage*
2. *Constant water content IC stage*
3. *Undrained shearing stage/constant water content shearing stage*
4. *Finishing stage*
5. Optionally repeat steps 1 to 4 using soil specimens under different consolidation pressures for as many soil specimens as required

Obtainable parameters:

1. Basic soil properties: void ratio, water content, density, degree of saturation
2. Main parameters: E_u, s_u

17.6.3 CU test

Test suitability and requirement:

1. Specific gravity (G_s) for basic soil properties calculation
2. Suitable for fine-grained soils or materials with low permeability
3. Soil specimen must be fully saturated prior to the *saturated IC stage*
4. Control of back and cell pressures is required
5. Water pressure transducer is required to measure pore-water pressure of the soil specimen
6. Porous stone must be properly saturated.
7. Several consolidation pressures (either single or multi-stage) are required if compression curve parameters, i.e., λ, κ, p'_c and R_0, are desired.

The CU test is required when it is desired to obtain the undrained compression behaviour of soil under specific consolidation pressure. The application of consolidation pressure helps to reduce the effect of soil disturbance on the soil specimen. BS 1377-8 (1990) recommends that the effective consolidation pressure (σ'_3) must be related to the effective vertical in situ stress (σ'_v) and must cover the effective stress range that will be experienced by the soil in situ. BS 1377-8 (1990) recommends that cell pressure of $0.5\sigma'_v$, σ'_v and $2\sigma'_v$ might be generally appropriate.

As pore-water pressure is measured throughout the test, it is possible to correlate the undrained compressibility parameters with the effective mean stress of the soil. The test can be done either in a single stage (multiple specimens, one consolidation and shearing stage) or multi-stage (one specimen, multiple consolidation and shearing stages). The sequence of the test procedures for the single-stage test is as follows:

1. *Preparation stage*
2. *Saturation with back pressure stage*
3. *Saturated IC stage*
4. *Undrained shearing stage/constant water content shearing stage*
5. *Finishing stage*
6. Repeats steps 1 to 5 by using different soil specimens for different consolidation pressures for as many consolidation pressures as required.

The sequence of the test procedure for the multi-stage test is as follows:

1. *Preparation stage*
2. *Saturation with back pressure stage*
3. *Saturated IC stage*
4. *Undrained shearing stage* without failing the specimen
5. Repeats steps 3 and 4 for the same soil specimen at higher consolidation pressure for as many consolidation pressures as required
6. *Finishing stage*

Obtainable parameters:

1. Basic soil properties: void ratio, water content, density, degree of saturation
2. Main parameters: $E_u(p')$
3. Additional parameters: $K'(p')$, p'_c, R_0, κ, λ, c, ϕ, c', ϕ'

17.6.4 CD test

Test suitability and requirement:

1. Specific gravity (G_s) for basic soil properties calculation
2. Suitable for coarse-grained soils. Fine-grained soils or materials with low permeability may require excessive testing duration
3. Soil specimen must be fully saturated prior to the *saturated IC stage*
4. Control over back and cell pressures is required
5. Water pressure transducer is required to measure the specimen's pore-water pressure
6. Porous stone must be properly saturated
7. Several consolidation pressures (either single or multi-stages) are required if compression curve parameters, i.e., λ, κ, p'_c and R_0, are desired

The CD test is conducted when effective stress parameters are required. However, the test duration can be excessively long when it is used to test fine-grained soils. For coarse-grained soils, it is quite common for the soil specimen to be reconstituted, while for fine-grained soils, the soil specimen is trimmed using a soil lathe before placing it into the triaxial apparatus.

BS 1377-8 (1990) recommends that the effective consolidation pressure (σ'_3) must be related to the effective vertical in situ stress (σ'_v) and must cover the effective stress range that will be experienced by the soil in situ. BS 1377-8 (1990) recommends that cell pressure of $0.5\sigma'_v$, σ'_v and $2\sigma'_v$ might be generally appropriate.

The strain rate must be slow enough to avoid excess pore-water pressure build-up during the test. As pore-water pressure is measured throughout the test, it is possible to correlate the drained compressibility parameters with the effective mean stress of the soil. The test can be conducted as either a single stage (multiple specimens, one consolidation and shearing stage) or multi-stage (one specimen, multiple consolidation and shearing stages). The sequence of the test procedures for the single-stage test is as follows:

1. *Preparation stage*
2. *Saturation with back pressure stage*
3. *Saturated IC stage*
4. *Drained shearing stage*
5. *Finishing stage*
6. Repeats steps 1 to 5 by using different soil specimen for different consolidation pressures for as many consolidation pressures as required

The sequence of the test procedures for the multi-stage test is as follows:

1. *Preparation stage*
2. *Saturation with back pressure stage*
3. *Saturated IC stage*
4. *Drained shearing stage* without failing the specimen
5. Repeats steps 3 and 4 for the same soil specimen at higher consolidation pressure for as many consolidation pressures as required
6. *Finishing stage*

Obtainable parameters:

1. Basic soil properties: void ratio, water content, density, degree of saturation.
2. Main parameters: $E'(p')$
3. Additional parameters: $K'(p')$, p'_c, R_0, κ, λ, c', ϕ'

17.6.5 IC test

Test suitability and requirement:

1. Specific gravity (G_s) for basic soil properties calculation
2. Suitable for any type of soils
3. Soil specimen must be fully saturated prior to the *saturated IC stage*
4. Control of back and cell pressures is required
5. Water pressure transducer is required to measure the specimen's pore-water pressure
6. Porous stone must be properly saturated

The purpose of conducting the IC test is to obtain a compression curve in terms of e-ln(p'). The sequence of the test procedures is as follows:

1. *Preparation stage*
2. *Saturation with back pressure stage*
3. *Saturated IC stage*
4. Repeat step 3 with different consolidation pressures
5. *Finishing stage*

Obtainable parameters:

1. Basic soil properties: void ratio, water content, density, degree of saturation
2. Main parameters: $K'(p')$, p'_c, R_0, κ, λ

17.6.6 At rest (k_0) consolidation test (k_0C)

Test suitability and requirement:

1. Specific gravity (G_s) for basic soil properties calculation
2. Suitable for any type of soils
3. Soil specimen must be fully saturated prior to the *saturated IC stage*
4. Control of back and cell pressures is required
5. Lateral strain measurement is required
6. Cell pressure must have a feedback system to maintain zero lateral strain
7. Water pressure transducer is required to measure the specimen's pore-water pressure
8. Porous stone must be fully saturated

While the oedometer test is the easiest k_0 consolidation test, it is difficult to measure the lateral stress of the soil specimen. Thus, without a lateral stress measurement, it is not possible to obtain k_0. For the triaxial apparatus, it is possible to control the lateral stress through the cell pressure. Thus, provided lateral strain is measured, it is possible to implement a feedback system such that the cell pressure is increased to maintain zero lateral strain (Piriyakul and Haegeman 2005). The k_0 value can then be determined using Equation 17.13.

$$k_0 = \frac{\sigma'_x}{\sigma'_z} = \frac{\sigma_3 - u}{\sigma_1 - u} \tag{17.13}$$

The sequence of the test procedures is as follows:

1. *Preparation stage*
2. *Saturation with back pressure stage*
3. *Saturated IC stage* (optional)
4. k_0 *consolidation stage*
5. *Finishing stage*

Obtainable parameters:

1. Basic soil properties: void ratio, water content, density, degree of saturation
2. Main parameters: k_0
3. Additional parameters: $K'(p')$, p'_c, R_0, $E_c(p')$, κ, λ

17.6.7 CCSS test

Test suitability and requirement:

1. Specific gravity (G_s) for basic soil properties calculation
2. Suitable for any type of soils
3. Control of back and cell pressure is required
4. Control of air pressure is required
5. Tensiometer is required to measure the specimen's pore-water pressure
6. High-air entry (HAE) ceramic disk is required to replace the porous stone at the bottom of the specimen
7. CU or CD test is required to complement the test result
8. Several consolidation pressures (either single or multi-stage) are required if compression curve parameters, i.e., λ, κ, p'_c and R_0, are required

The CCSS test is one type of triaxial test that is developed to measure unsaturated soil properties. The drawback of this method is an excessively long test duration due to the difficulties in maintaining the constant suction condition during the shearing stage and the long equilibrium time for the soil to reach the desired matric suction. In the CCSS test, the matric suction is varied under a constant IC stage in order to investigate the effect of matric suction at that particular IC stage. The number of tests required depends on the assumption of the soil behaviour such as the following:

1. The effect of matric suction at different consolidation pressures. The CCSS test under different constant matric suction conditions needs to be carried out for more than one consolidation pressure. For instance, if three different matric suctions and three consolidation pressures are needed, for each consolidation pressure, three single-stage CCSS tests or one multi-stage CCSS test with three different matric suctions are required. Hence, there will be a total of nine CCSS single-stage tests or three CCSS multi-stage tests.
2. The effect of constant matric suction at different consolidation pressures. The CCSS test under different consolidation pressures needs to be carried out at one matric suction. For instance, if there are three consolidation pressures, three single-stage CCSS tests or one multi-stage CCSS test for the same matric suction is required.

The use of three matric suctions and three consolidation pressures in the previous example is to illustrate the number of tests required when testing unsaturated soils for different matric suctions and consolidation pressures. As the time required for three single-stage CCSS tests is much longer than a multi-stage CCSS test, a multi-stage CCSS test is recommended.

The sequence of a single-stage CCSS test is typically conducted as follows:

1. *Preparation stage*
2. *Saturation with back pressure stage*
3. *Saturated IC stage*
4. *Suction increase/decrease stage*
5. *Constant suction shearing stage*
6. *Finishing stage*
7. Repeat steps 1 to 6 using different specimens for each consolidation pressure or suction.

The sequence of the multi-stage CCSS test at one consolidation pressure is typically conducted as follows:

1. *Preparation stage*
2. *Saturation with back pressure stage*
3. *Saturated IC stage*
4. *Suction increase/decrease stage*
5. *Constant suction shearing stage* without failing the specimen
6. Change the suction in step 4 and do step 5 for as many matric suctions as required
7. *Finishing stage*

Obtainable parameters:

1. Basic soil properties: void ratio, water content, density, degree of saturation
2. Main parameters: $E(p, s)$, $H(s)$
3. Additional parameters: $K'(p')$, p'_c, R_0, κ, λ, ϕ^b

17.6.8 ICSC test

Test suitability and requirement:

1. Specific gravity (G_s) for basic soil properties calculation
2. Suitable for any type of soils
3. Control of back and cell pressures is required
4. Control of air pressure is required
5. Tensiometer is required to measure the soil specimen's pore-water pressure

6. HAE ceramic disk is required to replace the porous stone at the bottom of the specimen
7. IC test is required to complement the test result

The purpose of the ICWCC test is to obtain a compression curve in terms of e-$\ln(p - u_a)$ under a different constant suction condition. The sequence of the test procedures is as follows:

1. *Preparation stage*
2. *Saturation with back pressure stage*
3. *Saturated IC stage*
4. *Suction increase/decrease stage*
5. *Constant suction IC stage*
6. Repeat step 5 with a different cell pressure
7. *Finishing stage*

Obtainable parameters:

1. Basic soil properties: void ratio, water content, density, degree of saturation
2. Main parameters: $K'(p')$, $p_{c,a}$ (s), $\kappa(s)$, $\lambda(s)$

17.6.9 CCWCS test

Test suitability and requirement:

1. Specific gravity (G_s) for basic soil properties calculation
2. Suitable for any type of soils
3. Control of back and cell pressures is required
4. Control of air pressure is required to consolidate the soil specimen to the initial matric suction
5. HAE ceramic disk is required to replace the porous stone at the bottom of the specimen if the pore-water pressure measurement is desired
6. CU or CD test is required to complement the test result

As a CS triaxial test requires a very long test duration, it is often not practical to conduct a CS triaxial test. An alternative method is to conduct CCWCS. In a CWC test, air pressure is drained. As the specimen is sheared under a constant water content condition, the degree of saturation increases as the void ratio decreases. This phenomenon is referred to as mechanical wetting and will cause an increase in pore-water pressure or reduction in matric suction (Wijaya 2017). The test can be done very fast, as there is no drainage of excess pore-water pressure and the compression of unsaturated soil is immediate.

The limitation of the CCWCS test is that interpretation of the parameters is more difficult as both net mean stress and matric suction change at the same time (Alonso et al. 1990) especially when no pore-water measurement is conducted during the test. Most constitutive models for unsaturated soils, such as Barcelona Basic Model (Alonso et al. 1990) and its modification (Wheeler and Sivakumar 1995), require CS tests to obtain the parameters.

The sequence of the single-stage CCWCS test procedures is as follows:

1. *Preparation stage*
2. *Saturation with back pressure stage*
3. *Saturated IC stage*
4. *Suction increase/decrease stage*
5. *Constant water content shearing stage*
6. Repeat steps 1 to 5 with different soil specimens for each cell pressure
7. *Finishing stage*

The sequence of the multi-stage CCWCS test procedures is as follows:

1. *Preparation stage*
2. *Saturation with back pressure stage*
3. *Saturated IC stage*
4. *Suction increase/decrease stage*
5. *Constant water content shearing stage* and stop the shearing stage before the specimen fails
6. Repeat steps 4 and 5 for as many matric suctions as required
7. *Finishing stage*

Obtainable parameters:

1. Basic soil properties: void ratio, water content, density, degree of saturation
2. Main parameters: $E(p, w)$, $H(w)$
3. Additional parameters: $K'(p')$, $K(s)$, p'_c, R_0, κ, λ, ϕ^b

17.6.10 Isotropic constant water content consolidation test (ICWCC)

Test suitability and requirement:

1. Specific gravity (G_s) for basic soil properties calculation
2. Suitable for any type of soils
3. Control of back and cell pressures is required
4. Control of air pressure is required if initial matric suction is applied
5. Tensiometer with HAE ceramic disk is required to replace the porous stone at the bottom of the specimen if pore-water pressure measurement is required
6. IC tests are required to complement the test result

The purpose of the ICWCC test is to obtain a compression curve in terms of e-$\ln(p - u_a)$ under a different constant water content condition. The sequence of the test procedures is as follows:

1. *Preparation stage*
2. *Saturation with back pressure stage*
3. *Saturated IC stage*
4. *Suction increase/decrease stage*
5. *Constant water content IC stage*
6. Repeat step 5 with different cell pressures
7. *Finishing stage*

Obtainable parameters:

1. Basic soil properties: void ratio, water content, density, degree of saturation
2. Main parameters: $K'(p')$, $K(p,w)$, $p'_{c,a}(w)$, $\kappa(w)$, $\lambda(w)$

REFERENCES

BS 1377-8 (1990). *Methods of test for soils for civil engineering purposes – part 8: shear strength tests (effective stress)*. UK: BSI.

Bao, C., Gong, B., & Zhan, L. (1998). Keynote lecturer, properties of unsaturated soils and slope stability of expansive soils. In *Proceedings of the 2nd International Conference on Unsaturated Soils (UNSAT 98)*. International Academic, Beijing, China, vol. 1, pp. 71–98.

Budhu, M. (2010). *Soil mechanics and foundations*. 3rd ed. Hoboken, NJ: Wiley.

Burrage, R., Anderson, J., Pando, M., Ogunro, V., & Cottingham, M. (2012). A cost effective triaxial test method for unsaturated soils. *Geotechnical Testing Journal*, 35(1): 50–59.

Escario, V. & Sáez, J. (1986). The shear strength of partly saturated soils. *Géotechnique*, 36(3): 453–456.

Farouk, A., Lamboj, L., & Kos, J. (2004). Influence of matric suction on the shear strength behaviour of unsaturated sand. *Acta Polytechnica*, 44(4): 11–17.

Fredlund, D. G., Morgenstern, N. R., & Widger, R. A. (1978). The shear strength of unsaturated soils. *Canadian Geotechnical Journal*, 15(3): 313–321.

Fredlund, D. G., Rahardjo, H., & Fredlund, M. D. (2012). *Unsaturated soil mechanics in engineering practice*. Hoboken, New Jersey: John Wiley & Sons, Inc.

Gan, J. K. M., Fredlund, D. G., & Rahardjo, H. (1988). Determination of the shear strength parameters of an unsaturated soil using the direct shear test. *Canadian Geotechnical Journal*, 25(3): 500–510.

Garven, E. A. & Vanapalli, S. K. (2006). Evaluation of empirical procedures for predicting the shear strength of unsaturated soils. In *Proceedings of 4th Int. Conf.*

of Unsaturated Soil, UNSAT 2006, ASCE Geotechnical Special Publication 147, ASCE, pp. 2570–2581.

Goh, S. G., Rahardjo, H., & Leong, E. C. (2010). Shear strength equations for unsaturated soil under drying and wetting. *ASCE Journal of Geotechnical and Geoenvironmental Engineering, 136*(4): 594–606.

Hilf, J. W. (1956). An investigation of pore-water pressure in compacted cohesive soils. PhD Thesis, Tech. Memo. No. 654. Denver, CO: U.S. Dept. of the Interior, Bureau of Reclamation, Design and Construction Div.

Khalili, N. & Khabbaz, M. H. (1998). A unique relationship for c for the determination of the shear strength of unsaturated soils. *Géotechnique, 48*(5): 681–687.

Lee, I. M., Sung, S. G., & Cho, G. C. (2005). Effect of stress state on the unsaturated shear strength of a weathered granite. *Canadian Geotechnical Journal, 42*(2): 624–631.

Marinho, F. A. M., Guzmán, G. G. C., & Orlando, P. D. G. (2016). Constant water content compression tests on unsaturated compacted soil with suction measurement using a HCT. *International Journal of Geomechanics, 16*(6): D4015008.

Oberg, A. & Sallfors, G. (1997). Determination of shear strength parameters of unsaturated silts and sands based on the water retention curve. *Geotechnical Testing Journal, 20*(1): 40–48.

Rahardjo, H., Lim, T. T., Chang, M. F., & Fredlund, D. G. (1995). Shear-strength characteristics of a residual soil. *Canadian Geotechnical Journal, 32*(1): 60–77.

Rassam, D. W. & Cook, F. J. (2002). Predicting the shear strength envelope of unsaturated soils. *Geotechnical Testing Journal, 25*(2): 215–220.

Rassam, D. W. & Williams, D. J. (1999). A relationship describing the shear strength of unsaturated soils. *Canadian Geotechnical Journal, 36*(2): 363–368.

Shen, Z. & Yu, S. (1996). The problems in the present studies on mechanics for unsaturated soils, atomic energy. In *Proceedings of Symp. on Geotechnical Aspects of Regional Soils*, pp. 1–9.

Tekinsoy, M. A., Kayadelan, C., Keskin, M. S., & Soylemaz, M. (2004). An equation for predicting shear strength envelope with respect to matric suction. *Computer and Geotechnics, 31*(7): 589–593.

Trinh, M. T., Rahardjo, H., & Leong, E. C. (2006). Shear strength and pore-water pressure characteristics during constant water content triaxial tests. *Journal of Geotechnical & Geoenvironmental Engineering, 132*(3): 411–419.

Vanapalli, S. K., Fredlund, D. G., Pufahl, D. E., & Clifton, A. W. (1996). Model for the prediction of shear strength with respect to soil suction. *Canadian Geotechnical Journal, 33*(3): 379–392.

Vilar, O. M. (2006). A simplified procedure to estimate the shear strength envelope of unsaturated soils. *Canadian Geotechnical Journal, 43*(10): 1088–1095.

Wijaya, M. (2017). Compression, shrinkage and wetting-induced volume change of unsaturated soils. In *Civil and environmental engineering (Ph.D.)* , Singapore: Nanyang Technological University.

Zhang, X. (2016) Limitations of suction-controlled triaxial tests in the characterization of unsaturated soils. *International Journal for Numerical and Analytical Methods in Geomechanics, 40*: 269–296.

FURTHER READING

Alonso, E. E., Gens, A., & Josa, A. (1990). A constitutive model for partially saturated soil. *Géotechnique*, 40(3): 405–430.

Trinh, M. T., Rahardjo, H., & Leong, E. C. (2006). Shear strength and pore-water pressure characteristics during constant water content triaxial tests. *Journal of Geotechnical & Geoenvironmental Engineering*, 132(3): 411–419.

Vanapalli, S. K., Fredlund, D. G., Pufahl, D. E., & Clifton, A. W. (1996). Model for the prediction of shear strength with respect to soil suction. *Canadian Geotechnical Journal*, 33(3): 379–392.

Chapter 18

Ring shear test

18.1 BACKGROUND

The ring shear apparatus was designed to obtain the residual strength of a soil at large displacement, mainly to investigate landslides with a pre-existing shear plane. The biggest advantage of the ring shear test over the direct shear test is that drained residual strength of clayey soils at large shear displacements can be determined without reversal in the direction of shear. There are two versions of the ring shear apparatus, the Bishop-type (Bishop et al. 1971) and the Bromhead-type (Bromhead 1979). The main difference between the Bishop-type and Bromhead-type of ring shear apparatus is that the Bishop-type ring shear apparatus has a split ring (upper and lower rings) equivalent to the split box in the direct shear test. The Bromhead ring shear apparatus is simpler, and it is easier to prepare a soil specimen in the Bromhead-type of ring shear apparatus as the thickness of the specimen is smaller about 5 mm. The advantages of the ring shear apparatus are that the area of the shear plane during the test remains constant and large displacement not possible in the direct shear test can be achieved in the ring shear apparatus to reach the residual shear stress condition. Although the failure shear plane in the Bromhead-type apparatus fails near the soil and porous element interface compared to the Bishop-type apparatus that fails at the centre of the soil specimen (between the upper and lower rings), the test results appear to be similar (Bromhead 1979). Comparison of the ring shear test with other tests (direct shear and triaxial) showed similar test results (Bromhead and Curtis 1983; Toyota et al. 2009). The Bromhead-type ring shear apparatus is adopted in BS 1377: part 7: 1990, clause 6, as a total stress strength test, while BS EN ISO 17892-10 and ASTM D6467-21 and D7608-18 allow the use of either Bishop-type or Bromhead-type of apparatus. The ring shear test has been modified to conduct tests on unsaturated soils (Vaunat et al. 2006, 2007; Infante Sedan et al. 2007; Merchán et al. 2011; Hoyos et al. 2011, 2014, 2020). This chapter will concentrate mostly on the Bromhead-type of ring shear apparatus, which is common to both BS, BS EN and ASTM.

DOI: 10.1201/b22304-18

18.2 RELATED STANDARDS

1. ASTM D6467-21, 2021. Test Method for Torsional Ring Shear Test to Determine Drained Residual Shear Strength of Fine-Grained Soils.
2. ASTM D7608-18. 2018. Standard Test Methods for Torsional Ring Shear Test to Measure Drained Fully Softened Shear Strength and Stress Dependent Strength Envelope of Fine-Grained Soils.
3. BS 1377:Part 7:1990 Clause 6 Determination of Residual Strength Using the Small Ring Shear Apparatus.
4. BS EN ISO 17892-10. 2018. Geotechnical Investigation and Testing. Laboratory Testing of Soil. Part 10: Direct Shear Test.

18.3 TEST METHODS AND APPARATUS

Typical ring shear apparatus is shown schematically in Figure 18.1. The ring shear test consists of two stages: consolidation stage and shearing stage. For the shearing stage, the test can be conducted under a constant normal load or constant volume condition. For unsaturated soils, the shearing stage can be conducted under a constant suction or constant water content condition. Hence, there are four possible test configurations that can be conducted for unsaturated soils (Infante Sedano et al. 2007): constant load with constant suction (CLS), constant volume with constant suction (CVS), constant load

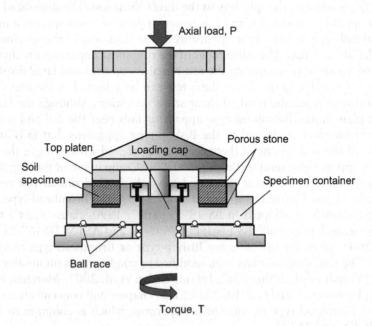

Figure 18.1 Schematic drawing of a typical Bromhead-type ring shear apparatus.

Table 18.1 Requirements of specimen container and soil specimen

Reference	Specimen container	Soil specimen
ASTM D6467-21	ID ≥ 50 mm ID/OD ≥ 0.6	Initial height ≥ 5 mm Max. particle size < 10% of initial height
ASTM D7608-18	ID ≥ 50 mm ID/OD ≥ 0.5	Initial height ≥ 5 mm Max. particle size < 10% of initial height
BS1377:Part 7: 1990, Clause 6	ID = 70 mm OD = 100 mm	Height = 5 mm Max. particle size < 1.18 mm
BS EN ISO 17892-10	OD ≥ 70 mm ID/OD ≥ 0.6 Height/width of annulus ≤ 1	Initial height ≥ 5 mm

with constant water content (CLW), and constant volume with constant water content (CVW). The four test configurations are useful for the development of constitutive models, but for practical purposes, probably only the CLS and CLW tests are more useful.

The requirements of the ring shear apparatus for the specimen container and soil specimen are given in the standards and summarised in Table 18.1.

For constant suction shearing, modifications need to be made to the ring shear apparatus to maintain the matric suction of the soil specimen constant during the test, and the shearing rate must be slow enough to allow the soil specimen to equilibrate with the applied matric suction constantly. For constant water content shearing, the conventional ring shear apparatus without modification can be used unless there is a need to change the matric suction of the soil at the beginning of the test and/or the need to measure matric suction during the test.

Two modifications need to be made: one to control the suction and the other to separate the air phase from the water phase. To control the suction, methods that have been employed include the vapour equilibrium method (Vaunat et al. 2006, 2007; Merchán et al. 2011) and axis-translation method (Infante Sedan et al. 2007; Hoyos et al. 2011, 2014, 2020). For the vapour equilibrium method, the top of the ring shear apparatus is covered (not necessarily airtight), and a vapour from a vessel of saline solution is circulated to the soil specimen and then re-circulated back to the vessel using an air pump (Vaunat et al. 2006, 2007; Merchán et al. 2011). A schematic drawing of a modified ring shear apparatus using the vapour equilibrium method is shown in Figure 18.2a. The actual suction applied is based on relative humidity and temperature reading within the cover. The equilibrium suction condition is based on observation of the rate of vertical deformation of the soil specimen strain to be less than 0.1%/day. The time for suction equilibrium takes several days to about two weeks. The suction of the soil specimen

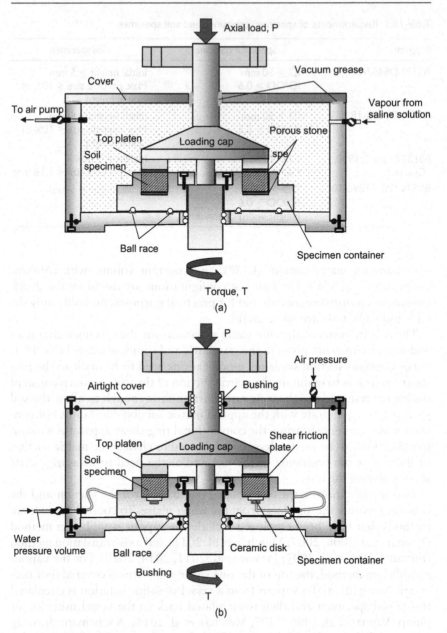

Figure 18.2 Schematic drawings of modified ring shear apparatus to test unsaturated soil. (a) Using vapour equilibrium method. (b) Using axistranslation method.

can be confirmed after the test using the chilled mirror hygrometer. This method is suitable for high suctions of 10–140 MPa (Merchán et al. 2011), where there is no need to control suctions to within 100 kPa and where shearing is conducted under the CLW condition. Using the axis-translation method, the modifications required are more extensive. The top of the ring shear apparatus has to be covered for airtightness to withstand the maximum air pressure that is applicable. A schematic drawing of a modified ring shear apparatus using the axis-translation method is shown in Figure 18.2b. Porous stone is not required in the top platen, shear friction plate can be used, or the surface of the top platen can be roughened to ensure that failure does not occur at the interface between the top platen and the soil. For the bottom of the specimen container, high-air-entry (HAE) ceramic disks need to replace the usual porous stone, with optional control of water pressure and volume below the HAE disks. Unlike the porous stone, the HAE ceramic disks need to be sealed using epoxy (Infante Sedan et al. 2007). Alternatively, HAE ceramic disks can be epoxy-sealed into steel rings with an O-ring on the circumference. The steel rings are then pressed into holes on the bottom platen (Hoyos et al. 2011, 2014, 2020). The ceramic disks occupy about 80% and 11.5% of the specimen area for Infante Sedan et al. (2007) and Hoyos et al. (2011, 2014, 2020) apparatuses, respectively (Figure 18.3). Because of the applied air pressure, a servo-load control of the normal load is needed to ensure that the net normal stress $(\sigma - u_a)$ is maintained constant during the CLS test and is adjustable during the CVS test. The axis-translation method is more suitable for the lower suction range where the highest matric suction is limited by the HAE value of the ceramic disks used.

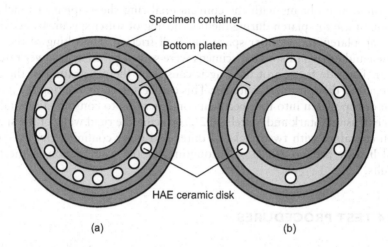

Figure 18.3 Arrangement of HAE ceramic disks used in (a) Sedan et al. (2007). (b) Hoyos et al. (2011, 2014, 2020).

The highest HAE of ceramic disks is 1500 kPa. However, the maximum air pressure that has been used so far in modified ring shear apparatus for an unsaturated soil test is 500 kPa (Infante Sedano et al. 2007; Hoyos et al. 2014, 2020).

The instrumentations for the ring shear apparatus include the original ring shear apparatus instruments which are the load transducers for the normal load, two load transducers for the torque arm or one torque transducer, one rotary potentiometer to measure angular rotation of the lower half of the soil specimen, one linear variable displacement transformer to measure deformation of the soil specimen, and additional instrumentations for the unsaturated soil test which includes one air pressure transducer and one digital pressure/volume controller to maintain constant water volume or constant water pressure. In addition, a servo control system with feedback is needed for the normal load to maintain constant load during the CLS test and to be adjustable during the CVS test.

The use of the axis-translation method to control matric suction is valid only when the water phase of the soil is continuous. Bocking and Fredlund (1980) suggest an upper limit of 85% degree of saturation should be applied when using the axis-translation technique because the presence of occluded air bubbles can result in an overestimation of the measured matric suction in the soil specimen. In addition, the use of elevated air pressure above the HAE ceramic disks may lead to diffused air accumulating below the ceramic disks (Padilla et al. 2006). Infante Sedano et al. (2007) incorporated a flushing system with air volume measurement in their apparatus, while Hoyos et al. (2011) flushed their apparatus of diffused air every 48 hours during a constant suction test.

Common problems with the conventional ring shear apparatus include tilting of the top platen during the test, oozing of softer specimens between the top platen and annular space and wall friction. The tilting of the top platen and oozing of softer specimens were solved by using a slower rate of shear. The side friction of the ring is caused by the penetration of the top platen into the specimen container. This can be solved by limiting the intrusion of top platen into the specimen container due to consolidation and/or soil extrusion (Stark and Vettel 1992), replacing the rigid walls of the specimen container with two stacks of outer and inner confining rings (Garga and Infante Sedano 2002) or accounting for the wall friction in the test results.

18.4 TEST PROCEDURES

The test procedures for the ring shear test of unsaturated soils are described next for the axis-translation method only. For the vapour equilibrium method, readers are asked to refer to Merchán et al. (2011). The test procedures can be divided into four stages: preparation stage, consolidation stage, shearing stage and finishing stage. The preparation of the apparatus is different for the vapour equilibrium.

The preparation stage involves the following steps:

1. Assemble the specimen container which contains the annular cavity for the soil specimen and ensure that all the water pressure/drainage lines are completely filled with water.
2. Saturate the HAE ceramic disk in the container by placing deaired water into the container to the specimen height and covering the ring shear apparatus with the airtight cover and introducing air pressure to near the air-entry value of the ceramic disk. Allow the water to drain through the ceramic disk and collect the water in a beaker or burette. Observe for air bubbles during the drainage process. The water should stop draining out once all the water above the ceramic disk has drained out if the ceramic disk is fully saturated. Repeat this step a few times. It may be necessary to couple this process with filling the specimen container with water, closing the drainage line and applying the air pressure overnight. The process of draining out the water is then repeated afterwards. This cycle of draining-pressuring-draining should be done at least once to ensure that the ceramic disks are fully saturated. The amount of water collected with time can be used to check the saturated permeability of the ceramic disks.
3. Prepare the soil specimen in the specimen container. The specimen can be reconstituted, compacted (static) or intact. For preparation of an intact/undisturbed specimen for the ring shear apparatus, please refer to Stark (2021). Hoyos et al. (2014) suggest placing a wet filter paper over the ceramic disks to ensure that the ceramic disks remain saturated when compacting the soil into the container.
4. Place the top platen over the top of the specimen. This may have to be done through the servo-controller if load control is available. A small seating load can be applied to ensure that the specimen is in good contact with the top and bottom platens, especially with the HAE ceramic disks.
5. Attach and adjust the vertical displacement transducers and zero the readings.
6. Zero the load or torque transducers.
7. Attach the top cover and connect the air pressure line.

The consolidation stage involves the following steps:

1. Start data recording and apply the net normal stress and matric suction in turn on the soil specimen in increments until the target values. Depending on the change in net normal stress and matric suction needed, the increments can be in two or more steps.

2. At the target net normal stress and matric suction, the consolidation stage is considered completed when there is negligible change in axial strain and negligible water flowing out of the soil specimen. ASTM D6467-21 specifies that the minimum soil specimen's height after consolidation should be 3 mm or 60% of its initial height.

The shearing stage depends on the type of tests that are conducted. The description of the following procedures follows the test configurations mentioned in Infante Sedano et al. (2007).

1. In the CLS test, both the net normal stress and matric suction are maintained constant. The shearing rate is selected such that the soil specimen maintains a constant matric suction. It is recommended to maintain air pressure constant but adjust the water pressure below the HAE ceramic disks to maintain constant matric suction. In this case, the normal load need not be adjusted, as the net normal stress is always constant as air pressure remains constant. The torque, volume and water volume of the specimen are continuously monitored until the completion of the CLS test.

2. In the CVS test, the matric suction is maintained constant as in the CLS test by keeping the air pressure constant but changing the water pressure below the HAE ceramic disks. In the CVS test, the soil specimen volume is maintained constant by changing the net normal stress. Since air pressure is maintained constant, the change can be affected by the normal load on the soil specimen. The shearing rate is selected such that the soil specimen maintains a constant matric suction. The normal load, torque, volume and water volume of the specimen are continuously monitored until the completion of the CVS test.

3. In the CLW test, the net normal stress is maintained constant since air pressure is constant. The water content of the soil specimen is maintained constant by not allowing water to drain through the ceramic disk (i.e., undrained with respect to water). The water volume below the ceramic disk is maintained constant, and the water pressure is monitored continuously. This can be easily achieved using a digital pressure/volume controller that is nowadays a common device in soil testing laboratories. The shearing rate in the CLW test is less stringent than in the other tests and probably could be sheared at the fastest rate compared to the other tests. The torque, volume and matric suction of the specimen are continuously monitored until the completion of the CLW test.

4. In the CVW test, the air pressure is maintained constant, and the normal load is adjusted to maintain a constant volume. The water content of the soil specimen is maintained constant by not allowing water to drain through the ceramic disk (i.e., undrained with respect to water). The water volume below the ceramic disks is maintained constant,

and the water pressure is monitored continuously to give the matric suction. The shearing rate is selected such that the soil specimen maintains a constant volume. The load, torque, volume and matric suction of the specimen are continuously monitored until the completion of the CVW test.

The previous steps can be repeated in a multi-stage ring shear test provided there is sufficient rotational displacement to carry out the multi-stages. Merchán et al. (2011) and Hoyos et al. (2014) conducted multi-stage suction-controlled ring shear tests.

Guidance on shearing rates for saturated soils in drained ring shear test is given by Stark et al. (2020) as a function of liquid limit. Stark et al. (2020) found that clogged or unsaturated porous disk requires a displacement rate as low as 0.018 mm/min. Guidance on the suitable shearing rate to be adopted for ring shear tests on unsaturated soils is lacking, as these depend on several factors. Merchán et al. (2011) tested Boom clay at high suction in CLS tests using a shearing rate of 0.089 mm/min. Hoyos et al. (2014) tested clay, silty-clayey sand and sand in CLS tests using a shearing rate of 0.025 mm/min.

The finishing stage involves the following steps:

1. Maintain the position of the normal load.
2. Release the air pressure and water pressure first. Drain the water from below the ceramic disk to prevent the soil specimen from imbibing water on the release of the normal load.
3. Release the normal load.
4. Remove the vertical displacement transducers and move the load transducers away from the torque arm.
5. Remove the top platen from the specimen. If any twisting action is needed to remove the top platen, do it by rotating along the shear direction and moving it up vertically.
6. Remove the specimen and record by photographing or sketching the failure surface.
7. Determine the final moisture content of the specimen.

18.5 CALCULATIONS

If the ring shear apparatus gives displacement in terms of a rotation angle β (in degrees), this is first converted to an apparent average linear displacement δ_a using Equation 18.1.

$$\delta_a = r.\beta \frac{\pi}{180^o} \tag{18.1}$$

where r is the average of the inner and outer radius of the specimen. The average linear displacement of the top platen δ given by Equation 18.2.

$$\delta = \frac{(\delta_1 + \delta_2)r}{L} \tag{18.2}$$

where
δ_1 and δ_2 = displacements at the torque arm load transducers where load is applied
L = distance between the two loading points on the torque arm

Using Equations 18.1 and 18.2, correct the average linear displacement δ' using Equation 18.3.

$$\delta' = \delta - \delta_a \tag{18.3}$$

The shearing rate of the shear test can be easily determined from a plot of δ' versus time. The average shear stress τ is calculated using Equation 18.4.

$$\tau = \begin{cases} \dfrac{3(F_1 + F_2)L}{4\pi\left(r_2^2 - r_1^2\right)} & \text{for torque arm with two load transducers} \\[4mm] \dfrac{3(T)}{2\pi\left(r_2^2 - r_1^2\right)} & \text{for torque arm with a torque transducer} \end{cases} \tag{18.4}$$

The net normal stress is calculated using Equation 18.5.

$$\sigma - u_a = \frac{P}{\pi\left(r_2^2 - r_1^2\right)} - u_a \tag{18.5}$$

For the consolidation stage, plot the vertical displacement versus time and matric suction versus time. For the shearing stage, plot τ versus δ', vertical displacement versus δ', τ versus $(\delta - u_a)$.

18.6 SUMMARY

The ring shear apparatus is more complex than the direct shear apparatus or triaxial apparatus. The correctness of the ring shear test depends on several factors. First-time users of the ring shear apparatus or a modified ring shear apparatus should always check the failure condition of the soil specimen and verify the test results with those reported in the literature.

REFERENCES

Bishop, A. W., Green, G. E., Garga, V. K., Andresen, A., & Brown, J. D. (1971). A new ning Shear apparatus and its application to the measurement of residual strength. *Géotechnique, 21*(4): 273–328.

Bocking, K. A. & Fredlund, D. G. (1980). Limitations of the axis translation technique. In *Proceedings of the 4th International Conference on Expansive Soils*, Denver, CO, pp. 117–135.

Bromhead, E. N. (1979). A simple ring shear apparatus. *Ground Engineering, 12*(5): 40–44.

Bromhead, E. N. & Curtis, R. D. (1983). A comparison of alternative methods of measuring the residual strength of London clay. *Ground Engineering, 16*(4): 39–41.

Fredlund, D. (2006). Unsaturated soil mechanics in engineering practice. *Journal of Geotechnical and Geoenviromental Engineering, 132*(3): 286–321.

Garga, V. K. & Infante, J. A. (2002). Steady state of sands in a constant volume ring shear apparatus. *Geotechnical Testing Journal, 25*(4): 1–8.

Hoyos, L. R., Velosa, C. L., & Puppala, A. J. (2011). A servo/suction-controlled ring shear apparatus for unsaturated soils: Development, performance, and preliminary results. *Geotechnical Testing Journal, ASTM, 34*(5): 413–423.

Hoyos, L. R., Velosa, C. L., & Puppala, A. J. (2014). Residual shear strength of unsaturated soils via suction-controlled ring shear testing. *Engineering Geology, Elsevier, 172*: 1–11.

Hoyos, L. R., Yepes, J. E., Velosa, C. L., & Puppala, A. J. (2020). Unsaturated shear strength of compacted clayey soil via suction-controlled ring shear testing. In *4th European Conference on Unsaturated Soils (E-UNSAT 2020), E3S Web Conference*, Volume 195, Article Number 03024. https://doi.org/10.1051/e3sconf/202019503024.

Infante, J. A., Vanapalli, S. K., & Garga, V. K. (2007). Modified ring shear apparatus for unsaturated soil testing. *Geotechnical Testing Journal, ASTM, 30*(1): 1–9.

Merchán, V., Romero, E., & Vaunat, J. (2011). An adapted ring shear apparatus for testing partly saturated soils in the high suction range. *Geotechnical Testing Journal ASTM, 34*(5): 433–444.

Stark, T. D. (2021). Constant volume ring shear specimen trimming and testing. *Geotechnical Testing Journal, 44*(5): 1279–1300. https://doi.org/10.1520/GTJ20200144

Stark, T. D., Cadigan, J. A., & Jafari, N. H. (2020). Drained shear displacement rates in fully softened strength torsional ring shear testing. *Geotechnical Testing Journal, 44*(5): 1350–1357. https://doi.org/10.1520/GTJ20200117

Stark, T. D. & Vettel, J. J. (1992). Bromhead ring shear test procedure. *Geotechnical Testing Journal, 15*(1): 24–32.

Toyota, H., Nakamura, K. Sugimoto, M., & Sakai, N. (2009). Ring shear tests to evaluate strength parameters in various remoulded soils. *Géotechnique, 59*(8): 649–659.

Vaunat, J., Amador, C., Romero, E., & Djeren-Maigre, I. (2006). Residual strength of a low plasticity clay at high suctions. In *Proceedings of Fourth International Conference on Unsaturated Soils*, Carefree, Arizona, vol. 1, pp. 1279–1289.

Vaunat, J., Merchán, V., Romero, E., & Pineda, J. (2007). Residual strength of clays at high suctions. In *Proceedings of the Second International Conference on Mechanics of Unsaturated Soils*, Weimar, Germany, vol. 2, pp. 151–162.

Chapter 19

Tension test

19.1 BACKGROUND

Soils in the vadose zone, especially at the ground surface, experience drying and wetting cycles to weather and climatic conditions. Shrinkage and formation of tensile cracks are the processes associated with drying. The tensile cracks affect the integrity of geotechnical structures such as slopes, dams, embankments and roads in terms of compressibility, permeability and strength.

The formation of tensile cracks is related to the tensile strength of soils. Research shows that the tensile strength of unsaturated soils is mainly influenced by matric suction of the unsaturated soils (De Souza Villar et al. 2009; Yin and Vanapalli 2018).

Tensile strength tests for soils are generally grouped based on the method of load application: direct and indirect tensile tests. Direct tensile strength tests usually involve constraining one of the boundaries of the test specimen and applying uniaxial tensile force to the opposite boundary. The direct tensile test is regarded as the only method where a specimen is subjected to true uniaxial tension and failure occurs along its longitudinal axis (Peters and Leavell 1988; Win 2006). In a direct tensile test, it is assumed that uniform stresses are applied to the specimen and torsional and bending stresses are absent. The tensile strength is computed as a ratio of the maximum load sustained by the specimen and the cross-sectional area upon which the load acts. Direct tensile strength tests have been conducted on soils (Tschebotarioff et al. 1953; Hasegawa and Ikeuti 1966; Ajaz and Parry 1975; Peters and Leavell 1988; Tang and Graham 2000; Trabelsi et al. 2010; Stirling et al. 2015; Araki et al. 2016; Tollenaar et al. 2017; Jhuo et al. 2019; Li et al. 2019; Murray and Tarantino 2019).

Despite the advantages of direct tensile strength tests, the validity of the tests has been questioned due to difficulties associated with the tests, such as difficulty in effectively clamping or holding the specimen at the ends, misalignment, stress concentration and eccentric loading (Kennedy and Hudson

DOI: 10.1201/b22304-19

1968). Creep and volume changes as a result of sustained loading during the test have also been reported (Win 2006).

To address some of the difficulties in direct tensile tests, indirect tensile tests have been developed. Although the failure mode in indirect tensile tests is a combination of compression and tension, indirect tensile tests have several advantages over direct tensile tests. It is relatively simple to conduct; the failure is in a region of uniform tensile stress; the specimens and equipment are the same as the compression tests; the surface conditions of the specimen do not affect the failure; there is less variation of the test results (Kennedy and Hudson 1968). Many studies have been conducted on soils using indirect tensile tests (e.g., Uchida and Matsumoto 1961; Kennedy and Hudson 1968; Khrishnayya and Eisenstein 1974; Dexter and Kroesbergen 1985; Das et al. 1995; Al-Hussaini 2009; Li and Wong 2013; Akin and Likos 2017; Murray et al. 2019; Pittaro 2019). Among the indirect tensile tests, the Brazilian tensile strength test is the most frequently used (Khrishnayya and Eisenstein 1974; Das et al. 1995; Vesga and Vallejo 2006; De Souza Villar et al. 2009; Beckett et al. 2015; Akin and Likos 2017; Wang et al. 2022). Comparison between direct and indirect tensile tests show no significant differences between the tensile strength (e.g., Vesga and Vallejo 2006; Fahimifar and Malekpour 2012; Kim et al. 2012).

In this chapter, one direct tensile test and two indirect tensile tests that have been more commonly used for testing unsaturated soils are described.

19.2 RELATED STANDARDS

1. ASTM D4543-08, 2018. Standard Practices for Preparing Rock Core as Cylindrical Test Specimens and Verifying Conformance to Dimensional and Shape Tolerances.
2. ASTM C496/C496M-17, 2017. Standard Test Method for Splitting Tensile Strength of Cylindrical Concrete Specimens.
3. ASTM D3967-18, 2016 Standard Test Method for Splitting Tensile Strength of Intact Rock Core Specimens.
4. ASTM D2216-10, 2010 Standard Test Methods for Laboratory Determination of Water (Moisture) Content of Soil and Rock by Mass.
5. ASTM D7012-14, 2014 Standard Test Methods for Compressive Strength and Elastic Moduli of Intact Rock Core Specimens under Varying States of Stress and Temperatures.
6. BS 1377-1 (2016). Methods of Test for Soils for Civil Engineering Purposes – Part 1:General Requirements and Sample Preparation.
7. ISRM Commission on Testing Methods, 1978. Suggest Methods for Determining Tensile Strength of Rock Materials. Int. J. Rock Mech. Min. Sci. Geomech. Abstr. 15, 101–103.

19.3 DIRECT TENSILE TEST

Direct tensile test for soils has not been standardised. The specimen shape used varies as shown in Figure 19.1. The dimensions of the specimens used are summarised in Table 19.1. Because of the specimen configuration, most of the soil specimens tested in direct tensile tests are reconstituted either by static or dynamic compaction. The specimen is prepared on a horizontal

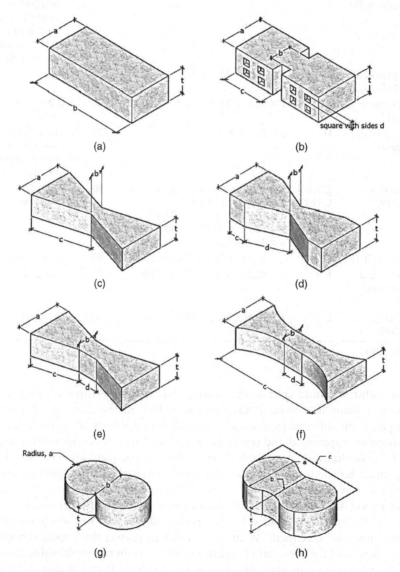

Figure 19.1 Shapes of specimens used in direct tensile tests.

Table 19.1 Dimensions of specimens

Reference	Shape	Dimensions (mm)	t^* (mm)	Loading rate
Nahlawi et al. (2004)	B	a = 100, b = 70, c = 85	100	0.5 – 3.9 mm/min
Stirling et al. (2005)	C	a = 50, b = 38, c = 54	50	0.24 – 1.22 mm/min
Tamrakar et al. (2005, 2007)	G	a = 50, b = 30	50	0.174, 0.342, and 0.882 mm/min, and 0.01, 0.14 and 1.10 mm/min, respectively.
Trabelsi et al. (2010, 2018)	E	a = 46, b = 20, c = 32.7, d = 20	30	0.1 mm/min
Tollenaar et al. (2017)	E	a = 90, b = 24, c = 95.9, d = 24	40	0.0005 – 1 mm/min
Wong et al. (2017)	H	a = 20.5, b = 25	25	Using dead weights on a hanger at 0.1 N increment.
Jhou et al. (2019)	C (10°)	a = 180, b = 148.3, c = 91.4	90	1 and 2.15 mm/min
	C (20°)	a = 180, b = 114.5, c = 95.8		
	C (30°)	a = 180, b = 76.1, c = 103.9		
	E (30°)	a = 180, b = 99.28, c = 80.8, d = 40		
	A	a = 45, b = 180		
Li et al. (2019)	D	a = 100, b = 60, c = 50, d = 53.4	50	0.2 mm/min
Murray and Tarantino (2019)	F	a = 45, b = 15, c = 120, d = 30	15	7 kPa/min
Cai et al. (2020)	E	a = 66, b = 26, c = 98, d = 20	~50	0.5 mm/s

*Thickness

bed within a mould split at the centre, and the tensile force is applied at one end while it is fixed at the other end. Due to the different specimen shapes, some differences in tensile strength between tests are expected. For triangular, trapezoidal and rectangular wedge-shaped specimens, Jhuo et al. (2019) concluded that triangular wedge-shaped specimens are most suitable for direct tensile tests. However, the tensile strengths of triangular-shaped specimens with different angles are different unless they have the same area ratio (Jhuo et al. 2019). The area ratio is defined as the cross-sectional area of the specimen divided by the cross-sectional area of the square base containing the specimen. As the area ratio increases, the tensile strength decreases until the area ratio is greater than a certain threshold value. Most of the apparatuses are specially designed or modified from existing soil testing apparatuses.

19.3.1 Test apparatus

A convenient test apparatus to modify is the direct shear test apparatus. Many researchers have customed built mould for the direct shear test apparatus (Nahlawi et al. 2004; Stirling et al. 2005; Trabelsi et al. 2010 2018; Murray and Tarantino 2019). These moulds consist of two identical boxes to hold the soil specimen. The bottom of the boxes has a guide which slides on roller bearings to reduce friction. One box is fixed while the other box is allowed to move as tension is applied. The tension can be applied by setting the drive motor of the direct shear apparatus to go in the reverse direction to exert a pull. A tension load cell is usually attached to the box that is fixed, and a horizontal LVDT to measure the relative separation of the two boxes is attached to the movable box while its tip rests on the fixed box. When preparing the soil specimen, the two boxes are held together at the sides. Depending on the design, the soil specimen can be prepared in the boxes outside of the direct shear apparatus and then transferred into the direct shear apparatus. The tension load cell and the LVDT are then positioned. The connectors connecting the two boxes are removed just before the start of the test. The tension load cell of 100 up to 300 N capacity (depending on the failure cross-section) and LVDT with a range of 5 mm–20 mm (depending on the specimen shape) are usually adequate for such tests.

19.3.2 Preparation of soil specimen

Procedures for preparing the soil specimens follow that described in Chapter 3. It is less common to conduct tensile tests on intact unsaturated soils, as there are difficulties in trimming the soil specimen. Even if the soil specimen can be trimmed successfully, gaps between the soil specimen and the specimen box need to be filled with a cementing agent to ensure that the soil specimen is tightly secured in the specimen box.

Increasing the cross-section area at the tension plane will generally improve the accuracy of the test. This can be done by either increasing the width of the tension plane and/or the thickness of the specimen. The width and thickness of the specimen used by others are shown in Table 9.1. Tamrakar et al. (2007) found that the tensile strength increases with the number of compacted layers and recommended three to four layers of compaction, each layer of 12.5 mm–16.5 mm thick. Specimen thickness of 25 mm–50 mm is generally suitable.

19.3.3 Test procedures

Test procedures for the test are as follows:

1. Attach the boxes together and prepare the soil specimen in the boxes as mentioned earlier.
2. Set up the boxes with the soil specimen in the direct shear apparatus.

3. Ensure that the unmovable box is securely fixed and move the lead screw of the stepper motor to secure it to the movable box.
4. Set up the tension load cell and the horizontal LVDT.
5. Check that the slack in the system is removed by moving the boxes backwards and forwards by observing the reading on the tension load cell.
6. Set the pulling direction and the predetermined displacement rate (motor speed). Stirling et al. (2015), Tollenaar et al. (2017) and Jhuo et al. (2019) found that the maximum tensile force is not affected by the displacement rate. A slow pulling rate may cause the moisture content of the soil specimen to change during the test.
7. Prepare for the data recording whether automatically or manually.
8. Remove the connectors connecting the two boxes and start the test.
9. The intervals of recording should be such that at least 10 to 15 readings are captured before tensile failure.
10. Continue the test until complete tensile failure where the soil specimen separates into two halves or where three to five readings after the maximum tensile force is recorded.
11. Stop the test and sketch/document the failed soil specimen.
12. Determine the moisture contents of the soil in the tension zone and the soil outside the tension zone.

19.3.4 Calculations

The tensile strength σ_t is given by Equation 19.1.

$$\sigma_t = \frac{P_{max}}{A_f} \qquad (19.1)$$

where
P_{max} = maximum force recorded during the test
A_f = cross-section area of failure plane. This is usually assumed as the cross-sectional area of the narrowed section of the soil specimen

19.4 INDIRECT TENSILE TESTS

Most of the laboratory test apparatuses for soil are commonly designed for compression tests. Hence, the indirect tensile test which uses compression mode of loading is more accessible and attractive to conduct. Two such tests which can be easily performed in the laboratory using existing apparatuses are described in the following sections.

19.4.1 Unconfined penetration test

The unconfined penetration test or the double punch test was developed by Fang and Chen (1971). The unconfined penetration test was originally developed to measure the tensile strength of concrete, but Fang and Chen (1971) showed that it is also applicable to soil. In the unconfined penetration test, a cylindrical specimen is positioned between two punches centred on the top and bottom of the specimen, as illustrated in Figure 19.2. A compression force is applied to the specimen through the punches which eventually punch through the soil when failure occurs. A conical wedge is formed at the top and bottom of the specimen where the punches are in contact with the soil specimen and the specimen is split into two or three pieces (Fang and Chen 1971), as shown in Figure 19.3.

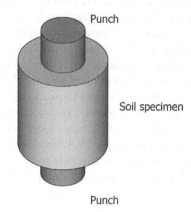

Figure 19.2 Unconfined penetration test.

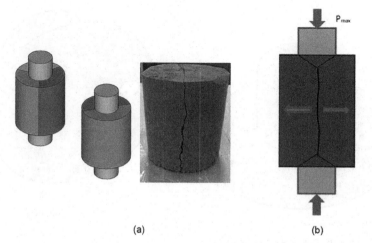

(a)   (b)

Figure 19.3 Failure condition of specimen in unconfined compression test. (a) Specimens failed into two or three pieces. (b) Failure condition.

19.4.1.1 Test apparatus and soil specimen

As the unconfined penetration test is not standardised, there is no specification of the specimen's dimensions. If compacted soils are used, the size of the specimen commonly follows the size of the Proctor mould (102 mm diameter × 113 mm height) or the California Bearing Ratio (CBR) mould (152 mm diameter × 176 mm height). However, specimens should typically have a height-to-diameter ratio of 0.46 to 1.0 (Fang and Fernandez 1981). Either the unconfined compression apparatus or the CBR compression apparatus may be used to provide compression loading on the specimens. The effect of loading rate on the tensile strength of the soil was found to be insignificant (Fang and Chen 1972, Kim et al. 2012). Hence, Fang and Chen (1972) recommend following the loading rate for the unconfined compression test specified in ASTM D2166-16 (2016) of 0.5%–2% per minute. The only additional apparatus that needs to complete the compression apparatus for the unconfined penetration test are the punches. The punches are cylindrical with a flat surface at the end that is in contact with the specimen. Tensile strength from the unconfined penetration test generally increases with the size of the punch (Fang and Chen 1972; Kim et al. 2012). Hence, Fang and Fernandez (1981) proposed that the ratio of the cross-sectional area of the punch to the cross-sectional area of the specimen should be between 0.2 and 0.3. A convenient set of punches that can be easily fabricated is shown in Figure 19.4. Small recesses can be made on the bottom of each punch for strong magnets to be fitted flush. The punch can be easily attached to the unconfined compression machine or CBR machine.

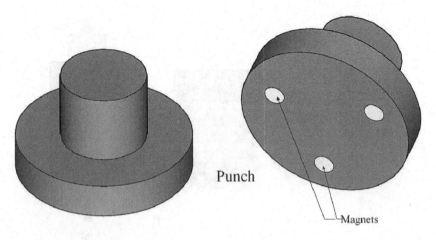

Punch

Magnets

Figure 19.4 Possible design of punch.

19.4.1.2 Test procedures

1. Place the soil specimen in the compression machine such that the soil specimen is centred with the punches directly along its central axis. An acrylic disk of the same diameter as the soil specimen with a central hole of a diameter slightly larger than the punch may be helpful as a guide to position the soil specimen (Figure 19.5).
2. Adjust the compression machine such that the upper punch just makes contact with the soil specimen.
3. Position the LVDT/dial gauge to record the axial deformation and record/zero the readings.
4. Apply the load at a strain rate of 0.5 to 2%/min. The strain rate should be chosen such that the soil fails within 15 minutes.
5. Record the load and deformation at sufficient time intervals to define the shape of the load-deformation curve.
6. Continue loading until the load starts to decline. Usually, three to five readings after the peak load value are sufficient.
7. Make a sketch or take a photo of the failed specimen.
8. Determine the water content of the soil specimen.

19.4.1.3 Calculations

The tensile strength σ_t is calculated using the maximum force P_{max} measured during the unconfined penetration test using Equation 19.2.

$$\sigma_t = \frac{P}{\pi \left(KrH - a^2 \right)}$$

(19.2)

where
K = constant
r = radius of specimen
H = height of specimen
a = radius of punch

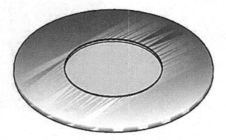

Figure 19.5 Guide to centre punch on soil specimen.

The K value in Equation 19.2 depends on friction angle, compression-tensile strength ratio, the angle of the conical wedge of soil at failure and the size of the punch. Fang and Fernandez (1981) recommend K values of 1.0 and 0.8 for the Proctor and CBR moulds, respectively.

19.4.2 Brazilian tensile strength test

The Brazilian tensile strength (BTS) test is standardised in ASTM D3967-18 (2018) to determine the splitting tensile strength of disk shape specimens by diametral line compression, as shown in Figure 19.5. The BTS test is intended for the testing of rock and concrete (ISRM 1978), but it has been used to test soils. The applicability of the BTS test for soil is often questioned as soil behaviour is more ductile (unlike rock and concrete, which exhibit brittle behaviour), and the specimen experiences much higher compression stresses. Since the Brazilian tensile test assumes that the applied loading is a line load and failure occurs along a diametrical line directly below the line load, tests where the failure departs greatly from the diametrical line are considered unreliable (Erarslan et al. 2012; Li and Wong 2013). The failure plane of the soil specimen at the end of the Brazilian tensile test should be checked for this condition.

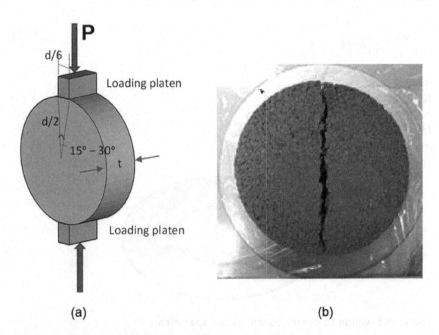

(a)   (b)

Figure 19.6 BTS test. (a) Schematic diagram. (b) Failed soil specimen.

19.4.2.1 Test apparatus and preparation of soil specimen

For soil which is more ductile than rock and concrete, curved bearing platens are more appropriate to limit the contact stresses. Guidelines for the dimensions of the curved bearing platens are given by ASTM D3967-18 (2018). The arc of contact of the curved loading platen with the specimen should be less than 15° or the width of contact should be less than 1/6 of the specimen's diameter. A smaller arc of contact will result in more uniform stresses in the central part of the specimen but increases the compressive stress relative to tensile stress. Gaspar and Jacobsz (2021) show that the ratio of compressive to tensile stresses is 18 for the loading arc of contact of 10° and decreases to 6.3 for the loading arc of contact of 30°. They recommended that the loading arc of contact of 30° be used, as the middle third of the soil specimen will still have relatively uniform tensile stress. For a loading arc of contact of 15°, the ratio of compressive to tensile stresses is about 12.1. Hence, a larger loading arc of contact of 15°–30° should be used when testing a soil specimen. Curved bearing platens can be easily designed for the unconfined compression machine or CBR compression machine.

The thickness (t) and diameter (d) of the cylindrical disk are specified to be in the t/d ratio of between 0.2 and 0.75 in ASTM D3967-16 (2016). The minimum diameter used should be at least ten times greater than the largest mineral grain constituent. This is usually satisfied for rock specimens, where the diameter normally used is 54 mm, the diameter of an NX core. Beckette et al. (2015) investigated the effect of specimen size on the tensile strength from the BTS test. They used 50 mm and 100 mm diameter specimens with a t/d ratio of 0.5 and found that the 100 mm diameter specimens gave more consistent results. Beckette et al. (2015) attributed the inconsistent results of the 50 mm diameter specimens to be due to the cracks forming at the large particles. The maximum particle size of the soil used by Beckette et al. (2015) was about 4–5 mm. This suggests that the diameter of the specimen should be more than 10 times the maximum particle size and 20 times seem to be a good ratio to avoid inconsistent tensile strength results. The t/d ratio of the specimen also affects the tensile strength. Specimens with higher t/d ratios have lower tensile strength (Rocco et al. 1999 and Akin and Likos 2017). Akin and Likos (2017) suggest that the BTS test for compacted clay be conducted on specimens with a t/d ratio of 0.3. Hence, the specimens used in the Brazilian tensile test should have a t/d ratio in the lower range of that recommended in ASTM D3967-16 (2016).

19.4.2.2 Test procedures

The Brazilian tensile test is conducted as follows:

1. Lightly mark a diametral line on the top surface of the soil specimen for use to centre the soil specimen in the compression machine.

2. Position the soil specimen in the compression machine such that the diametral line of the soil specimen is in line with the centre of the top and bottom curved platens.
3. Raise the bottom platen such that the top platen just contacts the soil specimen with little or no load on it.
4. Apply a constant rate of loading such that failure occurs within 1 to 10 min of loading and take at least 10 to 15 readings before failure.
5. Make a sketch or take a photo of the failed specimen.
6. Determine the water content of the soil specimen.

19.4.2.3 Calculations

The BTS test can be calculated using Equation 19.3.

$$\sigma_t = \frac{K'P_{max}}{\pi dt} \tag{19.3}$$

where
K' = constant
P_{max} = maximum load
d = diameter of specimen
t = thickness of disk specimen

The K' value in Equation 19.3 depends on the loading platen used. For flat platen, $K' = 2$ and for curved platen, $K' = 1.272$.

19.5 SUMMARY

Care should be exercised when comparing tensile strengths obtained from direct and indirect tensile tests. Even between different test methods within either direct or indirect tensile tests, the effects of sample preparation, shape and size of specimen cannot be accounted for fully. Tensile tests are a challenge for unsaturated soils which show ductile failure. Lutenegger and Rubin (2008) found that the unconfined penetration test is only applicable for soil compacted at a water content that is near the optimum. At very wet of optimum, the punches simply penetrate the soil without resulting in a true failure, and at very dry of optimum, failure occurs quickly, and as such, the specimen may not fail along a vertical failure plane as assumed. Such a conclusion should also apply to other tensile tests for soils.

Many studies have been carried out to investigate the correlation of the tensile strength obtained from different tests. Fang and Chen (1972) compared the tensile strength obtained from Brazilian tensile tests and unconfined penetration tests for soil and stabilised material and found them to be comparable. However, Kim et al. (2012) explained that a slightly lower tensile strength is obtained from the unconfined penetration test, as the failure plane is the weakest plane unlike the predetermined failure plane in the Brazilian tensile test.

Other tensile tests have been proposed by Wang et al. (2022).

REFERENCES

Akin, I. D., & Likos, W. J. (2017). Brazilian tensile strength testing of compacted clay. *Geotechnical Testing Journal*, 40(4): 608–617. https://doi.org/10.1520/GTJ20160180

Al-Hussaini, M. (1981). Tensile properties of compacted soils. In Yong, R. & Townsend, F. (Eds.), *Laboratory shear strength of soil*. West Conshohocken, PA: ASTM International, pp. 207–225.

Araki, H., Koseki, J., & Sato, T. (2016). Tensile strength of compacted rammed earth materials. *Soils and Foundations*, 56(2): 189–204. https://doi.org/10.1016/j.sandf.2016.02.003

Beckett, C. T. S., Smith, J. C., Ciancio, D., & Augarde, C. E. (2015). Tensile strengths of flocculated compacted unsaturated soils. *Geotechnique Letters*, 5(4): 254–260.

Cai, G., Shi, P., Kong, X. et al. (2020). Experimental study on tensile strength of unsaturated fine sands. *Acta Geotechnica*, 15: 1057–1065. https://doi.org/10.1007/s11440-019-00807-9

Das, B. M., Yen, S., & Dass, R. (1995). Brazilian tensile strength test of lightly cemented sand. *Canadian Geotechnical Journal*, 32(1): 166–171. https://doi.org/10.1139/t95-013

De Souza Villar, L. F., De Campos, T. M. P., Azevedo, R. F., & Zornberg, J. G. 2009. Tensile strength changes under drying and its correlations with total and matric suctions. In *Proceedings of the 17th International Conference on Soil Mechanics and Geotechnical Engineering*, vol. 1, pp. 793–796.

Dexter, A. R. & Kroesbergen, B. (1985). Methodology for determination of tensile strength of soil aggregates. *Journal of Agricultural Engineering Research*, 31(2): 139–147. https://doi.org/10.1016/0021-8634(85)90066-6.

Erarslan, N., Liang, Z. Z., & Williams, D. J. (2012). Experimental and numerical studies on determination of indirect tensile strength of rocks. *Rock Mechanics and Rock Engineering*, 45(5): 739–751.

Fahimifar, A. & Malekpour, M. (2012). Experimental and numerical analysis of indirect and direct tensile strength using fracture mechanics concepts. *Bulletin of Engineering Geology and the Environment*, 71(2): 269–283. https://doi.org/10.1007/s10064-011-0402-7.

Fang, H. Y. & Chen, W. F. (1971). New method for determination of tensile strength of soils. *Highway Research Record*, 345: 62–68.

Fang, H. Y. & Chen, W. F. (1972a). Further study of double-punch test for tensile strength of soils. In *Proceedings of the 3rd Southeast Asian Conference on Soil Engineering*, pp. 211–215.

Fang, H. Y. & Chen, W. F. (1972b). Further study of double-punch test for tensile strength of soils. In *Proceedings of the 3rd Southeast Asian Conference on Soil Engineering*, vol. 1, pp. 236–242.

Fang, H. Y. & Fernandez, J. (1981). Determination of tensile strength of soils by unconfined-penetration test. In Yong, R. & Townsend, F. (Eds.), *Laboratory Shear Strength of Soil*. West Conshohocken, PA: ASTM International, 10.1520/ STP28748S, pp. 130–144. https://doi.org/978-0-8031-4806-2

Gaspar, T. & Jacobsz, S. (2021). Brazilian tensile strength test conducted on ductile unsaturated soil samples. *Geotechnical Testing Journal*, 44. Published ahead of print, 01 May 2021. https://doi.org/10.1520/GTJ20190078.

Jhuo, Y.-S., Guan, Y., Ge, L., Xia, Z., & Kang, X. (2019). Assessment of direct tension tests on compacted sand-clay mixtures. *Journal of Materials in Civil Engineering*. https://doi.org/10.1061/(ASCE)MT.1943-5533.0002873

Kennedy, T. W. & Hudson, W. R. (1968). Application of the indirect tensile test to stabilized materials. *Highway Research Board*, 235: 36–48.

Khrishnayya, A. V. & Eisenstein, Z. (1974). Brazilian tensile test for soils. *Canadian Geotechnical Journal*, 11(4): 632–642. https://doi.org/10.1139/t74-064

Kim, T. H., Kim, T. H., Kang, G. C., & Ge, L. (2012). Factors influencing crack-induced tensile strength of compacted soil. *Journal of Materials in Civil Engineering, American Society of Civil Engineers*, 24(3): 315–320.

Li, D. & Wong, L. N. Y. (2013). The Brazilian disc test for rock mechanics applications: Review and new Insights. *Rock Mechanics and Rock Engineering*, 46(2): 269–287. https://doi.org/10.1007/s00603-012-0257-7

Li, H. D., Tang, C. S., Cheng, Q., Li, S. J., Gong, X. P. & Shi, B. (2019). Tensile strength of clayey soil and the strain analysis based on image processing techniques. *Engineering Geology*, 253(March): 137–148. https://doi.org/10.1016/j.enggeo.2019.03.017

Lutenegger, A. J. & Rubin, A. (2008). Tensile strength of some compacted fine grained soil unsaturated soils, *Advances in Geo-Engineering*, 608, 411–415.

Murray, I. & Tarantino, A. (2019). Mechanisms of failure in saturated and unsaturated clayey geomaterials subjected to (total) tensile stress. *Geotechnique*, 69(8): 701–712. https://doi.org/10.1680/jgeot.17.P.252

Nahlawi, H., Chakrabarti, S., & Kodikara, J. (2004). A direct tensile strength testing method for unsaturated geomaterials. *Geotechnical Testing Journal*, 27(4): 356–361. https://doi.org/10.1520/GTJ11767

Pittaro, G. (2019). Tensile strength behaviour of ground improvement and its importance on deep excavations using deep soil. In *XVII European Conference on Soil Mechanics and Geotechnical Engineering*, pp. 1–8.

Stirling, R. A., Hughes, P., Davie, C. T., & Glendinning, S. (2015). Tensile behaviour of unsaturated compacted clay soils—A direct assessment method. *Applied Clay Science*, 112–113: 123–133. https://doi.org/10.1016/j.clay.2015.04.011

Tamrakar, S. B., Mitachi, T., & Toyosawa, Y. (2007). Measurement of soil tensile strength and factors affecting its measurements. *Soils and Foundations*, 47(5): 911–918. https://doi.org/10.3208/sandf.47.911

Tamrakar, S. B., Toyosawa, Y., Mitachi, T., & Itoh, K. (2005). Tensile strength of compacted and saturated soils using newly developed tensile strength measuring apparatus. *Soils and Foundations*, 45(6): 103–111. https://doi.org/10.3208/sandf.45.103

Tang, G. & Graham, J. (2000). A method for testing tensile strength in unsaturated soils, *Geotechnical Testing Journal*, 23(3): 377–382. https://doi.org/10.1520/GTJ11059J

Tollenaar, R. N., van Paassen, L. A., & Jommi, C. (2017). Experimental evaluation of the effects of pull rate on the tensile behavior of a clay. *Applied Clay Science*, *144*: 131–140. https://doi.org/10.1016/j.clay.2017.04.026

Trabelsi, H., Jamei, M., Guiras, H., Hatem, Z., Romero, E., & Sebastia, O. (2010). Some investigations about the tensile strength and the desiccation process of unsaturated clay. *EPJ Web of Conferences*, *6*: 12005. https://doi.org/10.1051/epjconf/20100612005

Trabelsi, H., Romero, E., & Jamei, M. (2018). Tensile strength during drying of remoulded and compacted clay: The role of fabric and water retention. *Applied Clay Science*, 162: 57–68. https://doi.org/10.1016/j.clay.2018.05.032

Uchida, I. & Matsumoto, R. (1961). On the test of the modulus of rupture of soil sample. *Soils and Foundations*, *2*(1): 51–55.

Vesga, L. F. & Vallejo, L. E. (2006). Direct and indirect tensile tests for measuring equivalent effective stress in a kaolinite clay. In Miller, G. A., Zapata, C. E., Houston, S., & Fredlund, D. G. (Eds.), Proceedings of the Fourth International Conference on Unsaturated Soils April 2–6, 2006 | Carefree, Arizona, ASCE Geotechnical Special Publication No. 47, pp. 1290–1301.

Wang, R., Li, Y., Lv, D., Zhao, W., Zhang, C., Zachert, H., Eichhoff, G., & Beroya-Eitner, M. A. (2022). Comparison of test methods for determining the tensile strength of soil and weak rocks. *Frontiers in Earth Science*, *10*. https://doi.org/10.3389/feart.2022.835851

Wong, C. K., Wan, R. G., & Wong, R. C. K. (2017). Tensile and shear failure behaviour of compacted clay – hybrid failure mode. *International Journal of Geotechnical Engineering*. https://doi.org/10.1080/19386362.2017.1408242

Chapter 20

Wave velocities

20.1 BACKGROUND

Piezoelectric crystals were first used by Lawrence (1963, 1965) to generate one-dimensional compression waves through sand and glass beads, while piezoceramic bender elements were first used by Shirley (1978) to generate shear waves in laboratory tests. Since then, piezoelectric crystals and bender elements have been used in laboratory testing of soils and rocks. In a pulse transmission test, a wave is generated at one end of the soil specimen, and the time t required for the wave to travel through a soil specimen is recorded. The wave velocity V can be easily determined using Equation 20.1.

$$V = \frac{L}{t} \qquad (20.1)$$

where L = length of the soil specimen.

The measurement of the wave velocities provides the small-strain Young's modulus or shear modulus. For unsaturated soils, the compression wave velocity can also provide information about the degree of saturation. Pulse transmission test appears relatively easy to conduct, and many pulse transmission test kits are available commercially. Currently, two pulse transmission methods are used in soil tests: bender element test and ultrasonic test. The literature shows that the bender element test is the more common of the two pulse transmission methods for testing soils while the ultrasonic test is more common for testing rocks and concrete. Pulse transmission tests for soils are usually conducted on cylindrical specimens.

This chapter describes the bender element test and ultrasonic test used in unsaturated soil tests. The bender element test for soils has recently been standardised in ASTM D8295-19 (2019). The ultrasonic test for soils has not been standardised yet, but a standard exists for the testing of rocks (ASTM D2845-08) which was withdrawn in 2017.

20.2 RELATED STANDARDS

1. ASTM D2845-08 (Withdrawn in 2017). Standard Test Method for Laboratory Determination of Pulse Velocities and Ultrasonic Elastic Constants of Rock.
2. ASTM D8295-19 Standard Test Method for Determination of Shear Wave Velocity and Initial Shear Modulus in Soil Specimens Using Bender Elements.

20.3 BENDER ELEMENT TEST

The bender element is an electromechanical transducer and can act both as a transmitter and a receiver. Each bender element is a three-layer system: two sheets of a piezoelectric ceramic material such as lead zirconate titanate, barium titanate or lead titanate sandwich a centre shim of brass, stainless steel or other ferrous nickel alloys to give it strength. Hence, the bender element is a bimorph. When a bender element is given a voltage pulse, it changes its shape and generates a mechanical excitation, hence acting as a transmitter. When mechanically excited, a bender element converts the mechanical energy into an electrical output, hence acting as a receiver.

Two types of piezoceramic elements are available: x-poled and y-poled piezoceramic elements (Figure 20.1). The x-poled or y-poled refers to the directions of the polarisation for the two piezoelectric ceramic layers. If the polarisations of the two layers are pointing towards or away from each other, the bender element is x-poled. In the y-poled piezoceramic element, the polarisations for the two piezoelectric ceramic layers are pointing in the same direction. These piezoceramic elements can be wired in either a series or parallel configuration. In the series configuration, the excitation voltage is applied across the two piezoelectric ceramic layers, as shown in Figures 20.2(a) and (d). In the parallel configuration, the excitation voltage is applied to the two piezoelectric ceramic layers and the centre shim, as shown in Figures 20.2(b) and (c). In Figures 20.2(a) and (b), the piezoceramic element flexes and transmits a shear wave (S-wave). In Figures 20.2(c) and (d), the piezoceramic element extends and transmits a compression wave (P-wave). Hence, a single pair of bender elements can be used for both S-wave and P-wave velocity measurements by changing the wiring configuration (Lings and Greening 2001; Leong et al. 2009).

In terms of movement Δ per unit of applied voltage, the parallel configuration gives a larger movement by a factor of two as compared with the

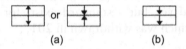

(a)   (b)

Figure 20.1 Types of piezoceramic elements used as bender elements. (a) x-poled. (b) y-poled.

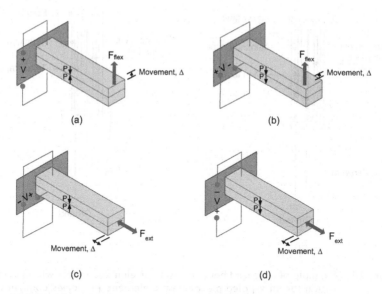

Figure 20.2 Bender element wiring configuration using x-poled and y-poled piezoceramic elements. (a) x-poled, series configuration. (b) y-poled, parallel configuration. (c) x-poled, parallel configuration. (d) y-poled, series configuration.

series configuration because of a reduction in impedance of 4:1 (Germano 2002). In other words, the parallel configuration requires only half of the voltage needed for the series configuration to achieve the same amount of movement. Hence, the parallel configuration is more suitable for the transmitter, as it gives a stronger mechanical motion for a given excitation voltage, whereas the series configuration is more suitable as a receiver, as it gives a higher output voltage for a given mechanical excitation.

Bender elements have been used in a number of soil tests, such as the triaxial test (Bates 1989; Brignoli et al. 1996; Pennington et al. 2001), oedometer test (Dyvik and Madshus 1985; Kawaguchi et al. 2001), direct shear test (Dyvik and Madshus 1985) and resonant column test (Dyvik and Madshus 1985; Gu et al. 2015; Hoyos et al. 2015). Recently, the bender element test has been standardised in ASTM D8295-19. However, the test standard is only for S-wave and describes bender element usage in triaxial type tests, although it allows for use in other types of laboratory tests such as the direct simple shear test, oedometer test and unconfined soil specimens.

20.3.1 Test set-up

Bender element test kits are now available commercially. Bender element test kits bought commercially usually come with the bender element embedded in platen and are ready for use once the platen is fixed into the

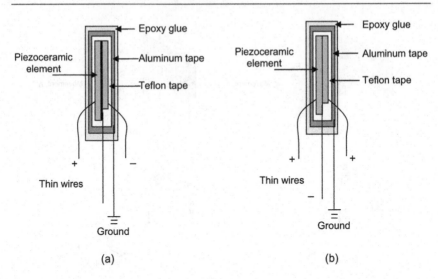

Figure 20.3 Example of prepared bender-extender element and its wiring configuration for an x-poled piezoceramic element. (a) Series configuration. (b) Parallel configuration.

testing apparatus. Bender elements can also be fabricated in-house. A schematic diagram of how a prepared bender element should look is shown in Figure 20.3. Wires are soldered to each of the three layers. To solder the wire to the centre shim, a small part of the piezoelectric ceramic is removed using a file to expose the centre shim. Once the wires are soldered, the resistance and capacitance of the bender element for pairs of wires should be checked with a multi-meter. The multi-meter should always show an open circuit, and the capacitance of the piezoelectric ceramic to metal shim is larger than that between the piezoelectric ceramic sheets. Once the check is completed, the bender element is then wrapped in a layer of Teflon tape followed by a layer of aluminium foil. The aluminium foil acts as an electromagnetic shield from background electrical noise. For better shielding, a grounding wire is attached to the aluminium foil, and epoxy is applied to coat the entire assembly. The importance of the grounding wire is shown in Figure 20.5. Figure 20.5 shows crosstalk reproducing the input signal in the receiver bender element's signal which was eliminated with grounding. Epoxy is applied to the bender element in thin layers, allowing sufficient time in between for the epoxy to set to ensure that it is completely waterproofed. The bender element is then ready to be fixed into a platen. To ensure that the bender element is completely insulated from the metal platen, silicone rubber is used to fix the bender element in the platen, as shown in Figure 20.4. In a triaxial set-up, the transmitter bender element is fixed to the bottom platen, which is rigidly attached to the triaxial cell, and

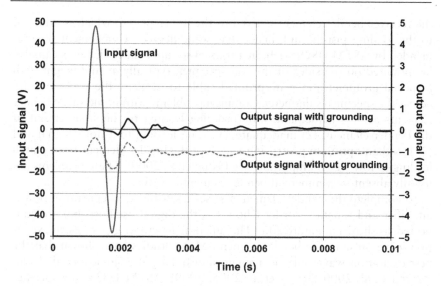

Figure 20.4 Typical output signals with and without grounding. (modified from Leong et al. 2009).

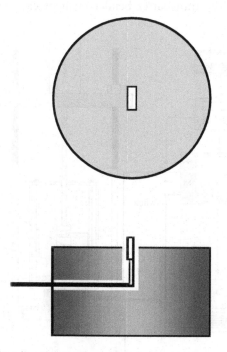

Figure 20.5 Fixing bender element in a platen.

the receiver bender element is fixed to the upper platen, which is attached to the loading piston, and, hence, has some degree of movement (recommended by ASTM D8295-19). In a triaxial set-up, bender elements can also be mounted on the sides of the soil specimen typically at mid-height with a special grommet and waterproofed cable to measure the wave velocity across the specimen's diameter. Common problems with home-made bender elements are (1) inadequate waterproofing leading to the short circuit of the bender element, (2) inadequate shielding leading to a noisy signal and (3) less than optimum protrusion length of the bender element from platen leading to low signal amplitude (see Leong et al. 2009). These problems are usually absent in commercial bender element test kits.

To complete the bender element test set-up, a function generator, power amplifier and a data recorder (shown as the digital oscilloscope) are connected, as shown in Figure 20.6. The function generator generates an excitation wave pulse to the bender element. Many studies have shown that the best excitation wave pulse is a single sinusoidal pulse (Jovicic et al. 1996; Blewett et al. 2000; Leong et al. 2005, 2009). ASTM D8295-19 specifies that the function generator should be capable of generating a single-period sine wave pulse with a frequency between 1 and 50 kHz. Usually, the wave pulse from the generator is amplified to a higher voltage through a power amplifier so that the transmitter bender element can produce a stronger

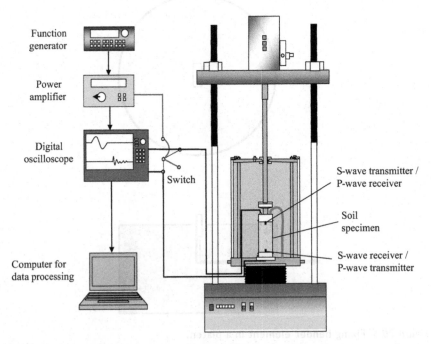

Figure 20.6 Typical bender element set-up in triaxial test apparatus.(from Leong et al. 2009).

movement. It is possible to add a power amplifier to the receiver bender element, but this may distort the signal and introduce errors in the interpretation. ASTM D8295-19 recommends avoiding the use of a power amplifier and signal filter between the bender element and the data recorder, and if necessary, the power amplifier should only be added between the function generator and the transmitter bender element before the connection to the data recorder. For the data recorder, ASTM D8295-19 recommends that it should have a time resolution of one microsecond (1 μs) and an amplitude resolution of at least 10 microvolts (10^{-5}V) for S-wave signal. Leong et al. (2009) found that a high-resolution oscilloscope ($\geq$ 12 bits) is needed to improve the resolution of the compression wave signal, as the compression wave signal has a smaller amplitude than the shear wave. In commercial bender element test kits, the function generator, power amplifier and data recorder (usually an oscilloscope interface) are generally integrated as a single unit.

20.3.2 Test procedures

As mentioned earlier, bender elements can be incorporated into a number of soil tests. The soil specimen is prepared in the usual manner, be it undisturbed or compacted for the target soil test. Determine the mass and dimensions of the soil specimen to calculate its density. Determine the moisture content of the leftover soil from the same sample or trimmings. For preparing a reconstituted or compacted soil specimen, see Chapter 3.

The selection of wave pulse frequency to the transmitter bender element and the resolution of the data recorder should consider the distance that the wave will travel from the tip of the transmitter bender element to the tip of the receiver bender element. This distance is shorter than the specimen length, as the bender elements protrude into the soil specimen. This distance is denoted as the tip-to-tip distance or L_{tt}. The alignment of the bender elements is important, and care needs to be taken that they are parallel and colinear at both ends of the soil specimen. A check can be made by bringing the tips of the two bender elements to be in contact and sending an excitation wave pulse to the transmitter bender element and recording the signal from the receiver bender element. There should not be any time delay between the two signals. If there is a noticeable delay, the delay should be noted and can be used to subtract from the measured travel time. The check should be repeated by interchanging the connections such that the transmitter bender element is acting as the receiver bender element, and the receiver bender element is acting as the transmitter bender element. When conducting such checks, avoid touching the bender elements or their attachments in case there is a short circuit, especially if a power amplifier is used to amplify the excitation wave pulse. ASTM D8295-19 recommends using a square wave pulse. A soft clicking sound could be heard if the bender elements are working properly. A functional check of the bender elements can also be done by lightly tapping the sides of the bender element and observing that a signal is registered on the data recorder.

After the functional check, place the soil specimen between the bender elements. Soft soil specimens should have no problem, as the bender elements should easily penetrate into the soil specimens. For stiffer soil specimens, a small slot needs to be made on the specimen face to allow the bender elements to be inserted. However, precaution needs to be exercised, as the quality of the bender element test relies heavily on good contact conditions between the bender elements and the soil specimen. ASTM D8295-19 recommends using "some material (soil particles, grout, and the like with similar stiffness to the test specimen)" to fill the slots before inserting the bender elements.

Once the soil specimen is set up, a bender element test check should be made under the unconfined condition that the bender elements are working properly before proceeding with the rest of the test set-up.

The test procedures in a bender element test are as follows:

1. Determine the specimen dimension across which the wave is travelling and subtract the protrusion lengths of the bender elements to obtain L_{tt}. The test set-up should allow measurement of any changes of this specimen dimension during the test.
2. Set the function generator to provide a single excitation sinusoidal wave to the transmitter bender element with a suitable frequency and amplitude. The amplitude of the excitation wave from the function generator can be amplified via the power amplifier, but the use of high voltages may deteriorate and shorten the life of the bender element (ASTM 8295-19).
3. Programme the data recorder to trigger recording on the initial rise of the transmitter bender element signal with a pre-trigger to record the signal prior to the start of the trigger so that the complete transmitter bender element signal is recorded.
4. Set up the display on the data recorder such that both transmitter and receiver bender element signals are shown on the same display with different vertical scales (voltage) and the same time scale which depends on the excitation wave frequency. The transmitter bender element signal has a much higher voltage range compared to the receiver bender element signal. Ensure that the voltage input to the data recorder is not exceeded when connecting the function generator through a power amplifier to the data recorder.
5. Start the function generator and record the bender elements' signals. If the receiver bender element signal is weak, increase the voltage of the excitation wave to the transmitter bender element by adjusting the power amplifier. The transmitter signal and the main part of the receiver signal should have about the same frequency. Adjust the

vertical and horizontal scales on the display of the data recorder such that the transmitter and receiver bender element signals can be clearly seen. If required, adjust the frequency of the function generator to achieve a receiver bender element signal with the main part that is easiest to read. Repeat the measurement thrice, and store the signals for data interpretation.

6. If the receiver bender element signal is noisy and no amount of adjusting the input wave frequency or voltage was able to produce a "clean" signal, some post-test signal processing without altering the characteristics of the signal can be considered. For this case, ASTM D8295-19 suggests using a time domain averaging function on the data recorder to average 10 to 100 records to "clean" the signal. The averaging function for 10 to 100 records can be performed by setting the function generator to generate a single sinusoidal wave pulse with a time delay between each of them. The time delay should be long enough that it does not interfere with each of the receiver bender element signals recorded. ASTM D8295-19 suggests that a conservative time delay for triaxial test specimen is 0.1s.

7. The quality of the bender element signals should improve with increasing effective or net confining pressure as the contact condition between the soil specimen and bender element improves.

8. If no receiver bender element signal is recorded despite increasing the voltage of the excitation wave to the transmitter bender element, there are two possibilities: (1) the transmitted wave does not have sufficient energy to travel to the receiver bender element, or (2) the receiver bender element is faulty. For case (1), a shorter specimen can be used, and the bender element test repeated if the test is solely to measure the wave velocities of the soil specimen. For case (2), the wires connected to the bender element can be checked for resistance and capacitance as mentioned earlier. In general, P-wave signals are weaker than S-wave signals, as the bender element movement acting as an extender is much smaller than acting as a bender (Leong et al. 2009).

20.3.3 Interpretations

Interpretation of bender element signals that are affected by poor contact, electronic drift and crosstalk (Lee and Santamarina 2005; Rio 2006) is difficult. Electronic drift and crosstalk can be largely resolved with good bender element manufacturing practices. Poor contact conditions may occasionally arise due to the stiffness of the soil specimen. However, there may still be ambiguity in determining the arrival time of waves for good quality bender element signals (Brignoli et al. 1996; Arulnathan et al. 1998; Arroyo et al. 2003b; Leong et al. 2009). The bender element signal can be interpreted either in the time domain or in the frequency domain. The time-domain methods determine the wave travel time t and the P- and S-wave velocities, V_p and V_s respectively, are calculated using Equation 20.2.

$$V_p \text{ or } V_s = \frac{L_{tt}}{t}$$ (20.2)

where
L_{tt} = distance between tip-to-tip of bender elements (m)
t = travel time of wave

The receiver bender element signal for the shear wave is affected by several factors, in particular, the parasitic P-wave giving rise to near-field effects and, hence, determining the travel time for S-wave is not straightforward. Most bender elements tests in the laboratory involve a small soil specimen, and the specimen's boundary reflects and refracts the wave as it travels from the transmitter bender element to the receiver bender element (Arulnathan et al. 1998). Each reflection or refraction may produce P-waves, and this interference distorts the receiver bender element's signal. A typical receiver bender element signal for the S-wave is shown in Figure 20.7.

According to Figure 20.7, the start-to-start travel time can be determined using different start points on the receiver signal, A, B or C. The method adopted by ASTM D8295-19 is to use point C to determine the travel time using the start-to-start method, t_{s-s}. The second method to determine travel time is to use the characteristic points (peak-to-peak method) t_{p-p} as indicated in Figure 20.7. This is the other method adopted by ASTM D8295-19. In fact, this is the preferred method of ASTM D8295-19, as there is less ambiguity in selecting the peak point of the main receiver signal compared to the wave arrival point due to the near-field effects in the receiver signal. To reduce the near-field effects, Leong et al. (2009) suggested checking the L_{tt}/λ ratio to be above 3.33 when determining the S-wave travel time using

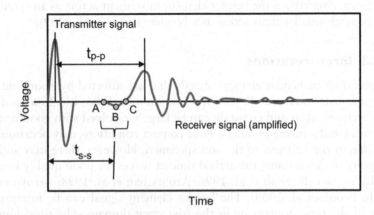

Figure 20.7 Travel time determination for S-wave.

the start-to-start method. The ratio can be calculated easily using the input excitation frequency f_{in} for the transmitter and the measured travel time Δt as given in Equation 20.3.

$$\frac{L_{tt}}{\lambda} = \left(\frac{L_{tt}}{V}\right) f_{in} = f_{in} \Delta t \tag{20.3}$$

where
λ = wavelength
V = wave velocity

There are two main frequency-domain methods to determine the wave travel time. These are the cross-correlation and cross-spectrum methods. In the cross-correlation method, the cross-correlation or similarities between the transmitter and receiver signals is calculated (Santamarina and Fratta 2005), and the time between the maximum amplitudes of the transmitter and the receiver signals is taken as the arrival time. The cross-spectrum method is an extension of the cross-correlation method. In the cross-spectrum method, phase analysis or "unwrapping" is conducted on the cross-power spectrum derived from the frequency spectra (Viggiani and Atkinson 1995). The arrival time is calculated from the slope of amplitude and phase angle with frequency. Implicit in the use of cross-correlation is that the receiver signal has the same shape or frequency as the transmitter signal (Jovicic et al. 1996; Santamarina and Fam 1997). This is only true when the excitation frequency exceeds the natural frequency of the bender/extender element (Rio 2006; Marjanovic and Germaine 2013). However, the transmitted signals are usually distorted by the reflected waves.

Although the frequency-domain methods are considered to be the most accurate (Viggiani and Atkinson 1995), they are more difficult to apply compared to the time-domain methods (Jovicic et al. 1996; Arroyo et al. 2003a, b). In addition, the comparison of S-wave velocities obtained from bender element tests using time- and frequency-domain methods and those obtained from resonant column tests show that the time-domain methods gave better agreement with the resonant column test results (Camacho-Tauta et al. 2013). This could be the reason why only the time-domain methods are adopted in ASTM D8295-19.

No issue has been reported on determining P-wave travel time, as it can be picked based on the first deflection of the receiver signal, i.e., a point equivalent to point A in Figure 20.7.

The P-wave and S-wave velocities are associated with the small-strain Young's modulus E_0 and small-strain shear modulus G_0 given in Equations 20.4 and 20.5, respectively.

$$E_0 = \frac{\rho V_s^2 \left(3V_p^2 - 4V_s^2\right)}{\left(V_p^2 - V_s^2\right)} \tag{20.4}$$

$$G_0 = \rho V_s^2 \qquad (20.5)$$

If both V_p and V_s are measured, other elastic constants such as Poisson's ratio ν (Equation 20.6), Lamé's constant (20.7) and bulk modulus K (20.8) can be determined.

$$\text{Poisson's ratio,} \nu = \frac{V_p^2 - 2V_s^2}{2\left(V_p^2 - V_s^2\right)} \qquad (20.6)$$

$$\text{Lam\'e's constant} = \rho\left(V_p^2 - V_s^2\right) \qquad (20.7)$$

$$\text{Bulk modulus,} K = \rho \frac{\left(3V_p^2 - 4V_s^2\right)}{3} \qquad (20.8)$$

20.4 ULTRASONIC TEST

The ultrasonic test makes use of piezoelectric crystals to generate compression or shear waves. The range of frequencies of ultrasonic waves is from 20 kHz to 1 GHz (Vary 1980). Unlike the bender element test, in the ultrasonic test, the piezoelectric crystals are bonded to the inside face of the platen and do not come into direct contact with the soil specimen. This arrangement avoids the problems commonly associated with the use of bender elements, such as difficulty to insert bender elements into stiff soil specimens, damage to bender elements due to insertion and waterproofing problems leading to short-circuiting of the bender elements. However, ultrasonic waves have smaller amplitude and high attenuation, and there is a time delay as the ultrasonic waves travel through the face of the platen. The ultrasonic test has been standardised for rocks to measure compression and shear wave velocities and elastic constants at small strain in ASTM D5285-08, but the standard was withdrawn in 2017.

For the ultrasonic test, there are three modes of operation (Kundu 2000): pulse-echo, through-transmission and pitch-catch. A single transducer used in pulse-echo mode acts as both transmitter and receiver, while two transducers are used in through-transmission and pitch-catch modes, where one transducer acts as the transmitter and the other transducer acts as the receiver. In the through-transmission mode, the transducers are placed on the opposite end of a specimen, while in the pitch-catch mode, the two transducers are placed on the same side of the specimen. The through-transmission mode is more suitable for measuring wave velocities and attenuations of geomaterials (Krautkrämer and Krautkrämer 1990) and hence only the through-transmission ultrasonic test is described here.

20.4.1 Test set-up

The ultrasonic test system is more commonly used to test specimens in the unconfined condition. For soils, the ultrasonic system is more commonly used in a triaxial test setting. A schematic diagram of such a system is shown in Figure 20.8. The ultrasonic test system consists of a pair of platens where the piezoelectric crystals are bonded to the inside face of the hollow platen, a pulse generator and receiver to generate and receive P or S-wave and a high-speed data acquisition system and a personal computer.

The P- and S-wave transducers consist of piezoelectric crystals that are bonded to the inside face of the platen. One possible arrangement of the P and S-wave piezoelectric crystals within the platens is shown schematically in Figure 20.9. There is one P-wave piezoelectric crystal and three pairs of S-wave piezoelectric crystals shown in Figure 20.9. Each pair of S-wave piezoelectric crystals is placed diametrically opposite, and they are excited

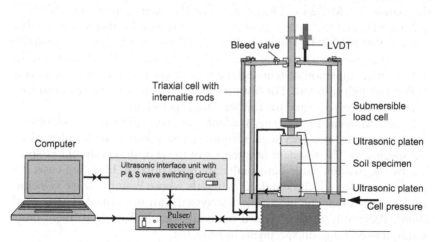

Figure 20.8 Typical ultrasonic test set-up in a triaxial apparatus(modified from Leong et al. 2004).

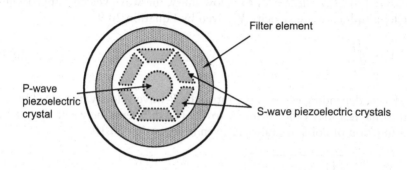

Figure 20.9 One possible arrangement of piezoelectric crystals for P- and S-waves.

in turn to generate a torsional S-wave. Other S-wave transducers may contain only two pairs of piezoelectric crystals (e.g., Nakagawa et al. 1996). The piezoelectric crystals change shape when a voltage pulse is applied and generates a voltage when they are mechanically excited. The P-wave piezoelectric crystal is a thickness-expander crystal, while the S-wave crystals are thickness-shear crystals. The P- and S-wave piezoelectric crystals have a natural frequency of between 200 and 1000 kHz. Depending on the dimensions and the way they are bonded to the platen, their resonant frequencies in the platen may be different and may vary from system to system. Therefore, either ultrasonic platen may act as the transmitter or receiver. The pulser is excited by a single voltage spike. A big difference between the bender element test and the ultrasonic test is that the P- and S-wave piezoelectric crystals have only one frequency of excitation at or near their natural frequencies. The excitation voltage used for the ultrasonic platens is usually higher than in the bender element test, as the waves have to transmit through the face of the platen. ASTM D2845-08 specifies that the excitation voltage should be at least 50V, and the data acquisition system should be able to sample data at rates of 0.2–50 kHz. Having a data recorder with a higher data sampling rate is desirable to increase the resolution of the signals. Usually, the settings of the data acquisition system and the type of wave to transmit are all controlled through software. For testing soil specimens, it is important to have at least the receiver signal for further data processing.

Many ultrasonic test systems available commercially are for rock or concrete specimens, although some ultrasonic test systems can be adapted for soil specimens. The resonant frequencies of the transducers should be specified by the manufacturer but usually are not available. In such a case, it is important to have some standard materials whose P- and S-wave velocities are known so that the ultrasonic test system can be verified (Leong et al. 2004). Preferably these standard materials should have a stiffness range within those of the soil specimens to be tested.

Leong et al. (2004) found that the ratio of the diameter of the specimen and wavelength of P-wave (D/λ_p) affects the type of P-wave velocity that is measured. As D/λ_p decreases to below unity, the wave velocity approaches the longitudinal wave velocity V_L given by Equation 20.9.

$$V_L = \sqrt{\frac{E}{\rho}} \tag{20.9}$$

Hence, depending on the D/λ_p ratio, the P-wave velocity measured by an ultrasonic test may range from V_L to V_p. The difference between V_p and V_L is a function of Poisson's ratio, v, given by Equation 20.10.

$$\frac{V_P}{V_L} = \sqrt{\frac{(1-v^2)}{(1+v)(1-2v)}} \tag{20.10}$$

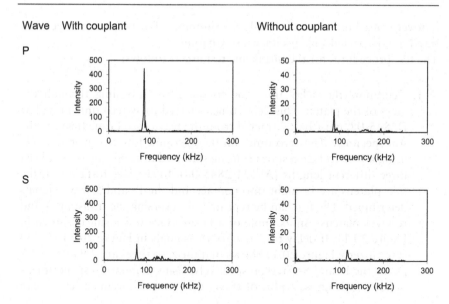

Figure 20.10 Frequency spectra of ultrasonic tests on saturated kaolin with and without acoustic couplant.

The ultrasonic platens are more difficult than the bender elements to incorporate in soil tests which require water or air drainage, as the area of the platen where the piezoelectric crystals are bonded below cannot have a filter element, as illustrated in Figure 20.8. The reduced area of the filter element will reduce the rate at which water can drain from the soil specimen.

It has also been found that signal quality in ultrasonic tests improves with the application of an acoustic couplant between the ultrasonic platens and the soil specimen. The acoustic couplant is a fluid such as water, glycerin, grease, petroleum jelly, phenyl salicylate or resin. Figure 20.10 shows the frequency spectra of P- and S-waves from an ultrasonic test on a saturated kaolin specimen. Figure 20.10 shows that the P-wave signal is stronger than the S-wave signal in the ultrasonic test with or without the acoustic couplant and that the signal intensity improves by at least ten times with the use of the acoustic couplant. In this case, water is not sufficient as an acoustic couplant. However, the use of an acoustic couplant may interfere with the other test that is being performed on the same specimen as the ultrasonic test.

20.4.2 Test procedures

The soil specimen is prepared in the usual manner, be it intact/undisturbed or reconstituted/compacted. Determine the mass and dimensions of the soil specimen to calculate its density. Determine the moisture content of the leftover soil from the same sample or trimmings. For preparing a reconstituted or compacted soil specimen, see Chapter 3.

The test procedures for the ultrasonic test are as follows:

1. Determine the delay time due to the wave travelling through the faces of the platen. This delay time, referred to as zero time in ASTM D2845-08, must be subtracted from the wave travel time through the soil specimen. The zero time can be determined by two methods: (1) using a face-to-face test or (2) using standard test specimens of at least three different lengths (ASTM 2845-08). In the face-to-face test, the two platens are brought into contact and the P- and S-wave signals determined. The test can be repeated by reversing the transmitter and receiver platens. An example of a face-to-face test result is shown in Figure 20.11. If delay time is zero, the signals in Figure 20.11 should start at zero. Figure 20.11 shows that the zero times are 2.9 μs and 4.8 μs for the P- and S-waves, respectively. Using standard test specimens of various lengths, a plot of travel time versus specimen length can

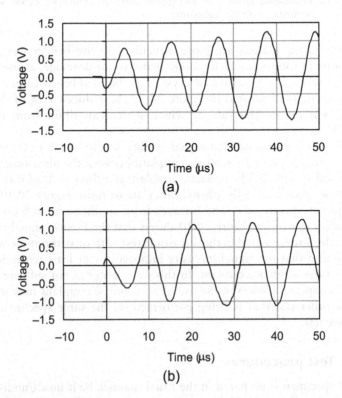

Figure 20.11 P- and S-wave signals from face-to-face tests(from Leong et al., 2004). (a) P wave. (b) S wave.

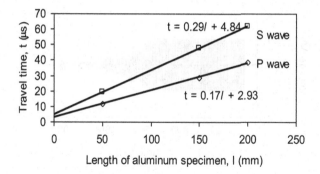

Figure 20.12 Travel times of P- and S-waves as a function of specimen length(from Leong et al. 2004).

be made, as shown for aluminium specimens of lengths 50, 150, and 200 mm in Figure 20.12. Figure 20.12 shows that the zero times for P-wave and S-wave are 2.93 µs and 4.84 µs for the P- and S-waves, respectively. Hence, either method will suffice. Both methods are mentioned in ASTM D2845-08.

2. Position the soil specimen on the bottom platen. For a triaxial test with confining pressure, place a rubber membrane over the soil specimen with the help of a split membrane stretcher and secure the bottom of the rubber membrane to the lower platen with an O-ring. Slowly bring the top platen to contact the top of the soil specimen and ensure that the top platen is contacting the soil specimen fully by checking the load on the piston. A trial ultrasonic test can be conducted to ensure that the wave signals are in order before securing the top of the rubber membrane to the top platen with an O-ring.

3. Carry on with the rest of the test set-up.

4. Input the necessary parameters (sample dimensions, output voltage of pulse generator, pulse type number of pulse, gain, sampling rate, filter, etc.) for the test into the software.

5. Conduct the ultrasonic test. Typical P- and S-wave signals are shown in Figure 20.13. If the signal is not clear enough to detect the first arrival of the wave, increase the voltage of the pulse generator and/or gain of the amplifier. Repeat the test to obtain at least three measurements of travel times. ASTM D2845-08 recommends that the travel times for P-wave should be measured to within 1% and the travel time for the S-wave should be measured to within 2%.

6. Repeat steps 4 and 5 whenever wave velocities are to be measured during the test, taking note of the length of the soil specimen whenever the ultrasonic test is carried out.

7. After the test, determine the mass and water content of the soil specimen.

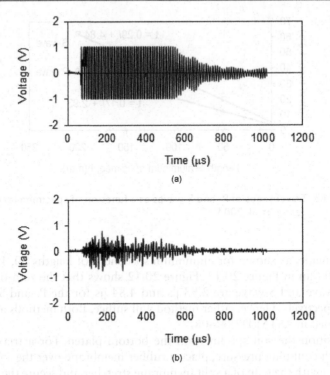

Figure 20.13 Typical P- and S-wave signals for a kaolin specimen from ultrasonic test. (a) P-wave signal. (b) S-wave signal.

20.4.3 Calculations

The P- and S-wave velocities (V_p and V_s) are calculated using Equation 20.11.

$$V_p \text{ or } V_s = \frac{L}{t - t_0} \qquad (20.11)$$

where t_0 = zero time.

Other elastic constants E_0, G_0, Poisson's ratio ν, Lamé's constant and bulk modulus K can be determined using Equations 20.4 to 20.8, respectively.

REFERENCES

Arroyo, M., Wood, D. M., & Greening, P. D. (2002). Source near-field effects and pulse tests in soil samples. *Géotechnique*, 53(3): 337–345.

Arroyo, M., Greening, P., & Muir Wood, D. (2003a). An estimate of uncertainty on current pulse test practice. *Italian Geotechnical Journal*, 37: 17–35.

Arroyo, M., Wood, D. M., & Greening, P. D. (2003b). Source near-field effects and pulse tests in soil samples. *Géotechnique*, *53*(3): 337–345.

Arulnathan, R., Boulager, R. W., & Riemer, M. F. (1998). Analysis of bender element tests. *Geotechnical Testing Journal*, *21*(2): 120–131.

Bates, C. R. (1989). Dynamic soil property measurement during triaxial testing. *Geotechnique*, *39*(4): 721–726.

Blewett, J., Blewett, I. J., & Woodward, P. K. (2000). Phase and amplitude responses associated with the measurement of shear-wave velocity in sand by bender elements. *Canadian Geotechnical Journal*, *37*(6): 1348–1357.

Brignoli, E. G. M., Gotti, M., & Stokoe, K. H., II. (1996). Measurement of shear waves in laboratory specimens by means of piezoelectric transducers. *Geotechnical Testing Journal*, *19*(4): 384–397.

Camacho-Tauta, J., Reyes, O., & Alvarez, J. (2013). Comparison between resonant-column and bender element tests on three types of soils. *Dyna (Medellin, Colombia)*, *182*: 163–172.

Dyvik, R. & Madshus, C. (1985). *Lab measurement of Gmax using bender elements*. Norwegian Geotechnical Institute Publication, No. 161, Oslo. pp. 186–196.

Germano, C. P. (2002). Flexure mode piezoelectric transducers. Morgan Electro Ceramics, Southampton, UK: Technical Publication TP-218.

Gu, X., Yang, J., Huang, M., & Gao, G. (2015). Bender element tests in dry and saturated sand: Signal interpretation and result comparison. *Soils and Foundations*, *55*(5): 951–962. https://doi.org/10.1016/j.sandf.2015.09.002.

Hoyos, L. R., Suescún-Florez, E. A., & Puppala, A. J. (2015). Stiffness of intermediate unsaturated soil from simultaneous suction-controlled resonant column and bender element testing. *Engineering Geology*, *188*: 10–28. https://doi.org/10.1016/j.enggeo.2015.01.014.

Jovičič, V., Coop, M. R., & Simič, M. (1996). Objective criteria for determining Gmax from bender element tests. *Geotechnique*, *46*(2): 357–362.

Kawaguchi, T., Mitachi, T., & Shibuya, S. (2001). Evaluation of shear wave travel time in laboratory bender element test. In *Proceeding of Fifteenth International Conference on Soil Mechanics and Geotechnical Engineering*, Istanbul. Balkema, pp. 155–158.

Krautkrämer, J. & Krautkrämer, H. (1990). *Ultrasonic testing of materials*. 4th ed. Berlin, Heidelberg: Springer.

Kundu, T. (2000). *Nondestructive testing techniques for material characterization: Modeling in geomechanics*. Chichester, UK: John Wiley and Sons Ltd.

Lee, J.-S. & Santamarina, J. (2005). Bender elements: Performance and signal interpretation. *Journal of Geotechnical and Geoenvironmental Engineering*, *131*(9): 1063–1070.

Leong, E. C., Cahyadi, J., & Rahardjo, H. (2009). Measuring shear and compression wave velocities of soil using bender – extender elements. *Canadian Geotechnical Journal*, *46*(7): 792–812.

Leong, E. C., Yeo, S. H., & Rahardjo, H. (2004). Measurement of wave velocities and attenuation using an ultrasonic test system. *Canadian Geotechnical Journal*, *41*(5): 844–860.

Leong, E. C., Yeo, S. H., & Rahardjo, H. (2005). Measuring shear wave velocity using bender elements. *Geotechnical Testing Journal*, *28*(5): 488–498.

Lings, M. L. & Greening, P. D. (2001). A novel bender/extender element for soil testing. *Geotechnique*, *51*(8): 713–717.

Marjanovic, J. & Germaine, J. (2013). Experimental study investigating the effects of setup conditions on Bender element velocity results. *Geotechnical Testing Journal*, 36: 20120131.

Nakagawa, K., Soga, K., & Mitchell, J. K. (1996). Pulse transmission system for measuring wave propagation in soils. *Journal of Geotechnical Engineering*, 122(4): 302–308.

Pennington, D. S., Nash, D. F. T., & Lings, M. L. (2001). Horizontally mounted bender elements for measuring anisotropic shear moduli in triaxial clay specimens. *Geotechnical Testing Journal*, 24(2): 133–134.

Rio, J. (2006.) Advances in laboratory geophysics using bender elements. PhD Thesis. London: University College London, University of London, p. 383.

Santamarina, J. & Fratta, D. (2005). *Discrete signals and inverse problems: An introduction for engineers and scientists*. Hoboken, NJ: John Wiley and Sons.

Santamarina, J. C. & Fam, M. A. (1997). Discussion: Interpretation of bender element tests. *Géotechnique*, 47(4): 873–877.

Vary, A. (1980) Ultrasonic measurement of material properties. In *Research techniques in nondestructive, testing IV*. (ed) R.S. Sharpe. New York: Academic Press, pp. 159–204.

Viggiani, G. & Atkinson, J. H. (1995). Interpretation of bender element tests. *Géotechnique*, 45(1): 149–154.

FURTHER READING

Cheng, Z. & Leong, E. C. (2014). A hybrid bender element – ultrasonic system for measurement of wave velocity in soils. *Geotechnical Testing Journal*, 37(3): 377–388. https://doi.org/10.1520/GTJ20120158. ISSN 0149-6115.

Chapter 21

Thermal conductivity

21.1 BACKGROUND

Heat transfer in soils and rocks is important in many geotechnical engineering applications related to climate change. One major climate change action is to reduce carbon dioxide emissions, especially through the reduced reliance on fossil fuels. The related geotechnical engineering applications where heat transfer is involved include the use of alternative energy sources such as nuclear energy (storage of radioactive waste, e.g., Sellin and Leupin 2013), geothermal energy (e.g., Jia et al. 2017), ground heat source pumps (e.g., Santa et al. 2020), energy piles (e.g., Laloui et al. 2006) and carbon dioxide sequestration (e.g., Vilarrasa and Rutqvist 2017). Dealing with these problems requires the thermal properties of soils and rocks for conditions ranging from dry to wet.

Heat flow q through soils is mainly by conduction and can be described by Fourier's law as given in Equation 21.1.

$$q = -\lambda \nabla T \tag{21.1}$$

where
λ = thermal conductivity of soil (W/mK or J/m°C)
T = temperature in °C

Measuring thermal conductivity λ is the subject of this chapter. The thermal properties of soils are heat capacity and thermal conductivity. Heat capacity can be defined in terms of mass as specific heat capacity or in terms of volume as volumetric heat capacity. Sometimes, thermal diffusivity is also used in heat flow problems. Thermal diffusivity is defined as the ratio of thermal conductivity and volumetric heat capacity. As soil consists of three phases (soil solids, water and air), the thermal properties of soils are a combination

Table 21.1 Thermal conductivity, specific heat capacity and volumetric heat capacity of common materials

Material	Density (kg/m³)	Thermal conductivity, λ (W/m.K)	Specific heat capacity (J/kg.K)	Volumetric heat capacity (MJ/m³K)
Air at 20°C	1.2	0.025	1,000	0.012
Water at 20°C	1,000	0.58	4,200	4.2
Clay minerals	2,660	2.9	800	2
Soil organic matter	1,300	0.25	2,500	2.7
Peat	-	0.4	-	3.8
Dry clay/silt	-	0.5	-	1.6
Wet clay/silt	-	1.7	-	3.4
Dry sand	-	0.4	-	1.6e6
Wet sand	1,600	1.8	1,700	2.7e6
Dry gravel	-	0.4	-	1.6e6
Wet gravel	-	1.8	-	2.4e6

of the thermal properties of each phase. De Vries (1963) proposed that the thermal conductivity of soil can be expressed using Equation 21.2.

$$\lambda = \frac{f_s\theta_s\lambda_s + f_w\theta_w\lambda_w + f_a\theta_a\lambda_a}{f_s\theta_s + f_w\theta_w + f_a\theta_a} \tag{21.2}$$

where
f_s, f_w, f_a = weighting factors for soil solids, water and air, respectively
$\theta_s, \theta_w, \theta_a$ = volumetric content for soil solids, water and air, respectively, i.e., volume of each phase with respect to the total volume
$\lambda_s, \lambda_w, \lambda_a$ = thermal conductivity for soil solids, water and air, respectively

The thermal conductivity of soils varies with water content, void ratio, mineralogical composition, organic content, and particle size (Abu-Hamdeh and Reeder 2000; Barry-Macaulay et al. 2013). Table 21.1 shows the thermal conductivity, specific heat capacity and volumetric heat capacity of common materials compiled from de Vries (1963), Hillel (1982) and Rosenberg (1983). Methods to measure the thermal resistivity of soils have been standardised in ASTM D5334 (2022) and IEEE-STD 442 (2017).

21.2 RELEVANT STANDARDS

1. ASTM D5334-22 (2022) Standard Test Method for Determination of Thermal Conductivity of Soil and Rock by Thermal Needle Probe Procedure.
2. IEEE-STD 442-17 (2017) IEEE Guide for Soil Thermal Resistivity Measurements of Soils and Backfill Materials.

21.3 THEORY

Test methods to determine the thermal conductivity of soils in the laboratory can be divided into steady-state and non-steady-state or transient methods. Steady-state methods for measuring thermal conductivity are well established for the measurement of thermal conductivity for construction materials and plastics. Steady-state methods require a constant temperature gradient within the soil specimen to be reached during the test to measure the thermal conductivity and take a long time. However, this can cause water movement in the soil in response to the applied temperature gradient and is generally considered not suitable for unsaturated soils (Bristow 2002). Examples of steady-state methods are the well-established guarded hot-plate method (van Donk et al. 2001, Bristow 2002) and the Pyrex heat flux meter (Côté and Konrad 2005). Generally, transient state methods for thermal conductivity measurements of unsaturated soils are preferred, as the tests do not require the establishment of a constant temperature gradient during the test, and, hence, the effects of water movement in the soil are smaller (de Vries 1952). Examples of transient methods are those employing line heat source (Bristow et al. 1994, Wu et al. 2014) and plane-heat source (Yao and Likos 2017).

For unsaturated soils, it is useful to determine the thermal conductivity of the soils over a wide range of saturation. It is possible to determine the soil-water characteristic curve and thermal conductivity of a soil at the same time (Dong et al. 2014; Yao et al. 2014); however, this chapter describes only the test for a single measurement of thermal conductivity for an unsaturated soil.

21.4 TEST METHOD

Both ASTM D5334-22 and IEEE-STD 442-17 use the line heat source for transient measurement of the thermal conductivity of soil. Hence, the line heat source is the only method described in this chapter. In ASTM D5334-22 and IEEE-STD 442-17, the line heat source is a needle probe. The requirement of the needle probe is that it is made of stainless steel and contains a linear heat source and a temperature sensor to measure the variation of temperature at a point along the line. The heat source and the temperature sensor should be electrically isolated from each other, as well as from the stainless-steel tube. Many types of needle probes for thermal conductivity measurement are available commercially.

ASTM D5334-22 provides instructions for constructing a thermal needle probe based on the design in Mitchell et al. (1977). The thermal needle probe suggested in IEEE-STD 442-17 is the same as ASTM D5334-22. Additionally, IEEE-STD 442-17 specifies that the stainless-steel needle should have a length-to-diameter ratio of 50. The heat source is provided by

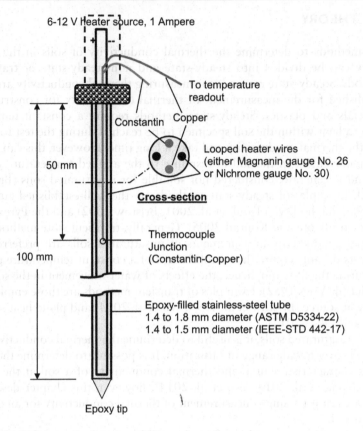

6-12 V Heater source, 1 Ampere

To temperature readout

Constantin   Copper

Looped heater wires
(either Magnanin gauge No. 26
or Nichrome gauge No. 30)

50 mm

Cross-section

Thermocouple
Junction
(Constantin-Copper)

100 mm

Epoxy-filled stainless-steel tube
1.4 to 1.8 mm diameter (ASTM D5334-22)
1.4 to 1.5 mm diameter (IEEE-STD 442-17)

Epoxy tip

Figure 21.1 Schematic drawing of thermal needle probe.

a constant current and a device to read the voltage to the nearest 0.01V and current in ampere is required. IEEE-STD 442-17 states that the power supply to the heater should be between 10 and 30 W/m. The temperature sensing element, either a thermocouple or thermistor, is connected to a recording device that is able to read temperature changes to a resolution of at least 0.1°C. A schematic drawing of the thermal needle probe suggested by ASTM D5334-22 and IEEE-STD 442-17 is shown in Figure 21.1.

21.5 TEST PROCEDURES

21.5.1 Calibration

The thermal needle probe should be calibrated before use using a material with a known thermal conductivity (typically $0.2 < \lambda < 5$ W/m·K). Suitable materials suggested by ASTM D5334-22 are dry Ottawa sand, Pyrex 7740, Fused Silica, Pryoceram 9606, glycerine (glycerol) at 25°C and water stabilised with 5g agar per litre at 25°C. The calibration should be done in a cylinder where the internal diameter is the greater of ten times the diameter of the needle probe or 40 mm, and the height should be at least 20% longer than the length of the needle. For solid materials, a hole can be drilled into the cylindrical solid specimen that will just tightly fit the probe, and the probe should be lightly thermal greased to improve thermal contact.

The calibration factor C_F is obtained using Equation 21.3.

$$C_F = \frac{\lambda_{known}}{\lambda_{measured}} \qquad (21.3)$$

where

λ_{known} = known thermal conductivity of calibration material

$\lambda_{measuredd}$ = measured thermal conductivity of calibration material (should be within one standard deviation of λ_{known})

The calibration is applied to all subsequent thermal conductivity measurements using the thermal needle probe.

21.5.2 Soil specimen

The soil specimen can be intact/undisturbed or compacted. There are some differences in the soil specimen requirements between ASTM D5334-22 and IEEE-STD 442-17, as listed in Table 21.2.

Table 21.2 Dimensions of specimens required for thermal conductivity measurement.

Specimen	ASTM D5334-22	IEEE-STD 442-17
Intact/undisturbed	Cut the sampling tube containing the sample to give a specimen length of 200 ± 30 mm. Minimum diameter of the specimen should be 51 mm	None
Compacted	Compact the specimen in a tube of a minimum internal diameter of 51 mm to a length of 200 ± 30 mm	Compact the specimen in a rigid cylinder of a minimum internal diameter of 100 mm in lifts of 25 mm to a length dependent on the probe. No guidance on what this should be. Hence the 20% clearance of probe length from the bottom of the specimen suggested in ASTM D5334-22 can be followed.

21.5.3 Test procedures

The test procedures for making thermal conductivity measurements are as follows:

1. Allow the specimen to come to equilibrium with the test temperature.
2. Weigh the specimen in the tube.
3. Insert the thermal needle probe down the centre axis of the specimen by either pushing the probe into the soil specimen or through a pre-bored hole so the whole length of the needle is embedded in the soil. A thin layer of thermal grease can be applied to the needle before inserting for better thermal contact.
4. Connect the heater wire of the thermal needle probe to the constant current source.
5. Connect the temperature wires to the readout unit.
6. Turn on the constant current source to the heater wire such that the temperature change is less than 10°C in 1,000 s.
7. Record the time and temperature throughout the heating time. At least 20 to 30 readings should be recorded at equal time intervals during the test. The total heating time depends on the size of the thermal needle probe; usually, 30 s–60 s heating time is needed for a needle probe that has a diameter of less than 2.5 mm. The heating time should not be so long that the heat reaches the boundary of the soil specimen.
8. Switch off the heater.
9. If cooling data is required, carry on recording the time and temperature at the same time interval as during heating for at least 20 to 30 readings.
10. After the test, determine the water content of the soil specimen.

21.5.4 Calculations

For a line source, the head conduction radially in a homogeneous medium is governed by Equation 21.4.

$$\frac{\partial T}{\partial t} = \lambda \left[\frac{\partial^2 T}{\partial r^2} + \frac{1}{r}\left(\frac{\partial T}{\partial r} \right) \right] \tag{21.4}$$

where r is the radial distance from the line source.

The solution to Equation 21.4 is given by Equation 21.5.

$$\Delta T = \begin{cases} -\dfrac{Q}{4\pi\lambda}\left[E_i\left(\dfrac{-r^2}{4Dt} \right) \right] & \text{for } t \leq t_h \\[4mm] -\dfrac{Q}{4\pi\lambda}\left[-E_i\left(\dfrac{-r^2}{4Dt} \right) + E_i\left(\dfrac{-r^2}{4D(t-t_h)} \right) \right] & \text{for } t > t_h \end{cases} \tag{21.5}$$

where
ΔT = change in temperature
Q = heat input (given as I^2R/L or VI/L where I is A current flowing through heater wire, R is the total resistance of heater wire, V is measured voltage and L is the length of needle probe)
E_i = exponential integral
D = thermal diffusivity
t_h = heating time

Solving Equation 21.5 is non-trivial, a simplified solution can be obtained by approximating the exponential integral in Equation 21.5 by the most significant term of its series expansion to give Equation 21.6.

$$\Delta T = \begin{cases} \dfrac{2.302Q}{4\pi\lambda}\log(t) & \text{for } t \leq t_h \\[2ex] \dfrac{2.302Q}{4\pi\lambda}\log\left(\dfrac{t}{t-t_h}\right) & \text{for } t > t_h \end{cases} \qquad (21.6)$$

Using Equation 21.6, temperature T is plotted against log(t). The plot of a good experiment is shown in Figure 21.2, and the plot of experiments that have some problems is shown in Figure 21.3. According to ASTM D5334-22, three portions of the plot can be identified: transient, quasi-steady and boundary effects. Figure 21.2 shows that the selected heating time is ideal, boundary effects are not observed in the heating data. A similar plot should

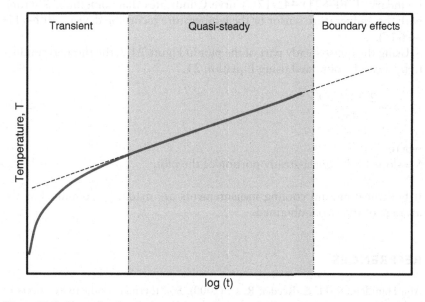

Figure 21.2 Plot of a good thermal conductivity measurement experiment with the thermal needle probe.

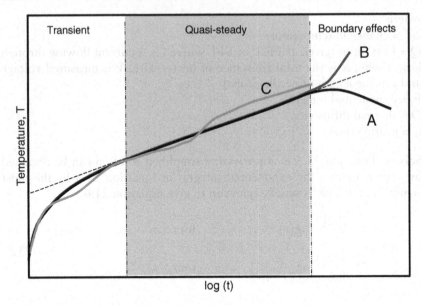

Figure 21.3 Plot of problematic thermal conductivity measurement experiments with the thermal needle probe.

be obtained for the cooling data. Figure 21.3 shows that curves A and B are affected by the boundary effects. Curve A represents heat loss during heating at the boundary, while curve B indicates heat gain during heating at the boundary (IEEE-STD 442-17). Curve C indicates that something is wrong with the temperature sensor or the temperature recording device (IEEE-STD 442-17).

Using the quasi-steady part of the plot in Figure 21.2, the thermal conductivity λ can be obtained using Equation 21.7.

$$\lambda = \frac{2.302 C_F Q}{4\pi S} \tag{21.7}$$

where
S = slope of the quasi-steady portion of the plot.

If both heating and cooling measurements are made, λ is obtained as the average of the two λ obtained.

REFERENCES

Abu-Hamdeh, N. H. & Reeder, R. C. (2000). Soil thermal conductivity effects of density, moisture, salt concentration, and organic matter. *Soil Science Society Of American Journal*, 64: 1285–1290. https://doi.org/10.2136/sssaj2000.6441285x

Barry-Macaulay, D., Bouazza, A., Singh, R. M., Wang, B., & Ranjith, P. G. (2013). Thermal conductivity of soils and rocks from the Melbourne (Australia) region. *Engineering Geology*, *164*: 131–138. https://doi.org/10.1016/j.enggeo. 2013.06.014

Bristow, K. L. (2002). Thermal conductivity. In Dane, J. H. & Topp, G. C. (Ed.), *Methods of soil analysis, part 4, physical methods*. Madison, WI: American Society of Agronomy, pp. 1209–1226.

Côté, J. & Konrad, J.-M. (2005). Thermal conductivity of base-course materials. *Canadian Geotechnical Journal*, *42*(1): 61–78. https://doi.org/10.1139/t04-081

Dong, Y., Lu, N., Wayllace, A., & Smits, K. (2014). Measurement of thermal conductivity function of unsaturated soil using a transient water release and imbibition method. *Geotechnical Testing Journal*, *37*(6): 1–11. https://doi.org/10.1520/ GTJ20140046. ISSN 0149-6115

Ghuman, B. S. and Lal, R. (1985). Thermal conductivity, thermal diffusivity, and thermal capacity of some Nigerian soils. *Soil Science*, *139*(1): 74–80.

Hillel, D. (1982). *Introduction to soil physics*. New York, NY: Academic Press.

Jia, G. S., Tao, Z. Y. Meng, X. Z., Ma, C. F., Chai, J. C., & Jin, L. W. (2019). Review of effective thermal conductivity models of rock-soil for geothermal energy applications. *Geothermics*, *77*: 1–11. https://doi.org/10.1016/j.geothermics.2018. 08.001.

Laloui, L., Nuth, M., & Vulliet, L. (2006). Experimental and numerical investigations of the behaviour of a heat exchanger pile. *International Journal for Numerical and Analytical Methods in Geomechanics*, *30*(8): 763–781. https://doi.org/10.1002/ nag.499

Lee, S.-J., Choi, J.-C., Baek, S., Kwon, T.-H., Ryu, H.-H., & Song, K.-I. (2016). Use of a pre-drilled hole for implementing thermal needle probe method for soils and rocks. *Energies*, *9*(10): 846. https://doi.org/10.3390/en9100846

Mitchell, J. K., Kao, T. C., & Abdel-Hadi, O. N. (1977). *Backfill Materials for Underground Power Cables*, Department of Civil Engineering, University of California at Berkeley, EPRI EL-506, June.

Santa, G. D., Galgaro, A., Sassi, R., Cultrera, M., Scotton, P., Mueller, J., Bertermann, D., Mendrinos, D., Pasquali, R., Perego, R., Pera, S., Di Sipio, E., Cassiani, G., De Carli, M., & Bernardi, A. (2020). An updated ground thermal properties database for GSHP applications. *Geothermics*, *85*. https://doi.org/10.1016/j. geothermics.2019.101758

Sellin, P. & Leupin, O. X. (2013). The use of clay as an engineered barrier in radioactive-waste management a review. *Clays Clay Miner 61*: 477–498. https:// doi.org/10.1346/CCMN.2013.0610601

Vilarrasa, V. & Rutqvist, J. (2017). Thermal effects on geologic carbon storage. *Earth-Science Reviews*, *165*: 2017. https://doi.org/10.1016/j.earscirev.2016. 12.011

Yao, J. & Likos, W. J. (2017). Transient plane and line source methods for soil thermal conductivity. *Geotechnical Testing Journal*, *40*(5): 858–870. https://doi. org/10.1520/GTJ20160333. ISSN 0149-6115

Yao, J., Oh, H., Likos, W. J., & Tinjum, J. M. (2014). Three laboratory methods for measuring thermal resistivity dryout curves of coarse-grained soils. *Geotechnical Testing Journal*, *37*(6): 1–12. https://doi.org/10.1520/GTJ20140120. ISSN 0149-6115.

FURTHER READING

Côté, J. & Konrad, J.-M. (2005). A generalized thermal conductivity model for soils and construction materials. *Canadian Geotechnical Journal*, 42(2): 443–458. https://doi.org/10.1139/t04-106

Haynes, W.M., Lide, D. R., Bruno, T.J. (2014). *CRC Handbook of chemistry and physics*. 95th ed. Boca Raton, FL: CRC Press.

Powell, R. W., Ho, C. Y., & Liley, P. E. (1996). *Thermal conductivity of selected materials, special publication, NSRDS-NBS 8*. Washington, DC: National Bureau of Standards.

Appendix A: Glossary of terms

Air-entry value (AEV) The potential or suction that air first enters a porous medium. Other names: bubbling pressure.

Available water (AW) The amount of water between in situ field capacity and the permanent wilting point. It is not the amount of water that can be absorbed by plant roots, which is plant specific. See also nonlimiting water range.

Bulk density In soil science, this is defined as the mass of dry soil per unit bulk volume. This is different from what is commonly used in geotechnical engineering where bulk density is total density. Bulk density is expressed in megagram per cubic metre, Mg m^{-3}.

Cavitation Refers to the formation of gas in a liquid volume when the pressure is reduced (tension is increased) to a critical level. According to Soil Science Society of America, cavitation typically occurs at about 0.08 MPa of water tension but may be lower in confined systems such as a tensiometer. In pressure measuring systems, cavitation can create discontinuity in a water system preventing the correct transmission of pressure in the system to the pressure measuring device e.g., vacuum gauge or pressure transducer.

Electrical conductivity (EC) When applied to a soil refers to the ability of soil to conduct electricity. More commonly, electrical conductivity of a solution or an extract of soil is commonly used to estimate the concentration of salt in a solution which can be used to estimate the osmotic potential of the solution.

Electrical resistivity A measure of the resistance of soil to conduct electricity and the inverse of electrical conductivity. It is used to estimate the soil-water matric potential from predetermined calibrations.

Field capacity (FC) Water content on a mass or volume basis that remains in a soil two or three days after having been wetted with water and after free drainage is negligible. See also available water.

Matric potential or matric head   Refers to the potential energy of soil water due to the attractive forces (adhesion and cohesion) between water and the soil matrix. Matric potential is expressed as energy per unit volume and is equal to the product of height of capillary rise, density of water and gravitational constant. Matric head is expressed as energy per weight and is equal to the height of capillary rise.

Matric suction   Commonly used in unsaturated soil mechanics as the difference between pore-air and pore-water pressures. However, it is no longer used in Soil Science Society of America (SSSA) publications, being replaced by matric potential.

Osmotic potential or pressure   Refers to the potential energy acting upon soil water due to the presence of solutes. This is the most misunderstood component of total suction. Osmotic potential or pressure requires a semi-permeable membrane where water containing solutes on one side will draw water from the other side which has a lower solute concentration. Omotic potential or pressure is important in plants where the semi-permeable membrane is provided by the cell walls through which water is being drawn into the plant. Osmotic potential, ψ_o, is given by the following equation:

$$\psi_o = CRT$$

where C is the total molar concentration, R is the universal gas constant, and T is the temperature.

Permanent wilting point (PWP)   The highest water content of a soil at which indicator plants growing in that soil cannot absorb the water and the plant wilts and fails to recover when placed in a dark and humid chamber. Often the water content at suction of 1500 kPa is taken as the PWP.

Pressure potential or head   Is the potential energy of soil water due to the weight of water (hydrostatic pressure) acting on that point of interest. Pressure head, h_p, is equal to the height of water between the point of interest and the free water surface (water table). It is also known as piezometric head and expressed as energy per unit volume and equals $\gamma_w h_p$ where γ_w is the unit weight of water.

Relative humidity   The ratio of the water vapour pressure to the saturated water vapour pressure at a given temperature. The relative humidity is used as a measure of water potential by meteorologists.

Salinity   Refers to the amount of soluble salts in water or in a soil (soil salinity). Soil salinity is usually determined by measuring the electrical conductivity of a saturation extract.

Saturation extract   Refers to the solution extracted from a soil at its saturation water content.

Soil-water characteristic   Also known as characteristic curve, water retention curve, water release curve or isotherm is the relationship between the soil-water content (by mass or volume) and the soil-water matric potential.

Vadose zone   Refers to the zone above the permanent water table where the soil is partially saturated.

Vapour flow   Flow of water vapour in soils from a moist or warm zone of higher potential to a drier or colder zone of lower potential.

Volumetric water content   Water content of soil expressed as a ratio of volume of water and volume of soil.

Appendix B: Formulas and conversion factors

FORMULAS

Relative humidity, R_h

$$R_h = \frac{u_v}{u_{vo}}(100\%) \tag{B.1}$$

where u_v is partial pressure of water vapour in the air and u_{vo} is saturated vapour pressure of water.

For practical application, the approximate saturated vapour pressure of water for temperature $T > 0°C$ can be obtained using a modified form of Buck (1981) equation:

$$u_{v0}(T) = 0.6112 \exp\left[\left(18.678 - \frac{T}{234.5}\right)\left(\frac{T}{257.14 + T}\right)\right] \tag{B.2}$$

where u_{v0} is in kPa. The error of Equation B.2 is less than 0.12% for 0°C to 100°C when compared with values from Kaye and Laby (1995).

More recently, Huang (2018) proposes more accurate equations to estimate the saturated vapour pressure of water for temperatures in the positive and negative ranges:

$$u_{v0}(T) = \begin{cases} \dfrac{\exp\left(34.494 - \dfrac{4924.99}{T + 237.1}\right)}{1000(T + 105)^{1.57}} & \text{for } T > 0°C \\[4mm] \dfrac{\exp\left(43.494 - \dfrac{6545.8}{T + 238}\right)}{1000(T + 868)^{2}} & \text{for } T \leq 0°C \end{cases} \tag{B.3}$$

The error of Equation B.3 is less than 0.1% for 0°C100°C when compared with values from Kaye and Laby (1995).

Total suction, π

$$\pi = -\frac{RT_k}{v_{w0}\omega_v}\ln\left(R_h\right) \qquad (B.4)$$

where R is universal gas constant = (8.31432 j/mol.K), T_k is temperature in Kelvin, v_{w0} is the specific volume of water or inverse of density of water (density of water is 998 kg/m³ at 293K or 20°C) and ω_v is molecular mass of water vapour = (18.016 kg/kmol).

CONVERSION FACTORS

Condition	J/kg	MPa	Bars	Rh	pF[1]
	−1	−0.001	−0.01	0.999993	1.01
Field capacity	−33	−0.033	−0.33	0.999756	2.53
Permanent wilting point	−1500	−1.5	−0.15	0.988977	4.18
Air dry	−100000	−100	−1000	0.477632	6.01
Oven dry	−1000000	−1000	−10000	0.000618	7.01

[1] pF = log base 10 of water tension in units of 10 cm of water.

REFERENCES

Buck, A. L. (1981). New equations for computing vapor pressure and enhancement factor. *Journal of Applied Meteorology*, 20: 1527–1532.

Huang, J. (2018). A simple accurate formula for calculating saturation vapor pressure of water and ice. *Journal of Applied Meteorology and Climatology*, 57(6): 1265–1272. Retrieved May 12, 2022, from https://journals.ametsoc.org/view/journals/apme/57/6/jamc-d-17-0334.1.xml

Kaye, G. W. C. & Laby, T. H. (1995). *Tables of physical & chemical constants*, 16th ed. NPL (National Physical Laboratory) Essex, England; New York: Longman Scientific & Technical.

Index

Note: Page numbers in **Bold** refer to tables; and page numbers in *italics* refer to figures